The Handbook of Neuropsychiatric Biomarkers, Endophenotypes and Genes

THE HANDBOOK OF NEUROPSYCHIATRIC BIOMARKERS, ENDOPHENOTYPES AND GENES

Volume 1: Neuropsychological Endophenotypes and Biomarkers

Volume 2: Neuroanatomical and Neuroimaging Endophenotypes and Biomarkers

Volume 3: Metabolic and Peripheral Biomarkers

Volume 4: Molecular Genetic and Genomic Markers

Michael S. Ritsner

Editor

The Handbook of Neuropsychiatric Biomarkers, Endophenotypes and Genes

Volume 3

Metabolic and Peripheral Biomarkers

 Springer

Editor
Michael S. Ritsner, M.D., Ph.D.
Associate Professor of Psychiatry, the Rappaport Faculty of Medicine
Technion - Israel Institute of Technology, Haifa and
Sha'ar Menashe Mental Health Center, Hadera, Israel

ISBN 978-1-4020-9837-6   e-ISBN 978-1-4020-9838-3

Library of Congress Control Number: 2008942052

Printed on acid-free paper

springer.com

Foreword

Common genetically influenced neuropsychiatric disorders such as schizophrenia spectrum disorders, major depression, bipolar and anxiety disorders, epilepsy, neurodegenerative and demyelinating disorders, Parkinson and Alzheimer's diseases, alcoholism, substance abuse, and drug dependence are the most debilitating illnesses worldwide. They are characterized by their complexity of causes and by their lack of pathognomonic laboratory diagnostic tests. During the past decade many researchers around the world have explored the neuropsychiatric biomarkers and endophenotypes implicated, not only in order to understand the genetic basis of these disorders but also from diagnostic, prognostic, and pharmacological perspectives. These fields have therefore, witnessed enormous expansion in new findings obtained by neuropsychological, neurophysiological, neuroimaging, neuroanatomical, neurochemical, molecular genetic, genomic and proteomic analyses, which have generated a necessity for syntheses across the main neuropsychiatric disorders. The challenge now is to translate these findings into meaningful etiologic, diagnostic and therapeutic advances.

This four volume collection of Handbooks offers a broad synthesis of current knowledge about biomarker and endophenotype approaches in neuropsychiatry. Since many of the contributors are internationally known experts, they not only provide up-to-date state of the art overviews, but also clarify some of the ongoing controversies, future challenges and proposing new insights for future researches. The contents of the volumes have been carefully planned, organized, and edited in close collaboration with the chapter authors. Of course, despite all the assistance provided by contributors and others, I alone remain responsible for the content of these Handbooks including any errors or omissions, which may remain.

The Handbook is organized into four interconnected volumes covering five major sections.

Volume 1 "Neuropsychological Endophenotypes and Biomarkers" contains 17 chapters composed of two parts emphasizing schizophrenia as a prototype. The first section serves as an introduction and overview of methodological issues of the biomarker and endophenotype approaches in neuropsychiatry and some technological advances. Chapters review definitions, perspectives, and issues that provide a conceptual base for the rest of the collection. The second section comprises chapters in which the authors present and discuss the neuropsychological, neurocognitive and neurophysiological candidate biomarkers and endophenotypes.

Volume 2 "Neuroanatomical and Neuroimaging Endophenotypes and Biomarkers", focuses on neuroanatomical and neuroimaging findings obtained for wide spectra of neuropsychiatric disorders.

Volume 3 "Metabolic and Peripheral Biomarkers", explores several specific metabolic and peripheral biomarkers, such as neuroactive steroid biomarkers, cortisol to DHEA molar ratio, mitochondrial complex, biomarkers of excitotoxicity, melatonin, retinoic acid, abnormalities of inositol metabolism in lymphocytes, and others.

Volume 4 "Molecular Genetic and Genomic Markers" contains chapters devoted to searching for novel molecular genetic and genomic markers in less explored areas. This volume includes an Afterword written by Professor Robert H. Belmaker.

Similarly to other publications contributed to by diverse scholars from diverse orientations and academic backgrounds, differences in approaches and opinions, as well as some overlap, are unavoidable. I believe that this collection is probably the first of its kind to go beyond the neuropsychiatric disorders and delve into the neurobiological basis for diagnosis, treatment, and prevention. The take-home message is that principles of the biomarker-endophenotype approach may be applied no matter what kind of neuropsychiatric disorder afflicts our patients.

The Handbook is designed for use by a broad spectrum of readers including neuroscientists, psychiatrists, neurologists, endocrinologists, pharmacologists, psychologists, general practitioners, geriatricians, graduate students, health care providers in the fields of neurology and mental health, and others interested in trends that have crystallized in the last decade, and trends that can be expected to evolve in the coming years. It is hoped that this collection will also be a useful resource for the teaching of psychiatry, neurology, psychology and mental health.

With much gratitude, I would like to acknowledge the contributors from 16 countries for their excellent cooperation. In particular, I am most grateful to Professor Irving Gottesman for his support of this project. His unending drive and dedication to the field of psychiatric genetics never ceases to amaze me. I wish to acknowledge Professor Robert H. Belmaker, distinguished biological psychiatrist, who was very willing to write the afterword for these volumes. I also wish to take this opportunity to thank my close co-workers and colleagues Drs. Anatoly Gibel, Yael Ratner, Ehud Susser, Stella Lulinski, Rachel Mayan, Professor Vladimir Lerner and Professor Abraham Weizman for their support and cooperation. Finally, I am forever indebted to my wife Galina Ritsner, sons Edward and Yisrael for their understanding, endless patience and encouragement when it was most required.

I sincerely hope that these four interconnected volumes of the Handbook will further knowledge in the complex field of neuropsychiatric disorders.

February, 2009   Michael S. Ritsner
                                                                                              Editor

Contents to Volume 3

Contributors to Volume 3

Pedro Abreu-Gonzalez Professor of Biochemistry, Department of Physiology, School of Medicine, University of La Laguna, La Laguna, Santa Cruz de Tenerife, Canary Islands, Spain

Galila Agam, Ph.D., Associate Professor, Psychiatry Research Unit and Department of Clinical Biochemistry, Faculty of Medicine, Ben Gurion University, Israel

Sarah J. Bailey Lecturer, Department of Pharmacy and Pharmacology, University of Bath, Claverton Down, UK
E-mail: S.Bailey@bath.ac.uk

Dorit Ben-Shachar, Ph.D., Head of Lab, Laboratory of Psychobiology, Department of psychiatry, B. Rappaport Faculty of Medicine, Rambam Medical Center, Technion IIT, Haifa, Israel
E-mail: shachar@tx.technion.ac.il

Yuly Bersudsky, M.D., Ph.D., Senior Lecturer, Faculty of Medicine, Ben Gurion University, Beersheva Mental Health Center, Beersheva, Israel

Yogesh Dwivedi, Ph.D., Associate Professor, University of Illinois at Chicago, Department of Psychiatry, Chicago, IL, USA
E-mail: ydwivedi@psych.uic.edu

Carlo Ferrarese, M.D., Ph.D., Professor of Neurology, Director of the Department of Neurology and of the Neurology Residency School, University of Milano-Bicocca, Ospedale San Gerardo, Monza, Italy
E-mail: carlo.ferrarese@unimib.it

Peter Gallagher Research Associate in Psychiatry, School of Neurology, Neurobiology and Psychiatry, Newcastle University, Leazes Wing (Psychiatry), Newcastle upon Tyne, UK
E-mail: peter.gallagher@ncl.ac.uk

Nadia De Giovanni Istituto Medicina Legale, Università Cattolica S. Cuore, Roma, Italy
E-mail: nadia.degiovanni@rm.unicatt.it

Andrea L. Glenn, M.A., Doctoral Student, University of Pennsylvania, Philadelphia, PA, USA
E-mail: aglenn@sas.upenn.edu

Manuel Henry Professor of Psychiatry, Department of Internal Medicine,
Dermatology and Psychiatry, School of Medicine, University of La Laguna,
La Laguna, Santa Cruz de Tenerife, Canary Islands, Spain

Christian Humpel Associate Professor Dr., Laboratory of Psychiatry
and Exp. Alzheimer's Research, Department of General Psychiatry,
Innsbruck Medical University, Innsbruck, Austria
E-mail: christian.humpel@i-med.ac.at

Josef Marksteiner, M.D., Associate Professor Dr., Laboratory of Psychiatry and
Exp. Alzheimer's Research, Department of General Psychiatry, Innsbruck Medical
University, Innsbruck, Austria
E-mail: J.Marksteiner@i-med.ac.at

Peter McCaffery Professor, University of Aberdeen, Institute of Medical Sciences,
Foresterhill, Aberdeen, UK
E-mail: p.j.mccaffery@abdn.ac.uk

Armando L. Morera Professor of Psychiatry, Department of Internal Medicine,
Dermatology and Psychiatry, School of Medicine, University of La Laguna,
La Laguna, Santa Cruz de Tenerife, Canary Islands, Spain
E-mail: amorera@ull.es

A. Leslie Morrow, Ph.D., Professor of Psychiatry and Pharmacology,
Associate Director, Bowles Center for Alcohol Studies, University of
North Carolina School of Medicine, USA
E-mail: morrow@med.unc.edu

Ghanshyam N. Pandey, Ph.D., Professor, University of Illinois at Chicago,
Department of Psychiatry, Chicago, IL, USA
E-mail: gpandey@psych.uic.edu

Patrizia Porcu Assistant Professor of Psychiatry, Bowles Center for Alcohol
Studies, University of North Carolina School of Medicine, USA
E-mail: patrizia_porcu@med.unc.edu

Michael S. Ritsner, M.D., Ph.D., Associate Professor of Psychiatry and
Head of Cognitive and Psychobiology Research Laboratory, The Rappaport
Faculty of Medicine, Technion – Israel Institute of Technology, Haifa and Chair,
Acute Department, Sha'ar Menashe Mental Health Center, Hadera, Israel
E-mail: ritsner@sm.health.gov.il

Gessica Sala, Ph.D., Post-doctoral Research Associate, Department of
Neuroscience and Biomedical Technologies, University of Milano-Bicocca, Italy
E-mail: gessica.sala@unimib.it

Jon Sen Specialty Registrar in Neurosurgery; Wessex Neurological Centre,
Southampton University Hospitals, UK
E-mail: jsen@ion.ucl.ac.uk

Lucio Tremolizzo, M.D., Ph.D., Neurologist and Post-doctoral Research
Associate, University of Milano-Bicocca; Ospedale San Gerardo, Monza, Italy
E-mail: lucio.tremolizzo@unimib.it

Part III
Possible Metabolic and Peripheral Biomarkers

Chapter 28
Peripheral Biomarkers in Dementia and Alzheimer's Disease

Christian Humpel and Josef Marksteiner

Abstract Alzheimer's disease (AD) is a chronic progressive neurodegenerative disease, and is the most prevalent type of dementia. Dementia is usually preceded by a stage of mild cognitive impairment (MCI), with a mean prevalence of about 16%. After Alzheimer's disease, the most common forms of dementia are vascular dementia and Lewy body dementia. Frontotemporal dementia is less common. To date, the most advanced biochemical biomarkers include cerebrospinal fluid levels (CSF) of beta-amyloid(1-42), total tau, and phospho-tau proteins. Decreased levels of beta-amyloid(1-42) and increased levels of tau and phospho-tau are the most reproducible chemical biomarkers for Alzheimer disease. However, laboratories for testing these biomarkers are not readily available, and they also require lumbal puncture. The development and validation of biomarkers for prediction, diagnosis and tracking of progression of dementia and MCI are increasingly important. This chapter reviews the use of CSF biomarkers and of putative blood-related markers.

Keywords Cerebrospinal fluid · blood · plasma · Alzheimer · diagnosis · mild cognitive impairment

Abbreviations AD: Alzheimer's disease; CSF: Cerebrospinal fluid; MCI: Mild cognitive impairment; NGF: Nerve growth factor; PBMC: Peripheral blood mononuclear cells

C. Humpel
Department of General and Social Psychiatry, Innsbruck Medical University, Austria

J. Marksteiner
Department of Psychiatry and Psychotherapy, Landeskrankenhaus Klagenfurt, Austria

Introduction

The life expectancy of humans has increased within the last 100 years from about 40 to about 77 years. As age is the main risk factor for Alzheimer's disease (AD), the number of patients suffering from AD, and mixed forms of dementia will dramatically increase within the next 50 years. It is expected that there will be about 80 million AD patients worldwide in 2050. Looking at these enormously high numbers of presumed AD patients, we have to further establish reliable diagnostic surrogate markers to diagnose and monitor disease progression. It will be essential to delay and counteract symptoms in AD and to start therapeutic treatment as early as possible. A valid and easily accessible diagnostic procedure should be the basis for the treatment.

Diagnosis of Alzheimer's Disease

Progressive impairment in memory and cognition is a key clinical feature of AD. The disorder is morphologically characterized by extracellular beta-amyloid plaque deposition, intraneuronal tau pathology, neuronal cell death, vascular dysfunction and inflammatory processes. Definitive diagnosis of AD requires both a clinical diagnosis of the disease and post mortem detection of beta-amyloid plaques and tau-pathology.[1] A probable diagnosis of AD can be established with a confidence of >90%, based on clinical criteria, including medical history, physical examination, laboratory tests, neuroimaging and neuropsychological evaluation.[2,3] Accurate, early diagnosis of AD is still difficult because early symptoms of the disease are shared by a variety of disorders, including mixed forms of dementia and depression, possibly also reflecting common neuro-pathological features.[2,3]

Mild Cognitive Impairment and Mixed Forms of Dementia

Probable AD can be diagnosed with a high confidence, whereas the diagnosis of mixed forms of dementia, such as vascular dementia, frontotemporal dementia or Lewy Body dementia, is more difficult.[4–15] The transition stage between normal aging and dementia is also referred to as mild cognitive impairment (MCI). MCI is defined as an impairment in one or more cognitive domains (memory) or an overall mild cognitive decline that is greater than would be expected for an individuals age or education but that is insufficient to interfere with social and occupational functioning.[15] In this stage, the validity of diagnosis is more difficult. MCI subtypes may have different outcomes for progression to dementia, and all progressive dementias may have their own predementia states.[16] Vascular MCI, for instance, is thought to result from cerebrovascular disease and is proposed to describe a prodrome of vascular dementia.[17] Patients with Parkinson's disease and MCI may be at higher risk of progressing to dementia than cognitively intact Parkinson's disease patients.[18]

The expected conversion rates from MCI to AD have been shown to be 15% (after 1 year), 20% (after 3 years) and 50% (after 5 years). Approximately 42% of MCI patients develop AD, 15% develop other forms of dementia and 41% remain cognitively stable.[8] It is not yet clear, which prognostic capacity biological markers in MCI patients will have to predict the likelihood of conversion from MCI to AD. Vascular dementia is the second most common form of dementia after AD and is further classified as cortical or subcortical dementia. It constitutes a group of syndromes relating to different vascular mechanisms. Autopsy studies have shown the association between AD and vascular lesions.[19]

Biomarkers in CSF

A biological marker or biomarker[10,20] is objectively measured and evaluated as an indicator of normal biological processes, pathogenic processes or pharmacological responses to a therapeutic intervention. It serves as an indicator for health and disease or confirms the risk of developing a disease. The biomarker can be primarily related to the disease or can be simply epiphenomenal in nature. The sensitivity and specificity and ease of use are the most important factors that ultimately define the usefulness of a biomarker for diagnosis.

A promising area of research for biochemical diagnosis of AD and mixed forms of dementia is the analysis of cerebrospinal fluid (CSF).[2,5,10,21–30] Three biochemical markers have been well established for CSF to diagnose AD: beta-amyloid(1-42), total-tau and Phospho-tau-181.[7,21–24,27,31–33] We have recently established in our laboratory the detection of these three biomarkers and we can reliably distinguish AD patients from controls and other forms of dementia.[31,32]

Beta-amyloid(1-42)

AD is characterized by extracellular beta-amyloid plaque depositions. Beta-amyloid is cleaved from the large amyloid-precursor protein by different enzymes, called secretases. Processing of the 42-amino acid long peptide correlates with enhanced plaque deposition. However, beta-amyloid accumulation in plaques is insufficient to cause the cell death observed in patients suffering from of AD. It is well established that beta-amyloid(1-42) is so far a prominent CSF biomarker in AD.[5,10,28,30,31,33–38] Analysis of CSF beta-amyloid(1-42) shows a highly significant reduction in AD patients compared to controls (<500 pg/ml) and in patients suffering from mixed forms of dementia including MCI (Table 28.1). It is suggested that these reduced CSF levels of beta-amyloid(1-42) are caused by a reduced clearance of beta-amyloid from brain to blood/CSF and an enhanced aggregation and plaque deposition in the brain. Patients with vascular dementia display significantly higher beta-amyloid(1-42) CSF levels compared to healthy age-matched control subjects and can be differentiated from AD with white matter lesions with a cut off of <750 pg/ml.[13] While there is no change in CSF beta-amyloid(1-40), there is a marked decrease in the ratio of beta-amyloid(1-42)/(1-40), which is useful in diagnosing AD.[31] Beta-amyloid(1-42) CSF levels are also reduced in progressive supranuclear palsy and corticobasal degeneration.[12] Thus beta-amyloid CSF levels do not discriminate between these diseases and AD. While beta-amyloid(1-42) is decreased in CSF of AD patients, beta-amyloid(1-38) or –(1-40) levels are unchanged, suggesting that ratios of CSF beta-amyloid(1-42) to (1-38) or to (1-40) are not useful.[39]

Total Tau

Tau is a neuronal protein and plays a role in neuronal transport. Dysfunction of tau (called tauopathy) is a key pathological feature in AD and tau is significantly enhanced in CSF in neurodegenerative diseases and AD[34] (Table 28.1). In normal controls total tau CSF levels are increasing by age[36]: <300 pg/ml (21–50 years), <450 pg/ml (51–70 years) and <500 pg/ml (>71 years). A high number of studies has reported that total CSF tau is a valid biomarker for AD and other forms of neurodegeneration.[2,10,21,23,30,31,33,35] Levels of total tau are significantly enhanced in AD patients compared to age-matched control subjects (>600 pg/ml) and in patients suffering from mixed dementia including the MCI group. Interestingly total tau levels are dramatically enhanced in Creutzfeld Jacobs disease (>1,300 pg/ml). Tau levels may also be a prognostic marker with a good predictive validity for conversion from MCI to AD, since high CSF tau was found in 90% MCI cases that later progressed to AD, but not in cases with stable MCI.[21]

Table 28.1 CSF biomarkers in Alzheimer's disease and dementia

Beta-amyloid(1-42)	Decreased (<500 pg/ml)
Total tau	Increased (>600 pg/ml)
Phospho-tau-181	Increased (>50 pg/ml)
Ratio (phospho-tau-181/beta-amyloid)*100	Increased (>10)
Nerve growth factor	Increased
Monocyte chemoattractant protein-1	Increased/unchanged
Vascular endothelial growth factor	Increased/unchanged
Transforming growth factor-beta	Increased/unchanged
Interleukin-8	Enhanced in MCI and AD
Insulin-like growth factor binding protein-6	Enhanced in AD
Macrophage-Colony stimulating factor	Enhanced in AD
Tumor necrosis-factor-alpha	Enhanced/reduced in MCI/AD/vascular dementia
Neuroserpin	Enhanced in AD
24-OH-cholesterol	Increased/unchanged

Phospho-Tau-181 and "Ratio"

Tau protein is highly hyper-phosphorylated at different sites in AD. In particular, the detection of phospho-tau at the position 181 is specific for AD compared to control subjects.[5,21,26,28,31,40] Phospho-tau-181 is significantly enhanced in AD compared to controls (>50 pg/ml) and in patients suffering from mixed dementia including the MCI group (Table 28.1). However, while phospho-tau-181 levels show relatively high intragroup variability, the ratio between phospho-tau-181 and beta-amyloid(1-42) has a lesser variability. A ratio between phospho-tau-181 and beta-amyloid(1-42) that is higher than 10 specifically differentiates AD from controls. In our hands, ratios of up to 40–90 can be observed in AD patients compared to controls. The analysis of other phosphorylated forms of tau (phospho-tau-199, phospho-tau-231, phospho-tau-235, phospho-tau-396, phospho-tau-404) may offer novel significant improvements to diagnose early AD[15,21] Interestingly, slightly decreased phospho-tau-181 levels may in some cases point to frontotemperal dementia (<15 pg/ml).[31]

Combination of These Three CSF Biomarkers

It is now well established that the combination of all three biomarkers (beta-amyloid(1-42), total tau and phospho-tau-181 or the respective ratio) significantly increase the diagnostic validity for AD (Table 28.1). Combining these three biomarkers yields a sensitivity of 95% and a specificity of 83% for detection of incipient AD in patients with MCI.[8,31] We recommend also to routinely determine total protein levels, which, however, do not influence the diagnostic values.[41]

Other CSF Biomarkers

Despite strong efforts to characterize other potential biomarkers in CSF, up to now no biomarkers with a higher sensitivity and specificity could be identified. From all tested proteins only. Nerve growth factor (NGF) has been found to be increased in CSF of AD patients.[31,42,43] NGF is the most potent trophic factor supporting survival of cholinergic neurons, which degenerate early in AD, possibly due to defective retrograde transport of NGF to the basal nucleus of Meynert. Interestingly, the increase of NGF is specific for AD and depends on the extent of neurodegeneration as expressed by the ratio of phospho-tau181/beta-amyloid(1-42) (Table 28.1).[4] Although raw NGF data do not reveal a significant difference, the comparison of NGF in AD-patients (phospho-tau181/beta-amyloid(1-42) ratio > 10) with healthy control subjects (ratio phospho-tau181/beta-amyloid(1-42) < 6) reveals a significant difference.[4] This might suggest that NGF accumulates in neurodegeneration of Alzheimer's type possibly only at a certain stage of the disease.

Several other studies measured growth factors and cytokines in CSF, however, the data are very heterogeneous and inconsistent and do not point to a selective biomarker for AD (Table 28.1). Monocyte chomoattractant protein-1 is enhanced in AD patients,[44] however, our recent study[4] showed that monocyte chomoattractant protein-1 levels are age-dependent and could not detect significant difference between AD and age-matched control subjects. Hepatocyte growth factor and vascular endothelial growth factor have been reported to be increased in CSF of AD-patient,[45] however in our study[4] this increase was not statistically significant. In agreement with other studies, brain-derived neurotrophic factor or glial-cell line derived neurotrophic factor levels are very low in CSF of controls but not changed in AD.[4] No changes were seen in fibroblast growth factor-2, monocyte inhibitoring protein-1, tumor necrosis factor-alpha or transforming growth factor-beta, although others reported that transforming growth factor-beta1 or tumor necrosis factor-alpha were enhanced in CSF of AD patients.[4,46,47] Others reported that interleukin-8, macrophage-colony stimulating factor and insulin-like growth factor binding protein-6 were enhanced in CSF of AD patients.[47] Interestingly, recently it was suggested that 24-S-hydroxycholesterol could be a sensitive biomarker for MCI because this marker correlated with CSF total tau.[48] The CSF levels of neuroserpin are significantly enhanced in AD compared to controls.[49] We and others also measured both forms of cholinesterases, acetylcholineesterase and butyrylcholineesterase, in CSF and did not find any differences between controls and AD.[43] In summary, only NGF seems to be consistently enhanced in CSF of severe AD patients, while other growth factors or cytokines do not show consistent changes between controls and AD patients. No other factors could be identified in CSF specific for MCI, vascular dementia or other forms of dementia.

Biomarkers in Blood

The routine diagnosis of AD and mixed forms of dementia from CSF has several drawbacks: lumbal puncture and collection of CSF is an invasive treatment with potential side effects. Commercial ELISAs for beta-amyloid, tau and phospho-tau are extremely expensive and screening of patients is often not possible. Follow up analysis of the same patient over years is difficult because of the invasive CSF collection. Thus there is a clear need to search for blood biomarkers to diagnose AD and other forms of dementia.

Biomarkers in Plasma and Serum

Several authors measured the standard CSF biomarkers beta-amyloid and tau in plasma, but the data were all very heterogeneous and not useful for diagnosis. Most studies have shown that plasma beta-amyloid(1-42) and beta-amyloid(1-40) levels are not different in AD and controls.[15] A significant increase of beta-amyloid(1-42) plasma levels have been seen in women with MCI but not in men compared to 72 cognitively normal age-matched subjects.[50] Nevertheless, measurement of beta-amyloid plasma has several drawbacks,[10] because plasma levels for beta-amyloid(1-42) are unstable and influenced by different medications.[31] Recent longitudinal studies showed that high plasma beta-amyloid(1-42) levels are risk factors for developing AD, however, there is agreement that this factor is not sensitive and specific for early diagnosis.[15] A decrease of serum beta-amyloid(42) autoantibodies have been found in AD,[51] however, it was concluded that this parameter alone is not useful as a diagnostic biomarker. Immunoreactivity for the Tau-protein was detected in human plasma but there was no obvious increase in dementia.[52]

Most studies focused on single isolated inflammatory plasma/serum biomarkers in AD, however, so far the heterogeneity was very high and controversial and a single biomarker did not yield any significant improvement in diagnosing AD. Recently, it has been shown that different cytokines (interleukins-12, -16, -18, transforming growth factor beta1) were significantly enhanced in plasma of AD and vascular dementia patients, confirming the inflammatory process in the disease.[53,54] ELISA-studies showed that ALZAS immunoglobulins are present in plasma of AD patients and preliminary ELISA pilot studies confirmed ALZAS to elicit a specific anti ct-12 autoantibody,[55] suggesting an induced auto-immune reaction. Furthermore oxidative stress markers such as cystatin C were increased in plasma and cathepsin D was decreased in plasma of AD patients.[56] Free copper might correlate to cognitive decline and was higher in the ApoE4 individuals.[57] We recently showed that both cholinesterases, acetylcholineesterase and butyrylcholineesterase, were not changed in plasma of AD and MCI patients. Others found that plasma levels of serotonin may be linked to the pathogenesis and progression of vascular dementia.[58] In summary, the analysis of a single plasma biomarker does not yield any novel information for diagnosing AD.

Despite these different single biomarkers in plasma, a recent promising study showed that the combination of 18 selected biomarkers in plasma may allow the diagnosis of AD.[47] These factors are chemokines, cytokines, growth factors and binding proteins (Table 28.2), involved in inflammation, transmigration of cells, proliferation/differentiation/survival of blood cells and neurons, angiogenetic factors and factors regulating other proteins. These 18 biomarkers were selected from 120 signalling proteins by filter-based arrayed sandwich ELISA and may allow to diagnose MCI with close to 90% accuracy.

The following section shortly summarizes the role of the 18 plasma biomarkers for diagnosing AD and MCI.[47] Angiopoietin-2 binds to the endothelial cell specific receptor Tie2 and promotes angiogenesis, sprouting and tube formation and the formation of new blood vessels. CCL5 (also known as RANTES) is chemotactic for monocytes, memory T-cells, eosinophils and basophils and plays an active role in recruiting leukocytes into inflammatory sites. CCL7 (also known as monocyte chemoattractant protein-3) attracts and activates monocytes and regulates the function of macrophages. CCL15 (also known as monocyte inhibitoring protein-1-δ) chemoattracts T cells and monocytes. CCL18 (also known as monocyte inhibitoring protein-4 or PARC) is chemotactic for activated T cells and non-activated lymphocytes. CXCL8 (also known as

Table 28.2 Potential putative biomarkers in plasma[47]

Biomarker	Change	Role in	Chromosome
Angiopoietin-2	↑	Angiogenesis	8
CCL5	↓	Inflammation	17
CCL7	↓	Inflammation	17
CCL15	↓	Inflammation	17
CCL18	↑	Inflammation	17
CXCL8	↑	Inflammation	4
Epidermal growth factor	↓	Proliferation	4
Granulocyte-colony stimulating factor	↓	PDS	17
Glial cell line-derived neurotrophic factor	↓	Survival	5
Intracellular cell adhesion molecule-1	↑	Transmigration	19
Insulin-like growth factor binding protein-6	↑	Control of IGFs	12
Interleukin-1α	↓	Inflammation	2
Interleukin-3	↓	PDS	5
Interleukin-11	↑	Inflammation	19
Macrophage-colony stimulating factor	↓	PDS	1
Platelet-derived growth factor-BB	↓	Mitogenesis	22
Tumor necrosis factor-α	↓	Inflammation	6
TRAIL-R4	↑	Control of TRAIL	8

CCL = chemokine that contains a C–C motif, CXCL = chemokine that contains a C–X–C motif, PDS = proliferation-differentiation-survival, TRAIL-R4 = tumor necrosis factor receptor superfamily, member 10d, decoy with truncated death domain.

interleukin-8) mediates the immune reaction, is chemoattractant and recruits neutrophils at sites of inflammation. Epidermal growth factor stimulates the proliferation of various epidermal and epithelial cells, inhibits gastric acid secretion and is involved in wound healing. Granulocyte colony-stimulating factor stimulates the bone marrow to produce granulocytes and stem cells and stimulates survival, proliferation and differentiation and function of mature and neutrophil precursors and white blood cells. Glial cell line-derived neurotrophic factor stimulates the survival of several neurons, such as e.g. dopaminergic neurons or motorneurons. Intracellular adhesion molecule-1 (CD45) is continuously present in low concentrations in the membranes of leukocytes and endothelial cells and plays a role in transmigration of leukocytes through the blood-brain barrier. Insulin-like growth factor binding protein-6 controls the distribution, function and activity of insulin-like growth factors. Interleukin-1α has a broad range of activities, e.g. stimulation of proliferation and maturation of thymocytes and B-cells, it is involved in immune defense against infections, enhances cell adhesion molecule expression on endothelial cells and induces transmigration of cells. Interleukin-3 is a hematopoietic growth factor that promotes the survival, differentiation and proliferation of megakaryocyte, granulocyte-macrophage, erythroid, eosinophil, basophil and mast cell progenitors and enhances thrombopoiesis, phagocytosis and cellular cytotoxicity and plays a role in immune defense. Interleukin-11 regulates hematopoiesis, stimulates megakaryocytes, stimulates lymphocytes and regulates bone metabolism. It also inhibits production of pro-inflammatory cytokines. Macrophage colony stimulating factor regulates proliferation, differentiation and survival of blood monocytes, tissue macrophages and progenitor cells. Platelet-derived growth factor-BB is mitogenic for cells of mesenchymal origin. Tumor-necrosis factor alpha is secreted by various cells including adipocytes, activated monocytes, macrophages, B cells, T cells and fibroblasts, it induces apoptotic cell death and has also other numerous effects, such as e.g. septic shock, autoimmune disease, rheumatoid arthritis, inflammation and diabetes. Tumor necrosis factor receptor superfamily, member 10d, decoy with truncated death domain (TRAIL-R4) is a receptor with an extracellular TRAIL-binding domain, a transmembrane domain and a truncated cytoplasmic death domain. It does not directly induce apoptosis and inhibits TRAIL-induced apoptosis. However so far no quantitative measurements of these 18 biomarkers have been performed and the results have not been verified by other dementia centers yet.

Biomarker in Blood Cells

Biomarkers in Platelets

In platelets of AD patients an increased level of beta-amyloid was found (Table 28.3), increased activation of beta-secretase and decreased activation of alpha-secretase and a decreased ratio of amyloid-precusor protein (130 kDa/110 kDa).[59] It was suggested that amyloid-precusor protein processing in platelets may be useful for diagnosing AD.[59]

Biomarkers in Peripheral Blood Mononuclear Cells and Isolated Monocytes

The detection of parameters in peripheral blood mononuclear cells (PBMC) seems to be promising (Table 28.3). Glycogen synthase kinase-3 was significantly increased in white blood cells in AD compared to healthy subjects.[60] More importantly, glycogen synthase kinase-3 levels were also increased in MCI. Thus, it was concluded that glycogen synthase kinase-3 could be an important parameter for diagnosis of MCI and AD. Similarly, in phytohemmagglutinin-stimulated cultures of PBMC of patients with MCI a dramatically enhanced secretion of interleukin-6 was found compared to controls after 24 h.[61] However AD patients did not differ from controls. Others focused on the inflammatory pathway and found a significant increase in the percentage of monocytes producing different cytokines (interleukins-1β, -6, -12, tumor necrosis factor-α) under basal conditions and after exposure to inflammatory stimuli.[62] They also reported that these cells differentially responded to inflammatory challenges when compared to controls. Higher spontaneous production of interleukin-1 or tumor necrosis factor-alpha by PBMCs are associated with the risk of incident AD.[62]

Protein degradation by the ubiquitin-proteasome system is an essential cellular mechanism that has come into focus of aging research because its activity decreases during the aging process.[64] As cells age the defective activity of the major proteolytic system leads

Table 28.3 Suggested putative biomarkers in blood cells

Platelets	Amyloid-precursor protein ratio 130/110 kDa reduced
PBMC	Glycogen synthase kinase-3 enhanced
	Phytohemmaglutinin-induced interleukin-6 release enhanced
Lymphocytes	Telomere shortening in vascular dementia
Monocytes	Interleukin-1beta, -6,-12, tumor-necrosis-factor-alpha enhanced
Monocytes/fibroblasts	Mutant-like p53 enhanced

to proteasome overload and to the intracellular accumulation of damaged and unfolded protein products. Misfolded proteins often aggregate and accumulate in the cells through life. Insoluble ubiquitinated protein aggregates are present in the pathological hallmarks of AD, particularly neuritic plaques and neurofibrillary tangles. Failure of the ubiquitin-proteasome pathway function has been linked to beta-amyloid toxicity. It is well known that ubiquitin levels are increased many fold in the cerebral cortex of patients with AD and the increase strongly correlates with the degree of neurofibrillary changes in the tissue.[65] Interestingly, in CD45 T-lymphocytes the ubiquitin-proteasome pathway has been shown to be reduced during aging.[66] In addition ubiquitin levels were increased in CSF of AD patients.[67] A dinucleotide deletion in human ubiquitin B messenger RNA leads to formation of polyubiquitin, which has been implicated in neuronal cell death in AD and other neurodegenerative diseases.[63] This issue is of importance and we are recently under way to measure ubiquitin levels in PBMC and lymphocytes.

Cellular senescence is a stress response phenomenon resulting in a permanent withdrawal from the cell cycle and the appearance of distinct morphological and functional changes, such as e.g. telomere shortening.[68] Telomeres are high order structures formed by DNA and a complex array of specialized proteins that cap and stabilize the physical ends of chromosomes.[68] In mammals, telomeric DNA consists of a variable number of nucleotides, which extend over several 1,000 base pairs in length and end in G-rich single stranded overhang.[69] Short telomeres (tandem TTAGGG repeats) induce DNA damage responsive pathways and subsequently induce the permanent cell cycle arrest.[64] Telomere shortening was also found in white blood cells and interestingly short telomere length was associated in peripheral white blood cells in vascular dementia (Table 28.3).[70] We are interested in this issue and started to measure telomerase and telomeres in monocytes of dementia patients. Despite blood cells also other cells became of interest to diagnose AD. Recently it has been reported that fibroblasts from sporadic AD patients specifically express an anomalous and detectable conformational state of the senescent marker p53 (mutant-like p53) that allows to differentiate them from fibroblasts of age-matched non-AD subjects.[71] Interestingly, the same group also showed that mononuclear blood cells from AD patients express a higher amount of mutant-like p53.[72]

Conclusions and Future Directions

In order for a diagnostic biomarker to be useful, certain criteria need to be met. These criteria include the following (see Chapter 1 in this book, Ritsner, Gottesman):

1. The biomarker should reflect some basic pathophysiological processes, and detect a fundamental feature of the disease.
2. The biomarker should be specific for the disease compared with related disorders.
3. The biomarker should not reflect clinical symptomatology and consequences of the disease.
4. The biomarker can be measured repeatedly over time and should be reproducible.
5. The biomarker should be measured in noninvasive and easy-to-perform tests that can be done at the bedside or in the outpatient setting.
6. The biomarker should not cause harm to the individual being assessed.
7. The biomarker should be reliable in many testing environments/labs.
8. The biomarker should be cost effective.

Taken together, up-to-date only the analysis of beta-amyloid(1-42), total tau and phospho-tau-181 in CSF allows reliable, sensitive and specific diagnosis of AD in body fluids. Other forms of dementia cannot be diagnosed with these biomarkers so far and other CSF biomarkers do not add any novel information. Unfortunately, the use of CSF biomarkers is limited because of the invasive collection. Thus, the discovery of peripheral blood biomarkers has several advantages over CSF biomarkers.

So far a single peripheral blood "super-biomarker" for diagnosis of dementia has yet not been found, and possibly will not exist. It seems very likely that only the combination of several biomarkers will be successful: (1) either several biomarkers in plasma, as reported e.g. by Ray et al.[47] using 18 signaling proteins, or and more likely (2) the combination of some plasma biomarkers and some blood-cell-derived biomarkers, such as e.g. the differences in monocyte phenotype or different expression of proteins or DNA changes (such as e.g. telomere shortening) in PBMC. So far, some biomarkers are of interest, not yet fully established international, but some reflect the basic pathophysiological process of the disease (e.g. amyloid-precursor protein changes, inflammation, cerebrovascular damage, etc.). However, these biomarkers do not yet allow to diagnose different forms of dementia with high sensitivity and specificity. The major advantage of peripheral blood biomarkers is that blood can be easily collected from patients, allowing repeated measurements over time and screening of many patients, also of younger individuals. Such screening over years may allow to identify an "age-related biomarker", which may be stable over years and might change at the beginning of the disease. In addition, all methods (e.g. ELISA or Western Blots) are well established in several laboratories, allowing good reproducibility of the assays. However, the procedures of detection, blood handling, transport and stability of proteins needs to be tested and international standards need to be defined. Unfortunately, the analysis of several combined biomarkers will dramatically increase the costs for laboratory analysis. In conclusion early, fast and cheap diagnosis from body fluids will become extremely important in the future to diagnose different forms of dementia and to measure therapy improvements.

This work was supported by the Austrian Science Funds (L429-B05).

References

1. McKeel DW, Price JL, Miller JP, et al. Neuropathologic criteria for diagnosing Alzheimer disease in persons with pure dementia of Alzheimer type. J Neuropathol Exp Neurol 2004; 63: 1028–1037
2. Fradinger EA, Bitan G. En route to early diagnosis of Alzheimer's disease – are we there yet? Trends Biotechnol 2005; 23: 531–533
3. Desai AK, Grossberg GT. Diagnosis and treatment of Alzheimer's disease. Neurology 2005; 64: S34–S39
4. Blasko I, Lederer W, Oberbauer H, et al. Measurement of thirteen biological markers in CSF of patients with Alzheimer's disease and other dementias. Dement Geriatr Cogn Disord 2006; 21: 9–15
5. Andreasen N, Vanmechelen E, Vanderstichele H, et al. Cerebrospinal fluid levels of total-tau, phospho-tau and Abeta42 predicts development of Alzheimer's disease in patients with mild cognitive impairment. Acta Neurol Scand 2003; 107: 47–51
6. Buerger K, Teipel SJ, Zinkowski R, et al. Increased levels of CSF phosphorylated tau in apolipoprotein E epsilon4 carriers with mild cognitive impairment. Neurosci Lett 2005; 391: 48–50
7. Solfrizzi V, D'Introno A, Colacicci AM, et al. Circulating biomarkers of cognitive decline and dementia. Clin Chim Acta 2006; 364: 91–112
8. Hansson O, Zetterberg H, Buchhave P, et al. Association between CSF biomarkers and incipient Alzheimer's disease in patients with mild cognitive impairment: a follow up study. Lancet Neurol 2006; 5: 228–234
9. Grossman M, Farmer J, Leight S, et al. Cerebrospinal fluid profile in frontotemporal dementia and Alzheimer's disease. Ann Neurol 2005; 57: 721–729
10. Frey HJ, Mattila KM, Korolainen MA, Pirttilä T. Problems associated with biological markers of Alzheimer's disease. Neurochem Res 2005; 30: 1501–1510
11. Mollenhauer B, Bibl M, Trenkwalder C, et al. Follow-up investigations in cerebrospinal fluid of patients with dementia with Lewy bodies and Alzheimer's disease. J Neural Transm 2005; 112: 933–948
12. Noguchi M, Yoshita M, Matsumoto Y, Ono K, Iwasa K, Yamada M. Decreased beta-amyloid peptide42 in cerebrospinal fluid of patients with progressive supranuclear palsy and corticobasal degeneration. Neurol Sci 2005; 237: 61–65
13. Stefani A, Bernardini S, Panella M, et al. AD with subcortical white matter lesions and vascular dementia: CSF markers for differential diagnosis. J Neurolog Sci 2005; 237: 83–88
14. Vanderstichele, H., de Meyer, G., Andreasen, N., et al. Amino-trunctaed beta-amyloid42 peptides in cerebrospinal fluid and prediction of progression of mild cognitive impairment. Clin Chem 2005; 51: 1650–1660
15. Borroni B, Di Luca M, Padovani A. Predicting Alzheimer dementia in mild cognitive impairment patients. Are biomarkers useful? Eur J Pharmacol 2006; 545: 73–80
16. Petersen RC, Morris JC. Mild cognitive impairment as a clinical entity and treatment target. Arch Neurol 2005; 62: 1160–1163
17. O'Brien JT. Vascular cognitive impairment. Am J Geriatr Psychiat 2006; 14: 724–733
18. Janvin CC, Larsen JP, Aarsland D, Hugdahl K. Subtypes of mild cognitive impairment in Parkinson's disease: progression to dementia. Mov Disord 2006; 21: 1343–1349
19. Snowdon DA, Greiner LH, Mortimer JA, Riley KP, Greiner PA, Markesbery WR. Brain infarction and the clinical expression of Alzheimer disease. The Nun Study. JAMA 1997; 277: 813–817
20. Henley SMD, Bates GP, Tabrizi SJ. Bionmarkers for neurodegenerative diseases. Curr Opin Neurol 2005; 18: 698–705
21. Blennow K. CSF biomarkers for Alzheimer's disease: use in early diagnosis and evaluation of drug treatment. Expert Rev Mol Diagn 2005; 5: 661–672
22. Blennow K. CSF biomarkers for mild cognitive impairment. J Internal Med 2004; 256: 224–234
23. Sjögren M, Vanderstichele H, Agren H, et al. Tau and Ab42 in cerebrospinal fluid from healthy adults 21–93 years of age: establishment of reference values. Clin Chem 2001; 47: 1776–1781
24. Zetterberg H, Andreasen N, Blennow K. Increased cerebrospinal fluid levels of transforming growth factor-beta1 in Alzheimer's disease. Neurosci Lett 2004; 367: 194–196
25. Galasko D, Chang L, Motter R, et al. High cerebrospinal fluid tau and low amyloid beta42 levels in the clinical diagnosis of Alzheimer disease and relation to apolipoprotein E genotype. Arch Neurol 1998; 55: 937–945
26. Hampel H, Buerger K, Zinkowski R, et al. Measurement of phosphorylated tau epitopes in the differential diagnosis of Alzheimer disease. Arch Gen Psychiat 2004; 61: 95–102

27. Kurz A, Reimenschneider M, Drzezga A, Lautenschlager N. The role of biological markers in the early and differential diagnosis of Alzheimer's disease. J Neural Transm 2002; 62: 127–133

28. Maddalena A, Papassotiropoulos A, Müller-Tillmanns B, et al. Biochemical diagnosis of Alzheimer disease by measuring the cerebrospinal fluid ratio of phosphorylated tau protein to beta-amyloid peptide42. Arch Neurol 2003; 60: 1202–1206

29. Shoji M, Matsubara E, Kanai M, et al. Combination asay of CSF tau, amyloid-beta1-40 and amyloid-beta1-42(43) as a biochemical marker of Alzheimer's disease. J Neurol Sci 1998; 158: 134–140

30. Ibach N, Binder H, Dragon M, et al. Cerebrospinal fluid tau and beta-amyloid in Alzheimer patients, disease controls and an age-matched random sample. Neurobiol Aging 2006; 27: 1202–1211

31. Blasko I, Kemmler G, Krampla W, et al. Plasma amyloid beta protein 42 in non-demented persons aged 75 years: effects of concomitant medication and medial temporal lobe atrophy. Neurobiol Aging 2005; 26: 1135–1143

32. Humpel C, Blasko I, Marksteiner J, et al. Diagnostik der Alzheimer Demenz und anderer Demenzen in der Cerebrospinalflüssigkeit. Neuropsychiatrie 2005; 19: 97–101

33. Hulstaert F, Blennow K, Ivanoiu A, et al. Improved discrimination of AD patients using beta-amyloid (1-42) and tau levels in CSF. Neurology 1999; 52: 1555–1562

34. Rapoport M, Dawson HN, Binder LI, Vitek MP, Ferreira A. Tau is essential to beta-amyloid-induced neurotoxicity. Proc Natl Acad Sci USA 2002; 99: 6364–6369

35. Olsson A, Vanderstichele H, Andreasen N, et al. Simultaneous measurement of beta-amyloid(1-42), total tau, and phosphorylated tau (Thr181) in cerebrospinal fluid by the xMAP technology. Clin Chem 2005; 51: 336–345

36. Sjögren M, Andreasen N, Blennow K. Advances in the detection of Alzheimer's disease-use of cerebrospinal fluid biomarkers. Clin Chim Acta 2003; 332: 1–10

37. Tapiola T, Pirttilä T, Mikkonen M, et al. Three-year follow-up of cerebrospinal fluid tau, beta-amyloid 42 and 40 concentrations in Alzheimer's disease. Neurosci Lett 2000; 280: 119–122

38. Sunderland T, Mirza N, Putnam KT, et al. Cerebrospinal fluid beta-amyloid1-42 and tau in control subjects at risk for Alzheimer's disease: the effect of APOEe4 allele. Biol Psychiat 2004; 56: 670–676

39. Schoonenboom NS, Mulder C, van Kamp GJ, Metha SP, Scheltens P, Blankenstein MA, Metha PD. Amyloid beta 38,40 and 42 species in cerebrospinal fluid: more of the same? Ann Neurol 2005; 58: 139–142

40. Lewczuk P, Esselmann H, Bibl M, et al. Tau protein phosphorylated at Threonine 181 in CSF as a neurochemical biomarker in Alzheimer's disease. J Mol Neurosci 2004; 23: 115–122

41. Marksteiner J, Hinterhuber H, Humpel C. Cerebrospinal fluid biomarkers for diagnosis of Alzheimer's disease: beta-amyloid(1-42), tau, phospho-tau-181 and total protein. Drugs Today 2007; 43: 423–431

42. Hock C, Heese K, Müller-Spahn F, et al. Increased CSF levels of nerve growth factor in patients with Alzheimer's disease. Neurology 2000; 54: 2009–2011

43. Marksteiner J, Michael P, Celine U, et al. Analysis of cerebraspinal fluid of Alzheimer patients: biomarkers and toxic properties. Pharmacology 2008; 82:214–220

44. Lue LF, Ryde IR, Brigham EF, Yang LB, Hampel H, et al. Inflammatory repertoire of Alzheimer's disease and nondemented elderly microglia in vitro. Glia 2001; 35: 72–79

45. Tsuboi, Y, Kakimoto K, Nakajima M, et al. Increased hepatocyte growth factor level in cerebrospinal fluid in Alzheimer's disease. Acta Neurol Scand 2003; 107: 81–86

46. Zetterberg H, Wahlund L-O, Blennow K. Cerebrospinal fluid markers for prediction of Alzheimer's disease. Neurosci Lett 2003; 352: 67–69

47. Ray S, Britschgi M, Herbert C, et al. Classification and prediction of clinical Alzheimer's diagnosis based on plasma signaling proteins. Nat Med 2007; 13: 1359–1362

48. Leoni V, Shafaati M, Salomon A, et al. Are the CSF levels of 24 S-hydroxycholesterol a sensitive biomarker for mild cognitive impairment. Neurosci Lett 2006; 397: 83–87

49. Nielsen HM, Minthon L, Londos E, et al. Plasma and CSF serpins in Alzheimer disease and dementia with Lewy bodies. Neurology 2007; 69: 1569–1579

50. Assini A, Cammarata S, Vitali A, et al. Plasma levels of amyloid beta-protein42 are increased in women with mild cognitive impairment. Neurology 2004; 63: 828–831

51. Brettschneider S, Morgenthaler NG, Teipel SJ, et al. Decreased serum amyloid beta1-42 autoantibody levels in Alzheimer's disease, determined by a newly developed immuno-precipitation assay with radiolabeled amyloid beta1-42 peptide. Biol Psychiat 2005; 57: 813–816

52. Ingelson M, Blomberg M, Benedikz E, et al. Tau immunoreactivity detected in human plasma, but no obvious increase in dementia. Dement Geriatr Cogn Disord 1999; 10: 442–445

53. Malaguarnera L, Motta M, di Rosa M, Anzaldi M, Malaguernera M. Interleukin-18 and transforming growth factor-beta1 plasma levels in Alzheimer's disease and vascular dementia. Neuropathology 2006; 26: 307–312

54. Motta M, Imbesi R, Di Rosa M, Stivala F, Malaguarnera L. Altered plasma cytokine levels in Alzheimer's disease: correlation with the disease progression. Immunol Lett 2007; 114: 46–51

55. Kienzl E, Jelliinger K, Janetzky B, Steindl H, Bergmann J. A broader horizon of Alzheimer pathogenesis: ALZAS – an early serum biomarker? J Neural Transm 2002; 62: 87–95

56. Straface E, Matarrese P, Gambardella L, et al. Oxidative imbalance and cathepsin D changes as peripheral blood biomarkers of Alzheimer's disease: a pilot study. FEBS Lett 2005; 579: 2759–2766

57. Squitti R, Ventriglia M, Barbati G, et al. "Free" copper in serum of Alzheimer's disease patients correlates with markers of liver function. J Neural Transm 2007; 114: 1589–1594

58. Ban Y, Watanabe T, Miyazaki A, et al. Impact of increased plasma serotonin levels and carotid atherosclerosis on vascular dementia. Atherosclerosis 2007; 195: 153–159

59. Tang K, Hynan LS, Baskin F, Rosenberg RN. Platelet amyloid precursor protein processing: a bio-marker for Alzheimer's disease. J Neurol Sci 2006; 240: 53–58

60. Hye A, Kerr F, Archer N, et al. Glycogen synthase kinase-3 is increased in white cells early in Alzheimer's disease. Neurosci Lett 2005; 373: 1–4

61. Magaki S, Mueller C, Dickson C, Kirsch W. Increased production of inflammatory cytokines in mild cognitive impairment. Exp Gerontol 2006; 42: 233–240

62. Guerreiro RJ, Santana I, Bras JM, Santiago B, Paiva A, Oliveira C. Peripheral inflammatory cytokines as biomarkers in Alzheimer's disease and mild cognitive impairment. Neurodegenerative Dis 2007; 4: 406–412

63. Tan ZS, Beiser AS, Vasan RS, et al. Inflammatory markers and the risk of Alzheimer disease: the Framingham study. Neurology 2007; 68: 1902–1908

64. Grillari J, Katinger H, Voglaue R. Aging and the ubiquitin-ome: traditional and non-traditional functions of ubiqutin in aging cells and tissues. Exp Gerontol 2006; 41: 1067–1079

65. Wang GP, Khatoon S, Iqbal K, Grundke-Iqbal I. Brain ubiquitin is markedly elevated in Alzheimer disease. Brain Res 1991 Dec 6; 566(1–2): 146–151

66. Ponnappan U. Ubiquitin-proteasome pathway is compromised in CD45RO+ and CD45RA+ T lymphocyte subsets during aging. Exp Gerontol 2002; 37: 359

67. Kudo T, Iqbal K, Ravid R, Swaab DF, Grundke-Iqbal I. Alzheimer disease: correlation of cerebro-spinal fluid and brain ubiquitin levels. Brain Res 1994; 639: 1–7

68. Baird DM, Telomeres. Exp Gerontol 2006; 41: 1223–1227

69. Erusalimsky JD, Kurz DJ. Cellular senescence in vivo: its relevance in ageing and cardiovascular disease. Exp Gerontol 2005; 40: 634–642

70. von Zglinicki T, Serra V, Lorenz M, et al. Short telomeres in patients with vascular dementia: an indicator of low antioxidative capacity and a possible risk factor? Lab Invest 2000; 80: 1739–1747

71. Lanni C, Racchi M, Mazzini G, et al. Conformationally altered p53: a novel Alzheimer's disease marker? Mol Psychiat 2008 Jun;13(6):641–647

72. Lanni C, Uberti D, Racchi M, Govoni S, Memo M. Unfolded 53: a potential biomarker for Alzheimer disease. J Alzheimer's Dis 2007; 12: 93–99

Chapter 29

S100B as a Potential Neurochemical Biomarker in a Variety of Neurological, Neuropsychiatric and Neurosurgical Disorders

Patrick Wainwright, Jon Sen, and Antonio Belli

Abstract S100B is a calcium-binding protein found in Schwann cells in the peripheral nervous system, and in astrocytes in the central nervous system. It has multiple functions including the inhibition of protein phosphorylation through interacting with kinase substrates, regulating enzyme activity and interacting with cytoskeletal elements. It is also involved in calcium homeostasis, and is believed to have a role in cytosolic calcium buffering.

In recent years it has been shown that S100B is elevated and released into the circulation in a wide variety of neuropathologic states. As such it has generated a great deal of interest as a surrogate biomarker for injury to the CNS. It has been shown to be raised in many organic brain disorders such as traumatic brain injury, subarachnoid haemorrhage, stroke, epilepsy, multiple sclerosis, Parkinson's disease and hydrocephalus. In addition to this there is now clinical and laboratory evidence that it is raised in neuropsychiatric disorders such as schizophrenia, depression, bipolar disorder, anxiety, post-traumatic stress disorder and neuropsychiatric systemic lupus erythematosus.

S100B has great potential to become a specific neurological screening tool that is predictive of outcome and reactive to treatments.

Keywords S100B · biomarker · neurochemistry · neuropathology · brain injuries

P. Wainwright, J. Sen, and A. Belli
Clinical Neurosciences, School of Medicine,
University of Southampton

Abbreviations FTD: Fronto-temporal dementia; LDH: Lactate dehydrogenase; MCAO: Middle cerebral artery occlusion; NPSLE: Neuropsychiatric systemic lupus erythematosus; SAH: Sub-arachnoid haemorrhage; TBI: Traumatic brain injury

Introduction

Chemical Biomarkers

A biomarker can be defined as a measurable indicator of a specific biological state.[1] It can be relevant to the stage of a disease process, the presence of disease or the risk of contracting a disease. A biomarker can take many forms, most commonly they are plasma measurements of specific proteins or molecules. Effective biomarkers can have a great variety of clinical uses. They can be used as screening tools or to diagnose and monitor disease activity. There is also great potential for biomarkers to guide molecularly-targeted rational therapies and to monitor response to treatments.[2]

Chemical biomarkers can be measured in the plasma or other fluids more proximal to the site of disease. Proximal fluids such as cerebrospinal fluid (CSF) can act as sinks for proteins or molecules secreted or leaked from diseased tissue and can provide a more accurate indication of disease activity. Measurements of protein levels are made more easily from plasma however, indeed laboratory assays to measure over 100 different proteins are currently used in routine clinical practice.[3] For a potential biomarker to be clinically useful it must be sensitive, specific, accurate and reliable. It must also provide a high predictive value.

M.S. Ritsner (ed.), *The Handbook of Neuropsychiatric Biomarkers, Endophenotypes and Genes,*
© Springer Science+Business Media B.V. 2009

Brain Specific Proteins

Brain specific proteins initially provoked interest in 1965, when a group separated proteins from the brain and liver of a variety of animals and found three proteins at consistently high concentrations in the brain samples, but at very low concentrations or absent in liver samples.[4] One of these proteins was soluble in 100% ammonium sulphate and was named S100 as a result. Since these proteins were identified, many more structural CNS-specific proteins have been discovered. Such proteins have attracted much attention in monitoring neuronal reactions to brain pathology as many are thought to be released into the circulation during such states. It has been suggested that the astrocyte is the predominant regulator of homeostasis within the brain and that it is affected in a range of neuropathological states. As such, brain specific proteins that indicate astrocytic activation are of particular interest when looking for potential neurochemical markers of CNS disease.[5]

S100B: An Introduction

A decade and a half after its initial discovery S100 was shown to consist of two distinct proteins, S100B and S100A1.[6] S100B is a 10.4 kDa protein and exists as a homodimer in cells. All S100 proteins contain two calcium binding domains of the EF-hand type that are interconnected by an intermediate hinge region.[8] At least 80–90% of S100B is located within the brain, with the remainder being found outside of the CNS.[7] It is located in astrocytes in the CNS and also in Schwann cells in the peripheral nervous system, and has been shown to have a variety of intracellular and extracellular activities.

S100B inhibits protein phosphorylation by interacting with kinase substrates and has been shown to have a great many protein targets.[8] S100B is also a regulator of enzymes involved in cellular energy metabolism and regulation of the cell cycle. Importantly, S100B is a calcium-binding protein and is believed to have a role in modulating intracellular signal transduction resulting from changes in calcium concentration.[7] It is also thought to be involved in calcium homeostasis and may play a role in cytosolic calcium buffering. S100B is also known to interact with cytoskeletal elements by inhibiting the assembly of microtubules. It has this effect by sequestering tubulin and altering the calcium-sensitivity of preformed microtubules.[9]

S100B is released from astrocytes in the CNS, although very little is currently known about the mechanism of action. However, it has been demonstrated that S100B secretion differs between brain regions. Astrocytes isolated and cultured from the cerebellum released S100B at a higher rate that those from hippocampal and cortical regions.[18] The physiological significance of the secretion of S100B from glia is poorly understood, although it appears to exert both neurotoxic and neuroprotective effects depending on concentration. S100B can stimulate neurite outgrowth at nanomolar concentrations.[10] It can also enhance the survival of neurons after injury[11] and during development.[12] It has been reported that S100B can cause apoptosis,[13] suggesting a physiological role for secreted S100B as a neurotrophic factor that could be of importance during development and in response to injury. This hypothesis is further supported by the observation that S100B release is increased during reactive synaptogenesis and lesion-induced sprouting.[19] Nanomolar levels of S100B have also been shown to have proliferative effects on astrocytes, possibly by increasing astrocytic ERK1/2 phosphorylation.[14] Other groups have reported data which support the suggestion that S100B may protect the CNS after injury. For example, very low S100B concentrations have been shown to protect hippocampal neurons from excitotoxic brain injury in vitro by activating expression of the receptor of advanced glycation products (RAGE)[15], and it has also been shown to attenuate the neurotoxic effect of trimethyltin on microglial and astrocytes.[16]

There is now a wealth of data to show that S100B is elevated and released into the circulation in a large variety of acute and chronic CNS disorders. In more general terms, S100B is released from astrocytes at times of cerebral metabolic stress. One group showed that primary astrocyte cultures subjected to oxygen and glucose deprivation for varying periods of time released S100B in a dose-dependant fashion implying an active mechanism of S100B release triggered by metabolic stress.[17] In many instances it is not clear whether S100B release is a cause or effect of a specific condition, but it is strongly implicated and so S100B is considered to be a strong candidate for a surrogate chemical biomarker of CNS injury deserving of further investigation.

The Role of S100B as a Biomarker in Neurological Conditions

Multiple Sclerosis

In 1868 Charcot described three steps in the pathophysiology of the condition we now recognise as multiple sclerosis. The first step was an initial microgial and astrocytic activation, followed by axonal degeneration and finally astrogliosis. The majority of research attention has been focused on damage occurring to axons, with relatively little attention paid to astrocytic activation despite evidence that this can precede axonal degeneration.[20] During this period of astrocytic activation, brain-specific proteins such as S100B are released into the CSF. Petzold et al. measured the levels of brain-specific proteins in the CSF of patients with three different subtypes of multiple sclerosis.[21] In this study the authors included 51 patients, 20 with relapsing remitting MS, 21 with secondary progressive MS and ten with primary progressive MS. They measured CSF levels of brain-specific proteins S100B, ferritin and glial-fibrillary acidic protein. Levels of S100B were significantly raised in the CSF of all clinical subtypes of MS when compared with control patients who had various neurological conditions, and ferritin levels were higher in secondary progressive MS than in control samples. The ratio of S100B to ferritin in the CSF was able to distinguish relapsing remitting MS from the other subtypes and control patients. The authors used the nine hole peg test, an examination of timed fine motor control, and showed that higher S100B CSF levels correlated with poorer performance in this test. The authors also examined post-mortem brain tissue from 12 patients with MS and eight control patients and found significantly higher S100B levels in the acute than in the subacute plaques, as opposed to ferritin levels which were elevated equally in all stages of the MS plaque. As such, they were able to show that S100B is an effective marker of the relapsing phase of the disease. A more recent study looking at CSF S100B levels in patients with relapsing remitting MS showed that S100B levels are increased during the acute relapsing phase of disease, but failed to show a correlation between concentration of S100B and clinical characteristics of disease.[22]

These findings are significant and have implications for future work which could influence clinical practice. These data suggest that S100B or combinations of several brain-specific proteins could be of diagnostic use in distinguishing between clinical subtypes of MS. This could then affect the future therapeutic decision-making process to optimise clinical management. There is also a suggestion that S100B could be of use in identifying the acute relapsing phase of relapsing remitting MS. Whether this could be of use in predicting clinical deterioration is currently unclear, but with more work this observation could prove useful both prognostically and for directing clinical management.

Data are less convincing in primary progressive MS. A recent study showed that serum S100B levels were not affected by intramuscular administration of interferon-beta-1a and there was also no correlation between levels of S100B and clinical disability scores of MRI pathology.[23] It is believed that the pathology of primary progressive MS differs from that of other forms of MS in that there is less inflammatory activity. This would fit with the finding that S100B is not an effective surrogate biomarker for this subtype as the pathophysiology would lead to less astrocytic activation and subsequently less release of S100B into the circulation.

Acute Ischaemic Stroke

The identification and effective implementation of an easily performed test that is both highly sensitive and specific for acute myocardial infarction has had dramatic effects in improving the care of patients with acute coronary syndromes. Serum analysis of troponin is now a routine laboratory investigation performed in hospitals throughout the world in suspected myocardial infarction. There is currently much interest in identifying a potential chemical biomarker of acute cerebral ischaemia that could have a similarly profound impact.

Initial animal studies have yielded interesting results. A recent laboratory study used *in vitro* brain slice preparations from rats and induced ischaemia by subjecting them to oxygen and glucose deprivation.[24] One hour of ischaemia increased S100B release by more than 50%, and this increase was even greater when the tissue slices underwent reoxygenation. This increase in S100B release was noted to occur in the absence of lactate dehydrogenase (LDH) release. LDH is an effective measurement of neuronal cell death,[25] therefore it is likely that the rise in S100B represents a specific neuronal response to ischaemia. A more recent study looked at the correlation between serum S100B levels and both cerebral oedema formation and neurological outcome in transient middle cerebral artery occlusion (MCAO) in the rat.[26] They demonstrated that MCAO resulted in increased serum S100B levels that peaked 48 h post-insult. MCAO also resulted in the formation of cerebral oedema in the ipsilateral cerebral hemisphere 24 h post-injury, becoming maximal 72 h post-injury. Interestingly, serum S100B concentration correlated not only with the extent of cerebral oedema formation but also with cerebral infarct volume. The authors also recorded neurological outcome up to 168 h post-injury by assessing postural reflex, visual placement, tactile placement, proprioceptive placement and proprioceptive adduction. They observed that serum S100B concentration correlated significantly with neurological outcome 168 h post-injury. These animal studies suggest that S100B could be a specific biomarker of cerebral ischaemia as opposed to neuronal cell death, and could also be an effective biomarker for predicting cerebral oedema formation and neurological outcome.

There have also been a number of human studies performed. Wunderlich et al. investigated 32 patients who were admitted to hospital within 6 h of acute ischaemic stroke.[27] Serum levels of S100B and neuron specific enolase (NSE) were assessed hourly between 1 and 6 h and then at multiple time points up to 120 h after stroke onset. Neurovascular status was assessed by transcranial and extracranial duplex sonography on admission and throughout hospital stay, and infarct volume was assessed in all patients. They observed that patients who had normal flow velocities throughout the circle of Willis had significantly lower serum levels of S100B and NSE than those that had main stem or branch occlusions. Serum S100B concentration also positively correlated not only with final infarct volume but also with the severity of the neurological deficit. S100B levels from 6 h onwards were associated with functional outcome, with levels above 0.2 μ/L 48 h post-injury being most strongly predictive of neurological outcome and functional status 3 months later. A further

study corroborated these findings, demonstrating that higher serum levels of S100B 24 h after acute ischaemic stroke correlated positively with baseline neurological deficit, computed tomography findings and functional outcome.[28] This study also looked at the effect of fibrinolytic therapy on serum levels of S100B. Although fibrinolysis was associated with a better functional outcome there was no effect on serum S100B concentration.

These studies suggest that S100B could potentially be an effective surrogate biomarker after acute ischaemic stroke that can predict outcome and deficit severity. Serum S100B levels become elevated within the first 24 h after stroke and can be of great value in terms of prognostication and acute management. However, the role of S100B in patients receiving fibrinolytic therapy needs further work to be fully defined.

Parkinson's Disease

Parkinson's disease (PD) is a progressive neurodegenerative disorder caused by a selective loss of dopaminergic neurons in the substantia nigra pars compacta leading to decreased striatal dopamine levels. There is growing evidence the glial cells may participate in the progression of this neurodegeneration through their activation and subsequent production of various compounds that could be neurotoxic or neuroprotective.[29,30] Post-mortem examination in patients with PD has demonstrated that dopaminergic neuronal loss in the substantia nigra is associated with significant astrogliosis and populations of activated microglial cells.[31] There is also a suggestion that glial activation persists after the initial neurotoxic insult has disappeared, and it is the glial cells that are responsible for the perpetuation of neurodegeneration. With this in mind it seems plausible that a biomarker of glial activation could prove useful diagnostically, prognostically and in terms of monitoring disease progress. In spite of this, there has only been one major study aimed at testing this hypothesis. Schaf et al. examined serum S100B levels in PD patients and compared them with controls.[32] They observed no significant difference in serum levels of S100B between patients with PD and controls. The natural history of neurodegeneration in PD is not known, and so there may well be peaks of S100B that were missed in this study, or it could simply be that S100B is not diagnostically useful in PD. However, the authors observed that serum

S100B levels correlated positively with the Hoehn and Yahr scale and negatively with the Schwab and England Activities of Daily Living (ADL) scale. The Hoehn and Yahr scale is a system used for describing how the symptoms of PD progress with a higher score indicating a higher level of disability experienced by the patient,[33] while the ADL scale estimates the abilities of PD patients with a lower score indicating a lower level of functioning and more severe disease.[34] Therefore, higher serum S100B levels indicated more severe disease and poorer functioning in PD patients. This suggests that S100B could potentially be an effective marker of disease progression in PD. Further work could clarify this role and investigate whether S100B levels were reactive to new treatments and subsequent alterations in clinical status.

Alzheimer's Disease

Alzheimer's disease (AD) is a common progressive neurodegenerative disorder and is the single largest cause of dementia today. It has been thought for some time that glial activation may underlie the development and progression of the neuropathological changes seen in AD. This idea was first suggested in 1989 when Griffin et al. detected increased microglial activation and astrogliosis in post-mortem brain of patients with AD neuropathology.[35] They also observed a very large increase in microglial immunoreactivity to IL-1 and astrocytic immunoreactivity to S100B. Subsequently, elevated levels of S100B have been detected in the brains of patients with AD.[36,37] It has since been suggested that overexpression of IL-1, S100B and other cytokines may initially be beneficial in AD but when occurring chronically will result in chronic glial activation, progressive neurodegeneration and the development of Alzheimer's type pathology.[38]

The development of Alzheimer's type neuropathology runs a chronic course with a long sub-clinical period where the disease is progressing without any noticeable clinical deterioration. There is much interest in finding potential biomarkers for detecting patients at this stage, particularly when coupled with the observation that treatments tend to become less effective when there is a more severe level of impairment. Due to the possibility of glial involvement and previous studies indicating a direct role for S100B in the generation of Alzheimer's

type pathology, a number of studies have looked at S100B levels in AD patients. One study compared CSF S100B concentrations in 68 patients with varying stages of AD with 25 healthy older subjects.[39] When considering all the AD patients together, there was no difference in CSF levels of S100B between the two groups. However, the authors went on to divide the AD group into mild/moderate clinical dementia and advanced stage clinical dementia as inferred by the Clinical Dementia Rating Scale criteria. They observed that CSF S100B levels were higher in the mild/moderate clinical dementia group than in the severe group, suggesting a role for increased S100B expression in the initial formation of neuritic plaques. A later study had similar aims and also investigated CSF levels of S100B in patients with a diagnosis of AD and compared them to those from patients with fronto-temporal dementia (FTD) and control patients with non-inflammatory neurological disease.[40] The authors here observed that CSF S100B levels were significantly elevated in the AD group, and that these levels correlated with the amount of cerebral atrophy present on MRI scan. These correlations were not observed in the FTD group.

These findings from clinical studies are interesting and suggest potential for S100B as a biomarker of some use in AD. It is also clear that S100B plays a role itself in the pathogenesis of AD. However, due to conflicting findings from clinical studies there is much work that needs to be done to clarify this potential role as a biomarker in the diagnosis and staging of AD.

The Role of S100B as a Biomarker in Neuropsychiatric Conditions

Schizophrenia

There is growing evidence to support a role for the astrocyte in the pathogenesis of schizophrenia. Early reports from post-mortem studies suggested that astrogliosis occurred in the brains of patients with schizophrenia.[41,42] However, this is not now accepted as being a feature of schizophrenia. More recent studies paint a more complex picture of astrocytic abnormality occurring in the brains of schizophrenic patients. One group investigated the anterior cingulated cortex and showed that schizophrenic patients had a 15–20% glial cell loss with preservation of neuronal density.[43] There are also reports of reduced glial cell numbers in the orbitofrontal and primary motor cortices, and dystrophic changes in glial cells of the prefrontal cortex.[44] It has also been reported that schizophrenic brains have reduced astroglial function after injury.[45] However, these findings are not entirely consistent as other groups have investigated this and reported no change in glial cell numbers in various brain regions. As such, it is currently unclear the exact nature of the astrocytic changes that occur in schizophrenia, but it is likely that the changes are region-specific and vary with patient group and disease severity.

Several studies have investigated S100B levels in patients with schizophrenia. The first study of this kind compared plasma S100B levels from schizophrenic inpatients taking neuroleptic medication with those from age- and gender-matched controls.[46] The authors observed that plasma S100B levels were significantly higher in schizophrenic patients than in control subjects and also that those patients with residual symptomatology and long-term psychiatric problems tended to have the highest levels. The same group then went on to investigate 26 unmedicated patients with acute paranoid schizophrenia, and showed that their admission plasma S100B levels were significantly higher than matched controls.[47] This difference had disappeared after 6 weeks of treatment. Interestingly, negative symptoms correlated positively with plasma S100B concentration, and those patients with more negative symptoms on admission had a smaller decrease in plasma S100B following treatment. Another group were able to show that a selection of schizophrenic patients that were either medication-free, outpatients or inpatients again had higher serum S100B levels than matched controls.[48] It has been demonstrated more recently that patients in whom negative symptoms persist have high plasma S100B levels that do not fall after an acute episode.[49] This was observed in contrast to patients with a lesser preponderance of negative symptoms whose S100B levels returned to levels similar to healthy age-matched controls following treatment.

To summarize these findings, it appears that serum S100B levels are raised in acute schizophrenic psychoses and this measurement could be used in conjunction with clinical findings to aid in the diagnostic process. It also seems that those patients who develop negative symptoms that persist after their initial acute psychotic episode has resolved have persistently high serum S100B levels whereas those without negative symptoms have

S100B levels that return to baseline. It remains unclear whether or not serial S100B levels could be useful in providing information about disease progress or severity, or response to treatment and as such further work is required to clarify this.

Major Depression

It has been known for some time that a degree of volume loss occurs in different brain regions in patients with major depression. One study demonstrated that depressed patients had reductions in volumes of the caudate and putamen when compared with age- and gender-matched controls.[50] As such, it appears that structural damage to neurons occurs at some stage in the development of major depression and that this damage could be assessed using a surrogate marker of brain injury.

Rothermundt et al. looked at plasma S100B levels in patients diagnosed with major depression.[51] Twenty-eight patients were compared with 28 matched controls and their symptoms were graded using the Hamilton Depression Rating Scale (HDRS). Patients were also sub-divided into having either melancholic or non-melancholic depression. Melancholic depression is often thought of as the most biological form of depression, and is characterised by pronounced anhedonia, excessive feelings of guilt, psychomotor retardation, affective symptoms that are distinct from feelings of bereavement and vegetative symptoms. The authors observed that before subgroup categorization into either melancholic or non-melancholic forms, the depressed patients had significantly higher plasma S100B levels. After categorization they showed that patients with non-melancholic depression in fact had similar plasma S100B levels to controls, and patients with melancholic depression had far higher levels. They did not, however, appreciate any correlation between S100B concentration and severity of symptoms. A further study looked at S100B levels in the CSF of patients with a current diagnosis of a mild or moderate depressive episode.[52] The authors observed that the presence of a depressive episode was associated with elevated CSF levels of S100B.

A further group then went on to investigate the possible use of S100B in predicting response to treatment with anti-depressant therapy.[53] They examined plasma S100B concentrations in 25 patients with major depression as defined in DSM-IV in comparison to those in matched controls. Mean plasma S100B levels were higher in depressed patients, and once again those patients with melancholic depression had higher levels than those with non-melancholic depression. Patients then underwent anti-depressant therapy for 4 weeks. Depressive symptoms experienced by the patients significantly were significantly reduced by this time. The relative response of patients to therapy in as assessed by their symptomatology was positively correlated with initial plasma S100B concentration. Those patients with higher initial S100B levels went on to experience a greater response to treatment. This positive correlation between initial S100B levels and magnitude of treatment response was more pronounced in patients with melancholic depression and, although present in those with non-melancholic depression, failed to reach statistical significance in this group. This is an interesting finding that at first glance appears somewhat unintuitive. As S100B is a marker of glial cell activation and general damage to the brain, one might expect higher levels to be associated with a poorer prognosis as this is what is seen in traumatic conditions. However, in such traumatic conditions S100B rise can be up to 100-fold, whereas the rises observed here and in other neuropsychiatric conditions are much lower, in the region of threefold. S100B can exert neuroprotective effects at such low concentrations[8] and is also known to act as a neurotrophic factor and neurite outgrowth promoter in serotonergic neurons and so could potentially be involved in the therapeutic response. These findings were replicated by another group, who also investigated visual evoked event-related potentials.[54] They observed similar findings in terms of S100B levels predicting treatment response, and also showed that only those patients with elevated S100B levels prior to treatment experienced improvements in attention and memory after anti-depressant treatment, as assessed through visual evoked event-related potentials.

A recent study addressed the question of whether or not increased S100B levels are only increased during the depressive episode or if they remain elevated during periods of remission.[55] The authors again used visual evoked event-related potentials to investigate attention and memory defects in 12 patients diagnosed with a recurrent major depressive disorder that was currently remitted for at least 3 months. They observed that six of these patients had elevated serum S100B levels by approximately three times, and the other six patients

had slightly raised levels but were not significantly different from matched controls. Interestingly, the authors observed that those patients who had significantly raised serum S100B concentrations had visual evoked event-related potentials in the normal range. This was in contrast to those patients with normal serum S100B levels who had marked attention deficits. These results appear to support previous findings in that depressed patients with raised serum S100B concentrations appear to have a more favourable prognosis in terms of treatment response and functioning.

It is clear that studies with higher patient numbers are required to further investigate the potential for S100B as a surrogate biomarker to predict treatment response and prognosis in major depression.

Bipolar Disorder

Bipolar disorder is a chronic psychiatric disorder with an estimated lifetime prevalence of 1.2%[56] and is characterised by instability of mood with recurrent episodes of depression and mania. As with major depression, there is growing evidence for an abnormality of glial cells in the pathogenesis of bipolar disorder. Initially, one group observed a marked reduction in glial number in the subgenual prefrontal cortex with a preservation of neuronal size and number in patients with bipolar disorder.[57] Other groups have subsequently reported similar findings. In a recent study Brauch et al. investigated glial cell number in post-mortem temporal cortical brain samples from patients with bipolar disorder.[58] The authors observed that the area occupied by glial cells was markedly reduced in samples from bipolar patients, as was the ratio of glial area to neuronal area. This reduction in size occupied by glial cells could represent glial dysfunction in the brains of these patients.

Machado-Viera et al. went on to investigate serum levels of S100B in unmedicated patients.[59] Serum S100B concentrations in 20 drug-free patients currently undergoing their first episode of mania were investigated and compared to healthy matched controls. The authors observed that average S100B levels were raised approximately threefold. This smaller increase in S100B concentration seen in bipolar patients is similar to that seen in depression and could potentially represent ongoing cerebral injury or aberrant cerebral functioning. The same group then investigated S100B in a rat model of mania using ouabain. Ouabain is a sodium-potassium ATPase inhibitor and has been shown to mimic the abnormalities observed during an acute manic episode.[60] Lithium, a mood stabilizer, has been shown to protect against the neurotoxic effects of ouabain administration to a certain extent.[61] The authors administered ouabain or a saline vehicle into the right lateral ventricle of rats and measured their behavioural response CSF levels of S100B as compared with baseline.[62] They observed an increase in activity as expected with this model of mania in those rats administered ouabain compared with controls and also a 30% increase in CSF levels of S100B. More recently, another group has investigated serum S100B levels in bipolar patients.[63] They investigated how serum S100B concentration varies in bipolar patients that are currently manic, depressed or euthymic. Patients that were euthymic had S100B levels similar to controls, while patients currently in a manic or depressive episode had elevated levels.

These data suggest that serum S100B is raised in the acute phase of bipolar disorder, whether that phase is depressive or manic, and remains at baseline during euthymic periods. The studies performed so far have only contained small amounts of patients, and larger studies are needed before any definite conclusions can be made. It would be useful to determine whether or not rises in S100B pre-empted clinical deterioration from euthymia and if so could management be optimised accordingly. However these preliminary data indicate that serum S100B levels could potentially become a useful adjunct in monitoring the patient with bipolar disorder.

Neuropsychiatric Systemic Lupus Erythematosus

Systemic lupus erythematosus (SLE) is a multi-system autoimmune disease that can frequently involve the nervous system. Neuropsychiatric SLE (NPSLE) can occur in up to 80% of patients with SLE.[64] NPSLE can have a variety of clinical presentations that can involve both the central and peripheral nervous system. Psychiatric symptoms can vary from mild depression to more severe disturbances. Neurological symptoms are also very varied and can include seizures, aseptic meningitis, cerebellar ataxia, cranial nerve lesions, migraines, cerebrovascular disease or polyneuropathy.

The pathogenesis of this cerebral involvement is poorly understood. It has been suggested that cerebral vasculopathy could occur due to immune complex deposition or that thrombosis occurs associated with antiphospholipid antibodies.[65]

NPSLE has proven difficult to diagnose due to the diversity of clinical syndromes and the frequent co-existence of other conditions. There are currently no specific markers or diagnostic tests that can reliably identify the existence of NPSLE or effectively monitor disease activity. As such, there is currently great interest in finding an effective biomarker to diagnose and monitor progression of disease in NPSLE. It is known that S100B levels are raised generally amongst SLE patients compared to controls, including those without nervous system involvement.[66] The group that observed this increase went on to look specifically at NPSLE. They were able to demonstrate that SLE patients with defined neuropsychiatric manifestations had higher serum levels of S100B than SLE patients with no involvement.[67] The authors also observed an association between higher serum S100B levels and the identification of anti-dsDNA antibodies. A further group investigated both serum and CSF S100B concentrations in NPSLE.[68] They observed higher S100B levels in serum and CSF compared to non-NPSLE patients, much like the previous study. The authors went on to assess S100B levels in different manifestations of NPSLE. They separated patients into an organic brain syndrome group, seizure group, cerebrovascular disease group, headache group, psychosis group and neuropathy group. The authors observed higher S100B levels in the organic brain syndrome, seizure, cerebrovascular disease and psychosis groups. S100B levels in the headache and neuropathy subgroups were not only lower than those in the other groups, but were not significantly higher than levels in non-NPSLE patients.

Increased S100B levels in patients with NPSLE could be potentially due to the presence of peri-vascular gliosis. Additionally alterations of the blood-brain barrier and antibodies directed against cerebral tissue could account for the observed differences. The two studies described above suggest that S100B could possibly be a complementary biochemical marker for the evaluation of NPSLE. However both studies were relatively small and further study is undoubtedly required to clarify the role of S100B in NPSLE, but it is clear that there is great potential to aid with the diagnosis of this difficult to diagnose complication.

The Role of S100B as a Biomarker in Neurosurgical Conditions

Traumatic Brain Injury

Traumatic brain injury (TBI) is a leading cause of death and disability amongst young adults.[69] When the brain experiences a traumatic insult, commonly due to a road traffic accident or a fall, astrocytes become activated and a reactive astrogliosis ensues.[70] This process leads to an upregulation of various glial markers, including S100B.[71] This response can be measured in the serum or CSF and as such S100B has become an established neurochemical biomarker of brain damage following TBI. Over the last decade many studies have looked at serum S100B levels in patients with TBI,[72–90] as well as CSF levels.[91,92] This large body of data demonstrates that serum levels of S100B tend to peak 1–2 days after injury, and then fall. Serum S100B levels correlate positively with severity of injury, and importantly can predict both short- and long-term neurological outcome, with higher values predicting a poorer outcome. There is also a positive correlation between serum S100B levels and volume of brain damage as assessed by computed tomography or magnetic resonance imaging. Recently it has been shown that early serum S100B levels are predictive of cognitive neuropsychological deficits in the long-term.[93]

Current management of acute TBI is reactive, in that treatments are instigated after an adverse clinical event, or secondary brain injury, has already occurred. A more preferable strategy would be to prevent such events from occurring. This would require a biomarker with predictive value, such as one able to predict acute increases in intracranial pressure. The studies described above suggest that S100B could be used in the neuromonitoring of patients with traumatic brain injuries. However, more recently it has become apparent that serum S100B levels can become significantly raised in trauma patients in the absence of cerebral injury.[94,95] These studies have shown that traumatically injured patients with acute fractures or burns injuries can have raised serum S100B concentrations even when the CNS is uninjured. Given that an isolated closed-head injury is a somewhat rare occurrence, this information casts doubt on the specificity of increased serum S100B levels as a marker of cerebral damage.

A more specific approach would be continuous monitoring of S100B levels in the extracellular fluid of the brain itself. This can be achieved by employing cerebral microdialysis techniques. Using this approach, S100B levels in the brain parenchyma can be continuously monitored to provide a constantly evolving picture that could potentially react to treatments as well as secondary cerebral insults. A temporal profile of extracellular activity can be built up during the various phases of injury. One small study used this method and the authors observed that peaks in extracellular levels of S100B were related to peaks in intracranial pressure.[96] This study had a sample size of two, and as such further work is required to fully elucidate the potential role of extracellular S100B in this setting. However, if the findings of this preliminary study can be replicated in larger trials, then the monitoring of extracellular S100B could assist in the management of patients with severe brain injuries.

Subarachnoid Haemorrhage

Subarachnoid haemorrhage (SAH) accounts for 5% of all strokes.[97] Within 1 month of the event half of the patients will have died and 40% of the survivors will be highly dependant.[98] There is much interest currently in developing tools to aid the predicting of clinical outcome in patients with SAH, and in particular developing a method of predicting episodes of potentially devastating cerebral vasospasm. It has been suggested that serum or CSF S100B could potentially act as a surrogate chemical biomarker for the severity of brain damage in patients with SAH.

A recent study published in 2006 investigated 74 patients admitted within 48 h of SAH who went on to have either surgical clipping or endovascular coiling within 2 days of admission.[99] The authors observed that serum S100B concentration correlates well with the initial severity of injury as assessed both clinically and radiologically. They also showed that the mean value of serial serum S100B levels taken during the first 8 days of admission was a strong independent predictive factor for 6 month outcome as assessed by the Glasgow outcome scale (GOS). They also noted that patients with aneurysms of the middle cerebral artery had higher mean S100B levels during the 8 day period which is consistent with data showing that aneurysms of this location are associated with a poorer prognosis.[100] However, this study reported no significant correlation between peaks in S100B levels and episodes of vasospasm although it is possible that there was not enough statistical power in the study to detect such a correlation.

Another group investigated 51 patients who underwent endovascular coiling within 4 days of SAH.[101] Consistent with the previously described study, they reported that the mean of daily S100B levels during the first 8 days of admission was a strong predictive factor for poor outcome at 12 months. They speculated that persistently high S100B values during this period could represent a higher severity of initial injury as well as complications of coiling and brain damage associated with raised intracranial pressure.

On the basis of these small studies it would seem that daily monitoring of serum S100B in the acute phase is a useful tool in predicting outcome in patients with SAH. It remains to be seen whether or not serum S100B is able to predict acute deteriorations in clinical condition as a result of cerebral vasospasm. As with traumatic brain injuries, it is possible that monitoring intraparenchymal levels of S100B using cerebral microdialysis techniques in neurological intensive care units to provide a constantly evolving biochemical picture may be a more appropriate way of achieving that in such a way that pre-emptive treatment could be initiated. However it is clear that much work needs to be done and many questions need to be answered before this could become a part of routine clinical practice.

Conclusion

It has been shown that the brain-specific protein S100B has potential as a surrogate biomarker of disease in a variety of different conditions involving the CNS.

It appears to be of most value in those conditions where the activation of astrocytes plays a prominent role in the pathogenesis. As such, it shows much promise in MS where it could either be used in isolation or as part of a panel of biomarkers in conjunction with ferritin and glial-fibrillary acidic protein. In this context it could be used to distinguish between disease subtypes before it is clinically obvious. It could also be used to biochemically identify an acute relapse of relapsing remitting MS.

S100B as a surrogate biomarker also shows particular promise in acute ischaemic stroke as it has been shown to accurately correlate with clinical outcome and severity of neurological deficit. There is a great clinical need for such a biomarker in this context and S100B could potentially fill this role either alone or in combination with neuron-specific enolase as part of a panel of biomarkers.

The evidence for using S100B in acute schizophrenic psychosis is also particularly compelling. Here it could be of considerable use in guiding early diagnosis and influencing treatment options. There also appears to be a correlation between S100B levels and negative symptoms which could prove useful clinically.

The role of S100B has also been well characterised in traumatic brain injury as an effective prognostic marker. S100B could be of most use here when continuously monitored in the extracellular fluid of the brain using cerebral microdialysis techniques on neurosurgical intensive care units.

This being said, there is clearly much work still to be done to fully characterise the place of S100B as a surrogate biomarker in the above conditions and all of the ones mentioned in this chapter. This additional work would be very worthwhile as data so far suggest that using S100B as a biomarker could potentially aid diagnostic and management decisions in a wide variety of neurological, neuropsychiatric and neurosurgical conditions.

References

1. Rifai N, Gillette M A and Carr S A. Protein biomarker discovery and validation: the long and uncertain path to clinical utility. Nat Biotechnol 2006;24:971–983
2. Etzioni R et al. The case for early detection. Nat Rev Cancer 2003;3:243–252
3. Burtis C A, Ashwood E R and Bruns D E (eds.). Tietz Textbook of Clinical Chemistry. Philadelphia, PA: Elsevier Saunders, 2005
4. Moore B W. A soluble protein characteristic of the nervous system. Biochem Biophys Res Commun 1965;19:739–744
5. Petzold A, Kier G, Lim D, Smith M and Thompson E J. Cerebrospinal fluid (CSF) and serum S100B: release and wash-out pattern. Brain Res Bull 2003;61:281–285
6. Cocchia D, Michetti R and Donato R. S100B antigen in normal human skin. Nature 1981;294:85–87
7. Sen J and Belli A. S100B in neuropathological states: the CRP of the brain? J Neurosci Res 2007;85:1373–1380
8. Donato R. S100: a multigenic family of calcium-modulated proteins of the EF-hand type with intracellular and extracellular functional roles. Int J Biochem Cell Biol 2001;33:637–668

9. Sorci A L, Agneletti R and Bianchi R. Association of S100B with intermediate filaments and microtubules in glial cells. Biochem Biophys Acta 1998;1448:277–289

10. Nishiyama H, Takemura M, Takeda T et al. Normal development of serotonergic neurons in mice lacking S100B. Neurosci Lett 2002;321:49–52

11. Barger S W, Van Eldick L J and Mattson M P. S100B protects hippocampal neurons from damage induced by glucose deprivation. Brain Res 1995;677:167–170

12. Van Eldick L J, Christie-Pope L M, Bolin E M et al. Neurotrophic activity of S100B in cultured dorsal root ganglia from embryonic chick and fetal rat. Brain Res 1991;542:280–285

13. Sorci G, Riuzzi F, Agneletti A L et al. S100B causes apoptosis in a myoblast cell line in a RAGE-independent manner. J Cell Physiol 2004;199:274–283

14. Goncalves D S, Lenz G and Karl J. Extracellular S100B modulates ERK in astrocyte cultures. Neuroreport 2000;11:807–809

15. Kogel D et al. S100B potently activates p65/c-Rel transcriptional complexes in hippocampal neurons: clinical implications for the role of S100B in excitotoxic brain injury. Neuroscience 2004;127:913–920

16. Reali C et al. S100B counteracts effects of the neurotoxicant trimethyltin on astrocytes and microglial. J Neurosci Res 2005;81:677–686

17. Gerlach R et al. Active secretion of S100B from astrocytes during metabolic stress. Neuroscience 2006;141:1697–1701

18. Pinto S S et al. Immunocontent and secretion of S100B in astrocyte cultures from different brain regions in relation to morphology. FEBS Lett 2000;486:203–207

19. McAdory B S, Van Eldick L J and Norden J J. S100B, a neurotrophic protein that modulates neuronal protein phosphorylation, is upregulated during lesion-induced collateral sprouting and reactive synaptogenesis. Brain Res 1998;813:211–217

20. Griffiths I et al. Axonal swellings and degeneration in mice lacking the major proteolipid of myelin. Science 1998;280:1610–1613

21. Petzgold A et al. Markers for different glial cell responses in multiple sclerosis: clinical and pathological correlations. Brain 2002;125:1462–1473

22. Rejdak K et al. Astrocytic activation in relation to inflammatory markers during clinical exacerbation of relapsing-remitting multiple sclerosis. J Neural Transm 2007;114:1011–1015

23. Lim E T et al. Serum S100B in primary progressive multiple sclerosis patients treated with interferon-beta-1a. J Negat Results Biomed 2004;3:4

24. Buyukuysal R L. Protein S100B release from rat brain slices during and after ischaemia: comparison with lactate dehydrogenase leakage. Neurochem Int 2005;47:580–588

25. Koh J Y and Choi D W. Quantitative determination of glutamate-mediated cortical neuronal injury in cell culture by lactate dehydrogenase efflux assay. J Neurosci Methods 1987;20:83–90

26. Tanaka Y, Koizumi C, Marumo T et al. Serum S100B indicates brain edema formation and predicts long-term neurological outcomes in rat transient middle cerebral artery occlusion model. Brain Res 2007;1137:140–145

27. Wunderlich M T, Wallesch C W and Goertler M. Release of neurobiochemical markers of brain damage is related to the neurovascular status on admission and the site of

arterial occlusion in acute ischaemic stroke. J Neurol Sci 2004;227:49–53

28. Jauch E C et al. Association of serial biochemical markers with acute ischemic stroke: the national institute of neurological disorders and stroke recombinant tissue plasminogen activator stroke study. Stroke 2006;37:2508–2513

29. Hirsch E C, Hunot S, Damier P and Faucheux B. Glial cells and inflammation in Parkinson's disease: a role in neurodegeneration? Ann Neurol 1998;44:115–120

30. Hirsch E C. Glial cells and Parkinson's disease. J Neurol 2000;247:58–62

31. Hirsch E C et al. The role of glial reaction and inflammation in Parkinson's disease. Ann N Y Acad Sci 2003;991:214–228

32. Schaf D V et al. S100B and NSE serum levels in patients with Parkinson's disease. Parkinsonism Relat Disord 2005;11:39–43

33. Hoehn M M and Yahr M D. Parkinsonism: onset, progression and mortality. Neurology 1967;17:427–442

34. Schwab R S and England A C. Projection technique for evaluating surgery in Parkinson's disease. In: Gillingham F J and Donaldson M C (eds.), Third Symposium on Parkinson's Disease. Edinburgh: Livingstone, 1969:152–157

35. Griffin W S T et al. Brain interleukin 1 and S-100 immunoreactivity are elevated in Down syndrome and Alzheimer disease. Proc Natl Acad Sci USA 1989;86:7611–7615

36. Marshak D R, Pesce S A, Stanley L C and Griffin W S T. Increased S100β neurotrophic activity in Alzheimer disease temporal lobe. Neurobiol Aging 1991;13:1–7

37. Sheng J G, Mrak R E and Griffin W S T. S100β protein expression in Alzheimer's disease: potential role in the pathogenesis of neuritic plaques. J Neurosci Res 1994;39:398–404

38. Mrak R E and Griffin W S T. Glia and their cytokines in progression of neurodegeneration. Neurobiol Aging 2005; 26:349–354

39. Peskind E R et al. Cerebrospinal fluid S100B is elevated in the earlier stages of Alzheimer's disease. Neurochem Int 2001;39:409–413

40. Petzgold A et al. Cerebrospinal fluid S100B correlates with brain atrophy in Alzheimer's disease. Neurosci Lett 2003;336:167–170

41. Bruton C J et al. Schizophrenia and the brain: a prospective cliniconeuropathological study. Psychol Med 1990;20:285–304

42. Stevens C D, Altshuler L L, Bogerts B et al. Quantitative study of gliosis in schizophrenia and Huntingdon's chorea. Biol Psychiat 1988;24:697–700

43. Cotter D et al. Reduced glial cell density and neuronal volume in major depression in the anterior cingulated cortex. Arch Gen Psychiat 2001;58:545–553

44. Cotter D R, Pariante C M and Everall I P. Glial cell abnormalities in major psychiatric disorders: the evidence and implications. Brain Res Bull 2001;55:585–595

45. Niizato K, Iritani S, Ikeda K et al. Astroglial function of schizophrenic brain: a study using a lobotomized brain. Neuroreport 2001;12:1457–1460

46. Wiesmann M et al. Elevated plasma levels of S-100b protein in schizophrenic patients. Biol Psychiat 1999;45:1508–1511

47. Rothermundt M et al. Increased S100B blood levels in unmedicated and treated schizophrenic patients are correlated with negative symptomatology. Mol Psychiat 2001; 6:445–449

48. Lara D R et al. Increased serum S100B protein in schizo-phrenia: a study in medication-free patients. J Psychiat Res 2001;35:11–14

49. Rothermundt M et al. S100B serum levels and long-term improvement of negative symptoms in patients with schizo-phrenia. Neuropsychopharmacology 2004;29:1004–1011

50. Parashos I A, Tupler L A, Blitchington T et al. Magnetic-resonance morphometry in patients with major depression. Psychiat Res 1998;84:7–15

51. Rothermundt M et al. S-100B is increased in melancholic but not in non-melancholic depression. J Affect Disord 2001;66:89–93

52. Grabe H J et al. Neurotrophic factor s100beta in major depression. Neuropsychobiology 2001;44:88–90

53. Arolt V et al. S100B and response to treatment in major depression: a pilot study. Eur Neuropsychopharmacol 2003;13:235–239

54. Hetzel G et al. The astroglial protein S100B and visually evoked event-related potentials before and after antidepres-sant treatment. Psychopharmacology 2005;178:161–166

55. Dietrich D E et al. Target evaluation processing and serum levels of nerve tissue protein S100B in patients with remit-ted major depression. Neurosci Lett 2004;354:69–73

56. Rush A J. Toward an understanding of bipolar disorder and its origin. J Clin Psychiat 2003;64:4–8

57. Ongur D, Drevets W C and Price J L. Glial reduction in the subgenual prefrontal cortex in mood disorders. Proc Natl Acad Sci USA 1998;95:13290–13295

58. Brauch R A et al. Glial cell number and neuron/glial cell ratios in post-mortem brains of bipolar individuals. J Affect Disord 2006;91:87–90

59. Machado-Viera R et al. Elevated serum S100B protein in drug-free bipolar patients during first manic episode: a pilot study. Eur Neuropsychopharmacol 2002;12:269–272

60. El-Mallakh R S et al. Intraventricular administration of ouabain as a model of mania in rats. Bipolar Disord 2003;5:362–365

61. Hennion J P, El-Masri M A, Huff M O et al. Evaluation of neuroprotection by lithium and valproic acid against oua-bain-induced cell damage. Bipolar Disord 2003;4:201–206

62. Machado-Vieira R. Increased cerebrovascular fluid levels of S100B protein in rat model of mania induced by ouabain. Life Sci 2004;76:805–811

63. Andreazza A C et al. Serum S100B and antioxidant enzymes in bipolar patients. J Psychiat Res 2007;41:523–529

64. Brey R L et al. Neuropsychiatric syndromes in lupus: prev-alence using standardized definitions. Neurology 2002;58:1214–1220

65. Bruyn G A. Controversies in lupus: nervous system involve-ment. Ann Rheum Dis 1995;54:159–167

66. Portela L V et al. Serum S100B levels in patients with lupus erythematosus: preliminary observation. Clin Diagn Lab Immunol 2002;9:164–166

67. Schenatto C B et al. Raised serum S100B protein levels in neu-ropsychiatric lupus. Ann Rheum Disord 2006;65:829–831

68. Yang X Y, Lin J, Lu X Y and Zhao X Y. Expression of S100B protein levels in serum and cerebrospinal fluid with different forms of neuropsychiatric systemic lupus erythe-matosus. Clin Rheumatol 2008;27:353–357

69. Fleminger S and Ponsford J. Long term outcome after traumatic brain injury. BMJ 2005;331:1419–1420

70. Eng L and Ghirnikar R. GFAP and astrogliosis. Brain Pathol 1994;4:229–237

71. Ghirnikar R S, Lee Y L and Eng L F. Inflammation in trau-matic brain injury: role of cytokines and chemokines. Neurochem Res 1998;23:329–340

72. McKeating E G, Andres P J and Mascia L. Relationship of neuron specific enolase and protein S-100 concentrations in systemic and jugular venous serum to injury severity and outcome after traumatic brain injury. Acta Neurochir Suppl 1998;71:117–119

73. Raabe A et al. Correlation of computed tomography find-ings and serum brain damage markers following severe head injury. Acta Neurochir (Wien) 1998;140:789–792

74. Rothoerl R D et al. S-100 serum levels after minor and major head injury. J Trauma 1998;45:765–767.

75. Ingebrigsten T et al. Traumatic brain injury in minor head injury: relation of serum S-100 protein measurements to magnetic resonance imaging and neurobehavioral outcome. Neurosurgery 1999;45:468–476

76. Raabe et al., Serum S-100B protein in severe head injury. Neurosurgery 1999;45:477–483

77. Woertgen C, Rothoerl R D, Metz C and Brawanski A. Comparison of clinical, radiologic and serum marker as prognostic factors after severe head injury. J Trauma 1999;47:1126–1130

78. Biberhaler P et al. Influence of alcohol exposure on serum S-100b levels. Acta Neurochir Suppl 2000;76:177–179

79. Elting J W et al. Comparison of serum S-100 protein levels following stroke and traumatic brain injury. J Neurol Sci 2000;181:104–110

80. Herrmann M et al. Release of glial tissue-specific proteins after acute stroke. A comparative analysis of serum con-centrations of protein S-100B and glial fibrillary acidic pro-tein. Stroke 2000;31:2670–2677

81. Ingebrigsten T et al. The clinical value of serum S-100 pro-tein measurements in minor head injury: a Scandinavian multicentre study. Brain Inj 2000;14:1047–1055

82. Jackson R G et al. The early fall in levels of S-100β in traumatic brain injury. Clin Chem Lab Med 2000;38:1165–1167

83. Mussack T et al. S-100b as a screening marker of the severity of minor head trauma (MHT): a pilot study. Acta Neurochir 2000;76:393–396

84. Otto M et al. Boxing and running lead to a rise in serum levels of S-100B protein. Int J Sports Med 2000;21: 551–555

85. Raabe A and Seifert V. Protein S-100B as a serum marker of brain damage in severe head injury: preliminary results. Neurosurg Rev 2000;23:136–138

86. Romner B, Ingebrigsten T, Kongstad P and Borgesen S E. Traumatic brain damage: serum S-100 protein measure-ments related to neuroradiological findings. J Neurotrauma 2000;17:641–647

87. Rothoerl R D, Woertgen C and Brawanski A. S-100 serum levels and outcome after severe head injury. Acta Neurochir Suppl 2000;76:97–100

88. Biberthaler P et al. Elevated serum levels of S-100B reflect the extent of brain injury in alcohol intoxicated patients after mild head trauma. Shock 2001;16:97–101

89. Herrmann M et al. Release of biochemical markers of damage to neuronal and glial brain tissue is associated with short and long term neuropsychological outcome

after traumatic brain injury. J Neurol Neurosurg Psychiat 2001;70:95–100

90. Nylen K et al. Serum levels of S100B, S100A1B and S100BB are all related to outcome after severe traumatic brain injury. Acta Neurochir (Wien) 2008;150:221–227

91. Pleines U E et al. S-100β reflects the extent of injury and outcome, whereas neuronal specific enolase is a better indicator of neuroinflammation in patients with severe traumatic brain injury. J Neurotrauma 2001;18:491–498

92. Berger R P et al. Neuron-specific enolase and S100B in cerebrospinal fluid after severe traumatic brain injury in infants and children. Pediatrics 2002;109:31

93. Watt S E et al. Protein S-100 and neuropsychological functioning following severe traumatic brain injury. Brain Inj 2006;20:1007–1017

94. Anderson R E et al. High serum S100B levels in paediatric patients undergoing corrective cardiac surgery with or without total circulatory arrest. Eur J Cardiothorac Surg 1999;16:32–37

95. Unden J et al. Raised serum S100B levels after acute bone fractures without cerebral injury. J Trauma 2005;58:59–61

96. Sen J et al. Extracellular fluid S100B in the injured brain: a future surrogate marker of acute brain injury? Acta Neurochir (Wien) 2005;147:897–900

97. Epidemiology of aneurysmal subarachnoid hemorrhage in Australia and New Zealand: incidence and case fatality from the Australasian Cooperative Research on Subarachnoid Hemorrhage Study (ACROSS). Stroke 2000;31:1843–850

98. Hop J W, Rinkel G J Algra A and van Gijn J. Case-fatality rates and functional outcome after subarachnoid hemorrhage: a systematic review. Stroke 1997;28:660–664

99. Weiss N et al. Prognosis value of plasma S100B protein levels after subarachnoid aneurysmal hemorrhage. Anesthesiology 2006;104:658–666

100. Kopera M, Majchrzak H and Kaspera W. Prognostic factors in patients with intracerebral hematoma caused by ruptured middle cerebral artery aneurysm. Neurol Neurochir Pol 1999;33:389–401

101. Pereira A R et al. Predictors of 1-year outcome after coiling for poor-grade subarachnoid aneurysmal hemorrhage. Neurocrit Care 2007;7:18–26

Chapter 30
Can the Cortisol to DHEA Molar Ratio be Used as a Peripheral Biomarker for Schizophrenia and Mood Disorders?

Peter Gallagher and Michael S. Ritsner

Abstract According to the vulnerability-stress concept patients with mood disorders and schizophrenia display increased sensitivity to stress. The hypothalamo-pituitary-adrenal (HPA) axis is one of the major hormonal systems mediating physical and psychological stress responses. Evidence for disturbances in HPA activation and abnormal HPA regulatory mechanisms in psychiatric illnesses is accumulating. A body of evidence has accrued demonstrating particularly dysfunction of the HPA axis leading to elevated peripheral cortisol levels in a proportion of patients. Recently there has been increased interest in the role of neurosteroids, particularly dehydroepiandrosterone (DHEA), which in its sulfated form (DHEAS) is one of the most abundant in humans. This chapter reviews the area around the study of these steroids in mood disorders and schizophrenia, particularly the evidence for the utility of the molar ratio of cortisol to DHEA and/or DHEAS as a peripheral biomarker.

Keywords Cortisol • neurosteroids • dehydroepiandrosterone • dehydroepiandrosterone-sulfate • schizophrenia • depression • bipolar disorder • peripheral biomarker

Abbreviations ACTH: Adrenocorticotropic hormone, beta-CL: Beta-cortolone, BPRS: Brief Psychiatric Rating Scale[1]; CL: Cortolone; CSF: Cerebrospinal fluid; DHEA: Dehydroepiandrosterone; DHEAS: Dehydroepiandrosterone sulfate; DHEA(S): Refers to both DHEA and DHEAS; DST: Dexamethasone suppression test; FGAs: First-generation antipsychotic agents; $GABA_A$: Gamma-aminobutyric acid type A receptor; GC: Glucocorticoid; GR: Glucocorticoid receptor; HDRS: Hamilton Depression Rating Scale; HPA axis: Hypothalamo–pituitary–adrenal axis; MR: Mineralocorticoid receptor; PANSS: The positive and negative syndrome scale[2]; ROC: Receiver operating characteristic; SGAs: Second-generation antipsychotic agents; THB: Tetrahydrocorticosterone

Introduction

The vulnerability-stress (or "diathesis-stress") models of mood disorders and schizophrenia have dominated theorizing about aetiology for over three decades.[3–6] More recently, with advances in our understanding of the biological processes mediating the effects of stress, these models have incorporated mechanisms to account for the adverse impact of stress on brain function.[7] Since the hypothalamic–pituitary–adrenal (HPA) axis plays an important role in stress regulation, the possible dysfunction of the HPA axis may be an important source in searching for biomarkers for mental disorders. In the present chapter we will first present a brief overview the HPA axis and methods of assessment of cortisol and DHEA(S)* followed by a review of evidence of abnormalities in blood

P. Gallagher
School of Neurology, Neurobiology & Psychiatry, Newcastle University, Leazes Wing (Psychiatry), Royal Victoria Infirmary, Newcastle upon Tyne, UK

M.S. Ritsner
Department of Psychiatry, The Rappaport Faculty of Medicine, Technion, Israel Institute of Technology, Haifa, and Sha'ar Menashe Mental Health Center, Hadera, Israel

* Throughout we use DHEA(S) to refer collectively to DHEA (dehydroepiandrosterone) and DHEAS (dehydroepiandrosterone-sulfate). The latter terms are used to refer to each specifically.

M.S. Ritsner (ed.), *The Handbook of Neuropsychiatric Biomarkers, Endophenotypes and Genes*,
© Springer Science+Business Media B.V. 2009

cortisol to DHEA molar ratio in mood disorders and schizophrenia.

In order to more systematically examine the evidence and utility of assessment of the cortisol to DHEA(S) molar ratio we performed an Ovid MEDLINE(R) and EMBASE database search (up to February 2008) using the following search-strategy: three general searches were performed using the terms: (i) "DHEA" or "DHEAS" or "DHEA-S" or "dehydroepiandrosterone" or "dehydroepiandrosterone sulfate"; (ii) "cortisol"; and (iii) "cortisol to DHEA ratio" or "cortisol to DHEA" or "cortisol to DHEAS ratio" or "DHEA to cortisol ratio". Then the diagnostic terms: (iv) "schizophrenia" or "major depression" or "bipolar disorder" or "manic depression" or "mood disorder" were entered. Finally, searches (i), (ii), and (iv) were combined; as were (iii) and (iv). In the resultant papers, duplicates were removed. Those of relevance are discussed in the section below.

It should first be noted that numerous studies have measured both cortisol and DHEA(S) in the same subjects but have not gone on to report the molar ratio of cortisol to DHEA(S). For the purposes of this chapter all studies are discussed, however more detail is presented for those examining the ratio.

The HPA Axis

The HPA axis is one of the major hormonal systems mediating physical and psychological stress responses. When activated, neurones in the paraventricular nucleus of the hypothalamus secrete corticotropin-releasing hormone which is transported via the hypothalamo–pituitary portal circulation to the anterior pituitary where adrenocorticotropic hormone (ACTH) is secreted through stimulation of pituitary corticotrophs. ACTH then stimulates the adrenal cortex to secrete glucocorticoids: corticosterone in rats and cortisol in humans.[8,9]

Cortisol

Under basal conditions, cortisol secretion exhibits a 24-h circadian rhythm in which concentrations are highest at waking and slowly decline to a nocturnal trough.[10] As with many hormones it is released in a pulsatile manner throughout this cycle.[11] A great deal of individual variation exists in the secretion of both ACTH and cortisol, but spontaneously occurring cortisol peaks are preceded by increases in ACTH levels, although secretion of the two hormones are not quantitatively linked throughout the day.[12] Indeed, analysis of ultradian variations within healthy individuals has shown a predominant periodicity in the oscillations of both hormones of between 55 and 140 min for ACTH and 95–180 min for cortisol, indicating that, on occasion, a single cortisol peak may be initiated by two ACTH peaks.[12] Levels also appear to exhibit seasonal variation with plasma cortisol being higher in winter, but where overall cortisol production rate may be reduced.[13–15]

Cortisol is involved in the regulation of fat, protein and carbohydrate metabolism, electrolyte balance, body water distribution, blood pressure and immunosuppressant anti-inflammatory action.[8] As discussed it is also a key regulator of the physiological stress response, through negative-feedback actions via corticosteroid receptors. Two distinct corticosteroid receptor subtypes have been identified; the mineralocorticoid receptor (MR; Type I) and the glucocorticoid receptor (GR; Type II). Both receptor types have been implicated in mediating glucocorticoid feedback,[16] however there are several differences in the distribution, occupancy and binding properties of the two receptors that affects their role physiologically. The MR is highly expressed in the limbic system whereas the GR is ubiquitous, being present in both subcortical and cortical structures, with a preferential distribution in the prefrontal cortex.[17]

Glucocorticoids bind to the MR with around a sixfold to tenfold greater affinity than to GR.[18] Consequently, at basal levels near complete occupation of MRs occurs. GRs are minimally occupied at this point and only during times of high cortisol secretion, such as the circadian peak or during stress, do MRs become saturated and GR occupancy increases.[16,19] A growing body of evidence indicates that alterations in HPA axis function may be a trait-marker of both mood disorders and psychosis and may exert significant causal and exacerbating effects on symptoms and neuropsychological functioning.

Dehydroepiandrosterone

Recently there has been increased interest in the role of other adrenal steroids such as dehydroepiandrosterone (DHEA) which, in its sulfated form (DHEAS) is the most abundant in humans.[20] DHEA is a naturally occurring excitatory neuroactive steroid (or *neurosteroid:* a term first proposed by Baulieu and colleagues in 1981[21] that applies to the steroids, the accumulation of which occurs in the nervous system independently, at least in part, of supply by the steroidogenic endocrine glands and which can be synthesized *de novo* in the nervous system[22]). The neurosteroidogenesis in the brain is independent of the peripheral production; brain DHEAS was not influenced by adrenal stimulation or inhibition with adrenocorticotropic hormone (ACTH) or dexamethasone, respectively, and increased 2 days after the stressful event of adrenalectomy and orchiectomy.[23] DHEA is a substrate for androstenedione and testosterone synthesis and may have a role as an adrenal androgen.[24] DHEA serves as a precursor of androstenedione, testosterone, as well as of approximately 50% of androgens in adult men, 75% of active estrogens in premenopausal women, and 100% of active estrogens after menopause.[25]

As with cortisol, DHEA levels have been shown to exhibit seasonal variation[26] although other studies have not found such changes and results seem far less consistent.[27,28] The diurnal rhythm of DHEA also appears to be less pronounced than that of cortisol.[29] Neurosteroids display multiple effects on the central nervous system (CNS) and may act as potential signaling molecules for neocortical organization during neuronal development.[30,31] In particular, neurosteroids can interact with various neurotransmitter systems to promote neuronal remodeling; they regulate growth of neurons, enhance myelinization and synaptogenesis in the CNS, affect synaptic functioning, and show neuroprotective properties.[32–34] Furthermore, these neurosteroids have been found in the mammalian brain at considerably higher concentrations than typically detected in serum or plasma.[23,35] There is evidence that neurosteroids may be involved in the vulnerability to developing neuropsychiatric disorders such as dementia, mood disorders, substance abuse and others (see for review).[36–40]

The precise mechanism of action of DHEA in the brain is less well known although it has been shown to have actions on membrane-bound receptors and is a gamma-aminobutyric acid type A ($GABA_A$) receptor antagonist[41] as well as a sigma-1 receptor agonist.[42,43] Recently it has been confirmed that neuroactive steroids (pregnenolone, DHEA, DHEAS, allopregnanolone) are present in human post-mortem brain tissue at physiologically relevant concentrations in the nanomolar range and that levels of pregnenolone and DHEA in posterior cingulate and parietal cortex are higher in subjects with schizophrenia and bipolar disorder compared to control subjects.[44] However, in addition to these neurosteroid properties, it is the putative role of DHEA(S) as a functional antagonist of cortisol's actions which have generated most interest in the study of patients with psychiatric illness.

Methods of Assessment

A number of methods are available for the assessment of basal steroid levels in humans. For example, for small, highly lipid-soluble molecules (such as cortisol) the unbound hormone can pass easily through the membranes of nucleated cells permitting 'free' steroid levels to appear in bodily fluids.[45] Levels can reliably be measured in urine, plasma and saliva with each having potential strengths and weaknesses.[46]

Urine

Urinary cortisol excretion results from glomerular filtration and is a useful index of integrated 24-h plasma free cortisol,[46] although steroid output over any fixed period of time can be reliably assessed.[47] Similarly reliable measurements of DHEA as well as many other steroid metabolites can be achieved.[48]

Saliva

Saliva sampling has certain advantages over plasma sampling, especially in patients whose HPA axes may be sensitive to stressful interventions such as venepuncture.[45,49–51] Due to the relatively small samples required to obtain steroid measurements, sampling can also be performed relatively frequently if necessary and can allow the circadian profile to be determined.[51] The analysis of the area-under-the-curve provides an estimate of the overall hormonal secretion over 24-h and – although not as precise – is more convenient than 24-h urine collection.

Importantly, several studies have examined the relationship between steroid levels in saliva compared to those in plasma. Cortisol and DHEA levels measured in saliva closely agree with free levels in the blood, due to the fact that cellular access and entry to the oral cavity are by a method of passive diffusion and therefore independent of saliva flow-rate and transport mechanisms.[45,52,53] However, it should be noted that the method of collection of saliva can have effects on the accuracy of this relationship.[53–55]

Plasma

The relationship between central and peripheral steroid levels is well established. Guazzo et al.[56] measured plasma and CSF levels of cortisol and DHEA(S) in a group of 62 subjects aged 3–85 years. Significant correlations in steroid-free subjects were observed between blood and CSF levels for DHEA (r = 0.65) and DHEAS (r = 0.88) but not for cortisol (r = 0.26). However, in the case of cortisol, there appeared to be some evidence of two distinct populations diverging at blood concentrations of 300–400 nmol, with a strong relationship evident in one of these. Also, a strong relationship between CSF and blood levels emerged in the in a sub-group of participants on exogenous steroid administration.[56] Brain DHEA(S) levels exceeded their respective concentrations in plasma.[23]

Many studies have examined aspects of HPA axis dysfunction using the variety of methodologies described above. We briefly discuss some of these in the following section before reviewing the evidence for the utility of examining the ratio of adrenal steroid secretion.

Alterations in HPA Axis

The first systematic studies of the abnormalities in steroid hormone secretion in psychiatric illnesses were carried out by Board and colleagues over half a century ago.[57,58] These initial findings were subsequently replicated by other groups and extended to show that levels reduced as patients recovered.[59–61] Many studies have replicated these findings using a variety of methodologies and collection methods. For instance, Walker and colleagues reviewed recent scientific findings on the role of the HPA axis in the expression of vulnerability for schizophrenia. The results indicate that psychotic disorders are associated with elevated baseline and challenge-induced HPA activity, that antipsychotic medications reduce HPA activation, and that agents that augment stress hormone (cortisol) release exacerbate psychotic symptoms.[62] In comparison to studies in patients with mood disorders there are relatively fewer that have assessed adrenal steroid secretion in schizophrenia.

Urinary Levels

In major depression, elevated urinary cortisol levels have been reported in many studies, e.g.[63–66] and may persist in recovery in some patients.[67] Although this pattern may not be evident in some sub-groups of patients and may even reverse with age/ illness chronicity.[68] Urinary DHEA levels have similarly been found to be elevated.[69] The psychotic sub-type of unipolar and bipolar disorders also appears to be associated with higher urinary cortisol levels.[70]

More recently, comprehensive analysis of multiple urinary steroid metabolites in medication-free patients with recurrent unipolar major depression revealed sex differences in some.[48] In male patients (compared to male controls) levels of DHEA, as well as tetrahydrocorticosterone (THB), allo-THB, beta-cortolone (beta-CL) were found to be significantly decreased. However, in female patients, DHEA levels did not significantly differ from their respective control group, although cortisol and allo-THB levels were significantly elevated, and etiocholanolone and beta-CL levels were significantly decreased.[48] Relationships between the ratio of cortisol and DHEA and their metabolites have also been examined in MDD in relation to symptom severity, with 11-beta-hydroxysteroid dehydrogenase (HSD) being correlated with severity in women, and 17-beta-HSD being positively correlated with severity in women but negatively correlated in men.[71]

Few studies have directly compared groups of patients with different diagnoses, although of those that have it has been found that 24-h urinary cortisol levels were higher in affective disorders compared with schizophrenia.[72,73]

Saliva

As discussed earlier, because of the ease with which samples can be collected, many studies have examined steroid levels in saliva (those assessing both cortisol and DHEA or the ratio in the same samples are presented in more detail subsequently in this chapter).

Recently there has been interest in the measurement of cortisol levels in saliva for the first hour after waking when cortisol levels are know to sharply rise. It should be noted however that free cortisol responses to awakening are influenced by the actual awakening time.[74]

Several studies have demonstrated that clear abnormalities can be observed in patients with mood disorders. Unmedicated depressed patients have been found to secrete up to 25% more cortisol in the first hour after waking than control subjects.[75] This increased cortisol response to waking (CRW) has been found to persist in remitted depressed patients.[76] Similarly, increased CRW has been observed in clinically well patients with bipolar disorder, with normal DST responses[77] and recently in young high-risk subjects who had never personally suffered from depression but who had a biological parent with a history of major depression.[78]

Abnormal diurnal variation of salivary cortisol levels has also been noted in chronic schizophrenia.[79] Blunted salivary cortisol responses following psychosocial stressors have also been noted, although overall cortisol levels did not differ between patients and controls.[80,81]

Plasma and CSF

Plasma sampling has often been adopted to take point-estimated of adrenal steroid secretion although as discussed, the pulsatile nature of release limits interpretation of findings. Of greater interest are those studies that have sampled at multiple time-points throughout the day to accurately profile the pattern of steroid secretion.

Wong and colleagues[82] performed a comprehensive assessment of plasma and CSF steroids every 30 min over 30 h in medication-free melancholic MDD patients. ACTH levels were not significantly different from healthy controls, however cortisol levels were significantly elevated as was the cortisol to ACTH ratio suggesting a relatively greater plasma cortisol response to a given simultaneous level of plasma ACTH.[82] CSF cortisol levels have been shown to be elevated in both unipolar and bipolar disorder, with even greater levels evident in patients with psychotic features.[83] Indeed, elevated afternoon cortisol levels have been found in first-episode, drug-naïve patients with schizophrenia.[84]

Those studies assessing DHEA(S) levels in affective disorders present a somewhat mixed picture (for an overview see ref.[85]), although many have looked at the effect on depressive symptoms rather than a clear diagnosis of mood disorder. Other differences likely arise due to methodological factors or through assessment of steroids in isolation rather than considering the relationship with other adrenal steroids (see below).

The Cortisol to DHEA Molar Ratio

It has been suggested that the assessment of cortisol or DHEA(S) alone may not be as informative as calculating the ratio of the two steroids – the cortisol to DHEA or DHEAS to cortisol molar ratio.[86] The notion is that DHEA(S) may maintain cortisol homeostasis by acting as a cortisol antagonist, particularly during periods of prolonged glucocorticoid hyperactivity. Several lines of evidence have shown that a variety of stressors result in a shift in the balance of cortisol and DHEA(S), in that there is an increase in cortisol synthesis and a decrease in androgen synthesis. In critical illness it has been demonstrated that not only do plasma levels of cortisol increase and DHEA decrease, but sensitivity of both to ACTH-stimulation is also correspondingly altered.[87] Similarly, during acute psychological stress, stimulation of adrenal steroid release is accompanied by a shift towards DHEA release.[88] This has also led to the recognition of the potent antiglucocorticoid properties of DHEA(S)[89] (and its active metabolites, see ref.[90]). In animals it has been demonstrated that DHEA protects hippocampal neurons against neurotoxin-induced cell death, possibly by decreasing nuclear GR levels.[91] DHEA(S) has also been shown to inhibit glucocorticoid-induced enzyme activity.[92] In healthy humans, acute administration of DHEA has been shown to rapidly reduce circulating cortisol levels[93] while reduction in 24h levels have been demonstrated with longer treatment trials in healthy older subjects.[94]

Since DHEA levels appear to have regulatory effects on glucocorticoid action in the brain, it has been argued that the ratio of cortisol to DHEA most accurately reflects the degree of 'functional' hypercortisolaemia.[95–97] Together, these studies highlight the importance of considering the somewhat symbiotic relationship between cortisol and DHEA(S) and suggests that examination of each in isolation may fail to be as informative as assessment of the ratio of the two.

Mood Disorders

An extensive series of longitudinal studies examining risk factors for the development of mood disorders in adolescents by Goodyer and colleagues[95,98–104] showed that the secretion of adrenal steroids is altered and of predictive utility. In saliva samples collected over 48h it was found that elevated evening cortisol and lower morning DHEA secretion were significantly, and independently, associated with major depression.[98] Different patterns of adrenal steroid secretion were associated with co-morbidity.[99]

Young and colleagues[105] assessed salivary cortisol to DHEA molar ratios over 2 consecutive days (at 8 am and 8 pm) in 44 medication-free major depressed patients compared to their matched controls. All patients were drug-free for at least 6 weeks although most were entirely medication-naïve (n = 26/44) and of the 18 who had previously received psychotropic medication, the time drug-free ranged from 6 to 336 weeks (median = 48 weeks). Depressive symptom scores in the patient group ranged from 15 to 30 (mean = 21) on the Hamilton Depression Rating Scale. Thirty patients (68%) were experiencing their first episode of depression. Although cortisol levels were elevated and DHEA levels decreased in the patient group, neither difference reached statistical significance however the molar cortisol to DHEA ratio was significantly elevated. It should be noted that saliva was collected using a salivette device

which, as discussed previously, studies have shown can affect the accuracy of DHEA measurement.[53–55]

In a comparison of depressed and remitted patients with major depressive disorder (the majority of whom were taking antidepressant medication) and matched controls, Michael and colleagues reported that salivary cortisol to DHEA ratios were significantly elevated, both at 8 am and 8 pm, compared to remitted patients and healthy controls who did not significantly differ. Furthermore, in a post hoc analysis, taking the 85th percentile morning (8 am) cortisol to DHEA ratio of the control group as a cut-off, 82.5% of the depressed group had cortisol to DHEA ratios that were equal to or greater than this value, while this occurred in only 15% of healthy controls.[106]

Using an intensive sampling methodology, Heuser and colleagues collected blood samples every 30 min over 24h in 26 depressed patients and 33 controls for assessment of cortisol and DHEA levels. Mean cortisol and DHEA levels, and minimum DHEA level was found to be elevated over the 24h period compared to controls.[107] An elevation in both cortisol and DHEA was also observed in a smaller group of female depressed patients.[108] However the cortisol to DHEA molar ratio was not calculated in either study. Interestingly, it was noted that these finding differed from a smaller earlier study which sampled blood at single time-points where cortisol levels were significantly elevated while DHEA did not differ.[109] Here the cortisol to DHEA ratio was also calculated and was found to be elevated in the morning (8 am) samples but not at 4 pm.[109] Elevated cortisol levels and cortisol to DHEA ratios have also been found in un-medicated female MDD patients with co-morbid borderline personality disorder[110] and in elderly depressed patients[111] although ageing itself is noted to significantly reduce DHEAS secretion.[112,113]

More discrepant results have been found in studies adopting single plasma-sampling methodology. In medication-free subjects (>4 weeks), Scott and colleagues[114] found evidence of increased ratios of both cortisol to DHEA, and cortisol to DHEAS. When assessed individually, neither cortisol nor DHEA levels differed significantly from controls although DHEAS levels were lower in the patients.[114] However, lower levels of both DHEA and cortisol with no difference in DHEAS have also been reported[115] although here the ratio was not calculated.

Other studies measuring the sulfated form have found elevated salivary DHEAS levels, even in the absence of abnormal cortisol levels in medicated patients with MDD. Although the sample size was somewhat modest, discriminant analysis indicated that 77% of subjects could be correctly classified by evening DHEAS levels.[116] In a preliminary study, Takebayashi and colleagues found DHEAS and cortisol levels to be significantly elevated compared to controls in plasma samples taken at baseline in an outpatient sample (aged <45 years). Following treatment, DHEAS had significantly decreased. There were no differences in the DHEAS to cortisol ratio of patients and controls at any point.[117] In one study of older depressed patients (>60 years) compared with matched controls, no differences in DHEA(S) to cortisol ratios were reported.[118]

Very little work has been carried out on assessing cortisol to DHEA ratios in patients with bipolar disorder. Using a repeated plasma sampling protocol, hypercortisolaemia has been observed in bipolar patients (with depressive symptoms) compared with controls, without alteration in DHEA levels or cortisol to DHEA molar ratio [119] (although see below for a comparison with schizophrenia patients and the effects of lithium treatment; also see Table 30.1).

Schizophrenia

A number of studies have assessed cortisol and DHEA(S) levels in patients with schizophrenia although fewer have examined these as the molar ratio. For example, an early study by Tourney and colleagues found a specific reduction in morning DHEA levels but no differences in cortisol, testosterone, and androstenedione.[120] Subsequently it was demonstrated that plasma DHEA diurnal rhythms could discriminate patients from healthy controls with high degrees of accuracy,[121] while other studies found no difference in levels compared with controls.[122]

Recent comparative studies of the blood DHEA(S) concentrations reported that serum DHEA and DHEAS concentrations range from 15.7 to 90.9 nmol/L, and from 4,928 to 12,777 nmol/L, respectively, among schizophrenia patients, as well as, from 24.0 to 68.8 nmol/L, and from 5,375 to 13,477 nmol/L among healthy subjects, respectively.[123–127] Meta-analysis of differences in mean concentrations of serum DHEA(S) between schizophrenia patients and control subjects show significant non-zero effect ($p < 0.001$), and significant heterogeneity of data ($p < 0.001$; ref.[40]).

In first episode, non-medicated patients, Strous and colleagues [128] found a selective increase in DHEA and DHEAS levels with no difference in cortisol or cortisol to DHEA(S) ratio. It was suggested that this may be a characteristic adaptive response in the acute stages of psychotic illness. However, this assumption has not been supported by a number of recent studies, which found elevated DHEA levels in chronic schizophrenia patients as well.[119,123,125] Cortisol levels and the cortisol to DHEAS ratio have also been found to be associated with duration of illness.[129]

Table 30.1 ROC analysis of cortisol, DHEA and the cortisol to DHEA ratio for bipolar and schizophrenia patients and healthy controls

	Cortisol	DHEA	Cortisol to DHEA ratio
Bipolar vs. control	AUC = 0.72 (95%CI = 0.53 to 0.90) Sensitivity = 0.90, Specificity = 0.60	AUC = 0.55 (95%CI = 0.35 to 0.80) Sensitivity = 0.95, Specificity = 0.20	AUC = 0.58 (95%CI = 0.38 to 0.78) Sensitivity = 0.90, Specificity = 0.45
Schizophrenia vs. control	AUC = 0.74 (95%CI = 0.56 to 0.91) Sensitivity = 0.75, Specificity = 0.70	AUC = 0.77 (95%CI = 0.59 to 0.94) Sensitivity = 0.90, Specificity = 0.55	AUC = 0.57 (95%CI = 0.37 to 0.76) Sensitivity = 0.90, Specificity = 0.40
Schizophrenia vs. Bipolar	AUC = 0.58 (95%CI = 0.38 to 0.77) Sensitivity = 0.45, Specificity = 0.85	AUC = 0.74 (95%CI = 0.56 to 0.92) Sensitivity = 0.90, Specificity = 0.50	AUC = 0.68 (95%CI = 0.49 to 0.87) Sensitivity = 0.75, Specificity = 0.60

Source: Data from ref.[119]

Other studies have replicated the finding of hypercortisolaemia in patients with chronic schizophrenia using repeated-sampling techniques.[119] Here, DHEA levels were found to be elevated compared to healthy controls and patients with bipolar disorder, with no differences between cortisol to DHEA molar ratios in the groups as a whole. Table 30.1 includes a receiver operating characteristic (ROC) analysis of the data from this study, comparing the patients with schizophrenia, bipolar disorder and healthy controls. DHEA levels and the cortisol to DHEA ratio best, and most reliably, discriminates patients with schizophrenia from healthy controls and patients with bipolar disorder. Whereas it is cortisol level that is the better discriminator in patients with bipolar disorder.

Some of the apparent inconsistencies in these findings are clarified in a series of studies by Ritsner and colleagues. In an initial assessment of 40 inpatients and 15 controls subjects,[130] plasma levels of cortisol and DHEA(S) were found not to significantly differ. However, molar cortisol to DHEA(S) ratios were significantly higher in schizophrenia patients than in healthy comparison subjects. In a later study in which both DHEA and DHEAS were measured[125] it was observed that while serum DHEA, androstenedione and prolactin levels were significantly increased in schizophrenia patients, DHEAS levels were significantly decreased, compared with healthy controls. Therefore some dissociation between sulfated and non-sulfated forms may occur.

High-Risk Studies

Evidence of abnormalities in steroid hormone levels in adolescents at high risk for psychopathologies has been described. High-risk was defined as having two or more risk factors from the following list: two or more moderately to severely undesirable life events in the previous 12 months; current parental marital disharmony or past marital breakdown; two or more lifetime exist events (bereavement and/or permanent separation) of personal significance (relative or friend) to the adolescent; high (>80th percentile) emotionality. In addition, the presence of a history of parental psychiatric disorder by itself qualified as high risk. Goodyer and colleagues[101] assessed 180 subjects aged 12–16 years in a longitudinal follow-up study and found that a greater instance of 'abnormal peaks' of cortisol and DHEA levels occurred in those experiencing a depressive episode within 12 months. In an extended follow-up, assessment over a 2 year period was also carried out in order to examine baseline predictors of persistent depression.[101] In this cohort, at 12 months, 30 participants had experienced an episode of major depression and 30 had not. Of the 30 who had, 19 went on to remit by 24 months while 19 remained depressed. Significantly lower baseline DHEA levels and an elevated 8 am cortisol to DHEA ratio was observed in those who remained depressed at 24 month compared to those who remitted. Persistent cases also had higher levels of self-reported depressive symptoms and ruminations at entry compared to never depressed, although logistic regression techniques showed that only the cortisol to DHEA ratio predicted persistence.[101]

Effects of Medication

Effects particularly of mood stabilisers and antipsychotics may influence the profile of steroid levels. Previous animal work has shown for example that chronic lithium administration can lower DHEA(S) levels in frontal cortex and hippocampus, and increase serum DHEA peripherally.[131] The atypical antipsychotic clozapine has been found to decreased rat brain cortical DHEA(S) levels.[132] It should however be noted that both antipsychotic treatment and elevated prolactin levels might stimulate the adrenal cortex to secrete DHEA. Raised prolactin levels are a common consequence of antipsychotic treatment. Indeed, elevated plasma levels of prolactin and DHEA have been found during sulpiride treatment in four out of five healthy male subjects compared with those of the controls.[133] However, the direct role of prolactin in the regulation of DHEA(S) is unclear, and the alterations in both prolactin and neurosteroids may be related to independent effects of antipsychotic treatment.

The majority of studies have examined peripheral cortisol and DHEA(S) levels in mood disorders and schizophrenia have done so in medicated patients, although there are some studies in medication-free patients.

Mood Disorders

In older depressed patients, Fabian and colleagues reported a reduction in DHEA(S) levels in those who remitted following 12 weeks of treatment.[118] As there were no baseline differences between groups and non-remitters had no change in hormone concentrations it was suggested that the data indicate that the decrease in DHEA and DHEAS in patients who remitted is related specifically to remission of depression rather than to a direct drug effect on steroid secretion.

In an exploratory analysis of the overall results,[119] after separating a group of patients with bipolar disorder into those taking lithium (n = 11) and those not (n = 9), compared to schizophrenia patients (n = 20; none of whom were taking lithium) and healthy controls (n = 20) it was found that cortisol and DHEA levels differed between the groups. For cortisol levels, bipolar patients not on lithium and schizophrenic patients exhibited higher cortisol levels than controls, while for DHEA levels, schizophrenia patients had higher levels than all other groups. Finally, although there was no overall group effect, a direct pairwise comparison revealed that bipolar patients not on lithium had significantly higher cortisol-DHEA ratios than schizophrenia patients.

Schizophrenia

In a series of recent studies, Ritsner and colleagues have examined the possible effects of antipsychotic treatment on DHEA(S) circulatory levels in schizophrenia patients. In the 2004 study[130] discussed previously cortisol to DHEA(S) molar ratios were significantly higher in schizophrenia patients than in healthy comparison subjects and this difference remained after separation by antipsychotic class (typical, atypical or both). Importantly, there was also no significant correlation between cortisol to DHEA(S) ratios and dosage of antipsychotic (converted to Defined Daily Dose - the average maintenance dosage as defined by the WHO Collaborating Center for Drug Statistics[134]).

In a more recent study it was similarly demonstrated that patients receiving first-or second-generation antipsychotic agents (FGAs or SGAs) had comparable serum DHEA(S) and their metabolite levels, although a step-wise regression analysis did suggest that a small proportion of variance in cortisol levels (<5%) was predicted by antipsychotic dose.[124] Other studies have also found no difference in cortisol or DHEAS levels.[135]

In the study of di Michelle,[123] secondary analyses were also performed to examine relationships with medication usage. There was no correlation between DHEA plasma levels and olanzapine equivalents in the entire group of patients. Also, when patients were divided into those who were treated with FGAs only (n = 9), those who were treated with SGAs only (n = 12) and those who were treated with a mixture of FGAs and SGAs (n = 2), DHEA plasma levels did not correlate with olanzapine equivalents in any of the three individual groups. Interestingly, a strong negative correlation (r = −0.83) existed between clozapine dosages and DHEA plasma levels in the group of patients treated with clozapine only (n = 3), although the issue of sample size should be noted.

Relationship to Treatment Response and Outcome

One important question to address is what specific effects changes in the ratio of adrenal steroids has on clinical symptoms or functioning. Few studies have examined the cortisol to DHEA(S) ratio as a predictor of outcome or response to treatment.

From the series of studies discussed previously, Goodyer and colleagues[95] conducted one of the first to look at cortisol to DHEA ratios and life events as predictors of outcome in 8- to 16-year-old subjects with first episode major depression, reassessed 12 months after presentation. Higher evening (20.00 or 24.00 h) cortisol to DHEA ratios at predicted persistent major depression although basal levels of either hormone alone or cortisol to DHEA ratios during daytime (08.00, 12.00 or 16.00 h) did not. Both high evening cortisol to DHEA molar ratio and one or more severely disappointing life event between presentation and follow-up predicted persistent major depression: 86% of subjects with both of these factors were still depressed at 36 weeks whereas 81% with neither factor were not.[95] Following the cohort out to 72 weeks, chronicity of depression was predicted by persistent elevated evening cortisol levels, as well as increasing depression-dependent life events over the follow-up period, and co-morbid obsessive-compulsive disorder (OCD) at presentation and at 36 weeks.[102]

In a mixed group of 17 inpatients (MDD with psychotic features: n = 2; schizophrenia with co-morbid depression: n = 10; schizoaffective with current depressive symptoms: n = 5) requiring ECT due to pharmacotherapy non-response, Maayan and colleagues reported that elevated baseline DHEAS concentrations may predict response to ECT. Patients had significantly higher levels of cortisol, DHEA, and DHEAS than did the healthy control subjects both at baseline and after the last session. In the patient group cortisol and DHEA plasma levels remained unaltered during the study period, but DHEAS levels rose significantly following treatment. After dividing the group post hoc using a cut-off of the mean baseline DHEAS level of controls plus 2 standard deviations, a greater proportion of patients in the elevated DHEAS group (n = 8 of 9) exhibited a clinical response (30% reduction in HDRS scores) compared to those in the lower DHEAS group (n = 1 of 8).[136]

Recently, Ritsner and colleagues[137] compared repeated measures of serum DHEA(S), and cortisol concentrations, as well as the molar ratios of cortisol to DHEA(S) at baseline, and 2 and 4 weeks later, between schizophrenia patients admitted for exacerbation of psychosis who responded and those who did not respond to antipsychotic treatment. Findings indicated that at baseline, treatment-responders (n = 20 of 43) had significantly higher serum levels of cortisol, DHEAS and cortisol to DHEA ratios compared to non-responders (n = 23 of 43). Across all three-assessment points the responders had significantly higher serum cortisol and cortisol to DHEA(S) ratios compared with non-responders, while DHEA(S) concentrations did not differ. Furthermore, using hierarchical logistic regression, the best-fit model (accounting for 42% of the variance and correctly classified 87.8% of the 43 patients as responder/non-responder) demonstrated an advantage of both cortisol to DHEA(S) molar ratios compared with serum cortisol concentrations for predicting response to treatment.[137]

In a small study (n = 10) lower DHEAS levels have been observed in patients with predominantly negative symptoms than those with positive symptoms.[138] In one study DHEAS levels were found not to differ in patients with low to moderate negative symptoms.[139]

In a correlation analysis in chronic patients with schizophrenia/schizoaffective disorder, DHEAS and DHEAS to cortisol ratio were correlated with semantic memory, while DHEA and DHEA to cortisol ratio were found to be positively correlated with attention and working memory. In addition DHEA to cortisol ratio was also positively associated with free recall performance and negatively with BPRS symptoms and Parkinsonism.[140] Another study utilising different tests has also found relationships between cognitive function and adrenal steroid ratios.[135] Here, both DHEAS levels and the DHEAS to cortisol ratio were shown to be positively associated with verbal memory, memory for faces and executive function in schizophrenia although no relationship was observed with symptom severity.[135]

Effect of DHEA Administration

DHEA administration has influences on circulating DHEA(S) and cortisol to DHEA(S) molar ratios among healthy elderly subjects.[141] Assessments of hormonal response to DHEA administration among schizophrenia patients have revealed varying findings[142–144]: administration of DHEA (100 mg/day) for *1 week* led to significant elevation in blood DHEAS compared to placebo. After *6 weeks* between-group differences were found on both DHEA and DHEAS levels, while *after 12 weeks* of DHEA treatment (150 mg/day) no significant between-group differences were detected. In the crossover study,[127,145] 55 schizophrenia patients exhibited at baseline *normal DHEA levels* (F = 1.8, $p = 0.19$), but significantly *decreased serum DHEAS levels* (F = 8.3, $p = 0.005$), and elevated levels of cortisol (F = 30.3, $p < 0.001$), prolactin (F = 9.3, $p < 0.001$), cortisol to DHEA (F = 7.2, $p = 0.009$) and cortisol to DHEAS (F = 13.3, $p < 0.001$) molar ratios when compared with 20 healthy subjects who were closely matched for gender, age, cigarette smoking, and body mass index (df = 1.75). DHEA administration (200 mg/day) *for a 6-week period* resulted in a significant increase in circulating DHEA ($p = 0.003$), DHEAS ($p < 0.001$), while cortisol to DHEA ($p = 0.004$) and cortisol to DHEAS ($p < 0.001$) molar ratios significantly decreased compared to the placebo period. Changes in these hormones and molar ratios remain significant even after Bonferroni correction ($p = 0.05/7 = 0.007$).

Methodological Issues

After reviewing the literature on cortisol and DHEA(S) levels in mood disorders and schizophrenia it is worth highlighting a number of methodological issues which may affect the interpretation of the levels of adrenal steroids; factors which are important to first consider.

- As discussed previously, the secretion of adrenal steroids exhibits significant diurnal variation. In psychiatric illness, the pattern of both these natural fluctuations has been shown to be altered. Subtle changes in the pattern and rhythm of cortisol secretion over 24 h have been demonstrated in depressive disorders[146,147] especially during the nocturnal quiescent period[148,149] compared with healthy individuals. Changes in pulsatility have also been demonstrated[150] (although also see[151,152] where rates of hypercortisolaemia were much lower and pulsatile and circadian components of the HPA axis normal, suggesting that abnormalities may be restricted to a sub-set of patients).
- Increased rates of cortisol suppression following the dexamethasone suppression test have been found in winter months compared to the rest of the year in MDD[153] whereas in chronic schizophrenia greater rates of cortisol non-suppression to the DST were found in February (Winter) as compared to November (Autumn/Fall) and May (Spring).[154] Thus, changes in sensitivity feedback regulation of the HPA axis appear to alter seasonally and to date, very few studies have included information on this potential confound.
- Iancu and colleagues examined cortisol, and DHEA(S) levels in 45 patients with chronic schizophrenia and found a significant decrease (33%) in DHEA levels in those who smoked cigarettes (defined as those who smoked at least 20 cigarettes a day for at least 5 years) compared to those who did not. Correlation analysis revealed that plasma DHEA levels correlated positively with cortisol levels and negatively with smoking, age and illness duration while DHEAS levels correlated negatively with age only. The only correlation with symptom levels was between plasma cortisol and PANSS-negative subscale score.[155]
- Early adverse life events have been linked to subsequent alterations in the extent and profile of HPA axis dysfunction in adults mood disorders and schizophrenia.[156,157]
- The direct effects of psychotropic medications are of course of clear importance to consider, especially in studies of patients likely to be taking combinations of such drugs.

Fundamental Questions

Having reviewed the above empirical data the core question is whether or not the cortisol to DHEA(S) ratio can be of utility as a biomarker in mood disorders and psychosis.

First, the definition of 'biomarker' has different meanings in different contexts. The Biomarkers Definitions Working Group (BDWG) of the National Institutes of Health Director's Initiative on Biomarkers and Surrogate Endpoints defines a biological marker (biomarker) as: *a characteristic that is objectively measured and evaluated as an indicator of normal biological processes, pathogenic processes, or pharmacologic responses to a therapeutic intervention.*[158]

In addition to their value in early efficacy and safety evaluations (in both animal and proof-of-concept clinical trials) other identified applications in disease detection and monitoring of health status included:

- Use as a diagnostic tool for the identification of those patients with a disease or abnormal condition
- Use as a tool for staging of disease
- Use as an indicator of disease prognosis
- Use for prediction and monitoring of clinical response to an intervention

Considering these points, the utility of the cortisol to DHEA molar ratio would appear to lie in the latter applications than the former. Past attempts at using tests of HPA axis as diagnostic tools have not proved fruitful, for example, the DST (see ref.[159]). However, with a better understanding of different factors that affect responses to the test[160,161] and refinement of the DST into more integrated tests of HPA function (such as the combined dex/CRH test[162–166]) has raised the possibility of using such measures as markers of early relapse and treatment response[167–169].

While a great deal of evidence has been reviewed on the alterations in peripheral cortisol and DHEA(S) levels and their ratio in different psychiatric illnesses the pattern and extent is somewhat variable, particularly with respect to whether DHEA(S) levels or the ratio is elevated, normal or decreased. Being cognizant of the effects of medication and careful controlling of other methodological factors (as discussed above) goes some way to reduce such discrepancies. However, differences in demographic/illness characteristics and their effects on steroid levels also create difficulties in producing a reliable diagnostic marker for general application, although some reviewed studies in very well defined populations have produced more reliable discrimination between patients and healthy subjects. These require replication in larger samples.

From the studies reviewed, more promising findings at present are those examining the cortisol to DHEA(S) ratio as an indicator of prognosis or for prediction and monitoring of clinical response to an intervention. In these areas, longitudinal studies have shown that assessment of peripheral steroid levels and particularly the cortisol to DHEA(S) ratio can have predictive utility in high-risk groups, while other studies have demonstrated utility in predicting treatment response, especially in schizophrenia. More work is needed to explore this hypothesis in mood disorders.

It has further been highlighted that *"biomarker measurements can help explain empirical results of clinical trials by relating the effects of interventions on molecular and cellular pathways to clinical responses. In doing so, biomarkers provide an avenue for researchers to gain a mechanistic understanding of the differences in clinical response that may be influenced by uncontrolled variables (for example, drug metabolism)"*.[158]

This is likely to be an important area of research with respect to the utility of the cortisol to DHEA(S) molar ratios. Several studies have explored the utility DHEA administration as a treatment in general with some encouraging results.[142–145,170–172] However, subpopulations of patients with elevated cortisol to DHEA(S) ratios or decreased DHEA(S) to neurosteroid levels may specifically benefit from augmentation strategies which improve treatment response and outcome. Understanding the complex modulatory effects of neurosteroids within the brain, for example the effects of DHEA(S) on GABA systems,[173,174] may also identify novel targets for, or predictors of, treatment.

Finally, we return to discuss the specific criteria that a diagnostic biomarker needs to meet (see Chapter 1 of this book for more extensive elaboration on these criteria). We will now discuss the different results presented in this chapter using these criteria:

1. *The biomarker should reflect some basic pathophysiological process, and detect a fundamental feature of the disease with high sensitivity and specificity*

As is seen in the preceding chapter, there are many studies that have examined the role of the HPA axis and (neuro-)steroid levels in the pathophysiology of severe mental illnesses. A limited number of studies have performed sensitivity and specificity analyses on the cortisol, DHEA(S) or their molar ratio. Whilst some have shown these to be quite favourable, heterogeneity in the samples with regard to diagnosis, clinical/illness factors, medication, and a number of other medicating factors can introduce variance into the results that reduces and limits the precision of such measures.

2. *The biomarker should be specific for the disease compared with related disorders*

Few studies have directly compared the cortisol to DHEA(S) ratio across different disorders, possibly due in part to the complexities of which factors to match diagnostic groups on (also, see above point 1). There is preliminary evidence to suggest that individuals with schizophrenia may be characterised by changes in DHEA levels compared to those with mood disorders. More work is needed to establish if it is possible to separate related (but often heterogeneous) disorders by cortisol to DHEA(S) ratio specifically. However, as discussed earlier, the utility of the cortisol to DHEA(S) ratio may be somewhat limited on this point.

3. *The biomarker should not reflect symptomatology of disorder*

Similarly, more work is needed to examine the cortisol to DHEA(S) ratio in patients in remission, especially longitudinal studies in the same individuals. There is some evidence of abnormalities in high-risk groups, however the majority of work has focussed on individuals while they are symptomatic and the proportion of patients that continue to exhibit changes in cortisol to DHEA(S) ratio outside of these periods is relatively unknown at this point.

4. *The biomarker can be measured repeatedly over time and should be reproducible*

On a practical level it is relatively straightforward to assess the cortisol to DHEA(S) ratio repeatedly over time. However, the reproducibility of such assessments is relatively unknown. One factor that should be considered is the alterations that can occur across seasons and within individuals (such as menstrual cycle effects, physical changes, etc.) that could have differential effects over time. Again this could be examined in a longitudinal study or with repeat sampling in the same individuals in different mood states (for example, see the case series for dex/CRH results[175]).

5. *The biomarker should be measured in non-invasive and easy-to-perform tests that can be done at the bedside or in the outpatient setting*

The ease with which samples can be obtained non-invasively is described earlier in this chapter (see 'Methods of Assessment'), studies having been conducted previously in both in- and out-patient samples. Single samples which provide a point-estimate of steroid levels are less accurate and useful than multiple sampling techniques, however this should be balanced with the notion of 'easy-to-perform'. More data are needed to establish the optimum sampling number/times as has been done previously with other HPA markers.[165,176,177]

6. *The biomarker should not cause harm to the individual being assessed*

As discussed in point 5, although steroid levels can be measured in CSF through more invasive procedures, it is also possible to assess levels in serum and even urine or saliva.

7. *The biomarker should be reliable in many testing environments/labs*

The availability of commercial assay kits and standardised published methods has meant that estimates can be reliably obtained. To enhance this, full information on inter- and intra-assay coefficients of variation across quality control runs should be included in all published reports along with other data on assay sensitivity.

8. *The biomarker should be inexpensive*

Finally, the issue of cost should be considered in relation to point 5 above, i.e. being related to how many samples are required. Assays for DHEA(S) tend to be more expensive than those for cortisol, but are comparatively inexpensive to perform on individual samples.

Conclusions

From the evidence examined it appears that measurement of the cortisol to DHEA(S) molar ratio is a promising biological marker in mood and psychotic disorders. Taken together findings suggest that patients with major depression do exhibit an increased cortisol to DHEA ratio when symptomatic. Alterations in DHEA(S) levels appear somewhat less consistent. Due to the limited number of studies it is less clear if such changes can be observed in bipolar disorder across different illness phases, although the occurrence of hypercortisolism in a significant proportion of patients is more clearly established. In schizophrenia, more consistent findings relate to predictors of clinical outcome/treatment response, particularly DHEA(S) levels; and in depression and adolescents at high-risk, cortisol to DHEA molar ratio. The application of these finding to treatment studies are a natural extension to this work and many are currently underway.

As highlighted in this chapter there are many factors that can affect the assessment of steroid levels, not least of which are sampling methodology and medium, timing, and clinical factors such as illness characteristics and medication status. There is a need to conduct further, well-controlled studies, particularly longitudinal analysis in drug-free cohorts followed through treatment. Such work may increase our understanding of the pathophysiology of these illnesses, establish reliable markers of outcome and response in sub-groups of patients and elucidate new avenues for treatment.

References

1. Overall JE, Gorham DR. The Brief Psychiatric Rating Scale. *Psychol Rep*. 1962;10:799–812.
2. Kay SR, Fiszbein A, Opler LA. The positive and negative syndrome scale (PANSS) for schizophrenia. *Schizophr Bull*. 1987;13(2):261–276.
3. Rosenthal D. *Genetic theory and abnormal behavior*. Washington, DC: McGraw-Hill; 1970.
4. Zubin J, Spring B. Vulnerability: A new view of schizophrenia. *J Abnorm Psychol*. 1977;86:103–126.
5. Norman R, Malla A. Stressful life events and schizophrenia: I. A review of the research. *Br J Psychiatry*. 1993; 162:161–166.
6. Norman RM, Malla AK. Stressful life events and schizophrenia. II: Conceptual and methodological issues. *Br J Psychiatry*. 1993 162:166–174.

7. Walker E, Mittal V, Tessner K. Stress and the Hypothalamic Pituitary Adrenal Axis in the Developmental Course of Schizophrenia. *Annu Rev Clin Psychol.* 2008;4:189–216.

8. Berne RM, Levy MN. *Physiology.* Fourth ed. St. Louis: Mosby; 1998.

9. Feldman S, Conforti N, Weidenfeld J. Limbic pathways and hypothalamic neurotransmitters mediating adrenocortical responses to neural stimuli. *Neurosci Biobehav Rev.* 1995;19(2):235–240.

10. Weitzman ED, Fukushima D, Nogeire C, Roffwarg H, Gallagher TF, Hellman L. Twenty-four hour pattern of the episodic secretion of cortisol in normal subjects. *J Clin Endocrinol Metabol.* 1971;33(1):14–22.

11. Young EA, Abelson J, Lightman SL. Cortisol pulsatility and its role in stress regulation and health. *Front Neuroendocrinol.* 2004;25(2):69–76.

12. Follenius M, Simon C, Brandenberger G, Lenzi P. Ultradian plasma corticotropin and cortisol rhythms: time-series analyses. *J Endocrinol Invest.* 1987;10(3):261–266.

13. Walker BR, Best R, Noon JP, Watt GC, Webb DJ. Seasonal variation in glucocorticoid activity in healthy men. *J Clin Endocrinol Metab.* 1997;82(12):4015–4019.

14. King JA, Rosal MC, Ma Y, Reed G, Kelly TA, Stanek EJ, 3rd, Ockene IS. Sequence and seasonal effects of salivary cortisol. *Behav Med.* 2000;26(2):67–73.

15. Hansen AM, Garde AH, Skovgaard LT, Christensen JM. Seasonal and biological variation of urinary epinephrine, norepinephrine, and cortisol in healthy women. *Clin Chim Acta.* 2001;309(1):25–35.

16. Reul JM, de Kloet ER. Two receptor systems for corticosterone in rat brain: microdistribution and differential occupation. *Endocrinology.* 1985;117(6):2505–2511.

17. Patel PD, Lopez JF, Lyons DM, Burke S, Wallace M, Schatzberg AF. Glucocorticoid and mineralocorticoid receptor mRNA expression in squirrel monkey brain. *J Psychiatr Res.* 2000;34(6):383–392.

18. de Kloet ER, Oitzl MS, Joels M. Stress and cognition: are corticosteroids good or bad guys? *Trends Neurosci.* 1999;22(10):422–426.

19. de Kloet ER, Reul JMHM. Feedback action and tonic influence of corticosteroids on brain function: A concept arising from the heterogeneity of brain receptor systems. *Psychoneuroendocrinology.* 1987;12(2):83–105.

20. Morfin R. *DHEA and the brain.* London/New York: Taylor & Francis; 2002.

21. Baulieu EE. Steroid hormones in the brain: Several mechanisms? In: Fuxe K, Gutafsson JA, Wetterberg L, eds. *Steroid Hormone Regulation of the Brain.* Oxford: Pergamon; 1981:3–14.

22. Baulieu E-E, Robel P. Dehydroepiandrosterone (DHEA) and dehydroepiandrosterone sulfate (DHEAS) as neuroactive neurosteroids. *PNAS.* 1998;95(8):4089–4091.

23. Corpechot C, Robel P, Axelson M, Sjovall J, Baulieu EE. Characterization and measurement of dehydroepiandrosterone sulfate in rat brain. *Proc Natl Acad Sci USA.* 1981;78:4704–4707.

24. Gurnell EM, Chatterjee VKK. Dehydroepiandrosterone replacement therapy. *Eur J Endocrinol.* 2001;145(2):103–106.

25. Regelson W, Kalimi M. Dehydroepiandrosterone (DHEA) – the Multifunctional Steroid II. Effects on the CNS, Cell Proliferation, Metabolic and Vascular, Clinical and Other Effects. Mechanism of Action? *Ann N Y Acad Sci.* 1994 719:564–575.

26. Garde AH, Hansen AM, Skovgaard LT, Christensen JM. Seasonal and biological variation of blood concentrations of total cholesterol, dehydroepiandrosterone sulfate, hemoglobin A1c, IgA, prolactin, and free testosterone in healthy women. *Clin Chem.* 2000;46(4):551–559.

27. Bjornerem A, Straume B, Oian P, Berntsen GKR. Seasonal Variation of Estradiol, Follicle Stimulating Hormone, and Dehydroepiandrosterone Sulfate in Women and Men. *J Clin Endocrinol Metab.* 2006;91(10):3798–3802.

28. Brambilla DJ, O'Donnell AB, Matsumoto AM, McKinlay JB. Lack of Seasonal Variation in Serum Sex Hormone Levels in Middle-Aged to Older Men in the Boston Area. *J Clin Endocrinol Metab.* 2007;92(11):4224–4229.

29. Hucklebridge F, Hussain T, Evans P, Clow A. The diurnal patterns of the adrenal steroids cortisol and dehydroepiandrosterone (DHEA) in relation to awakening. *Psychoneuroendocrinology.* 2005;30(1):51–57.

30. Baulieu EE, Robel P. Dehydroepiandrosterone and dehydroepiandrosterone sulfate as neuroactive neurosteroids. *J Endocrinol.* 1996;150 Suppl:S221–239.

31. Mao X, Barger SW. Neuroprotection by dehydroepiandrosterone-sulfate: role of an NFkappaB-like factor. *Neuroreport.* 1998;9(4):759–763.

32. Friess E, Schiffelholz T, Steckler T, Steiger A. Dehydroepiandrosterone – a neurosteroid. *Eur J Clin Invest.* 2000;30 Suppl 3:46–50.

33. Wen S, Dong K, Onolfo JP, Vincens M. Treatment with dehydroepiandrosterone sulfate increases NMDA receptors in hippocampus and cortex. *Eur J Pharmacol.* 2001;430 (2–3):373–374.

34. Johnson MD, Bebb RA, Sirrs SM. Uses of DHEA in aging and other disease states. *Ageing Res Rev.* 2002; 1(1):29–41.

35. Baulieu EE. Neurosteroids: of the nervous system, by the nervous system, for the nervous system. *Recent Prog Horm Res.* 1997;52:1–32.

36. Epperson CN, Wisner KL, Yamamoto B. Gonadal steroids in the treatment of mood disorders. *Psychosom Med.* 1999;61(5):676–697.

37. Sundstrom Poromaa I, Smith S, Gulinello M. GABA receptors, progesterone and premenstrual dysphoric disorder. *Arch Women Ment Health.* 2003;6(1):23–41.

38. Eser D, Schule C, Baghai TC, Romeo E, Rupprecht R. Neuroactive steroids in depression and anxiety disorders: clinical studies. *Neuroendocrinology.* 2006;84(4):244–254.

39. Girdler SS, Klatzkin R. Neurosteroids in the context of stress: implications for depressive disorders. *Pharmacol Ther.* 2007;116(1):125–139.

40. Ritsner MS, Gibel A, Ratner Y, Weizman A. Dehydroepiandrosterone and Pregnenolone Alterations in Schizophrenia. In: Ritsner MS, Weizman A (eds) *Neuroactive Steroids in Brain Function, Behavior and Neuropsychiatric Disorders: Novel Strategies for Research and Treatment.* Springer Science+Business Media, B.V; 2008:251–298.

41. Hansen SL, Fjalland B, Jackson MB. Differential blockade of gamma -aminobutyric acid type A receptors by the neuroactive steroid dehydroepiandrosterone sulfate in

posterior and intermediate pituitary. *Mol Pharmacol.* 1999;55(3):489–496.

42. Maurice T, Phan VL, Urani A, Kamei H, Noda Y, Nabeshima T. Neuroactive neurosteroids as endogenous effectors for the sigma1 ([sigma]1) receptor: pharmacological evidence and therapeutic opportunities. *Jpn J Pharmacol.* 1999;81(2):125–155.

43. Maurice T, Gregoire C, Espallergues J. Neuro(active)steroids actions at the neuromodulatory sigma1 ([sigma]1) receptor: Biochemical and physiological evidences, consequences in neuroprotection. *Pharmacol Biochem Behav.* 2006;84(4):581–597.

44. Marx CE, Stevens RD, Shampine LJ, Uzunova V, Trost WT, Butterfield MI, Massing MW, Hamer RM, Morrow AL, Lieberman JA. Neuroactive steroids are altered in schizophrenia and bipolar disorder: relevance to pathophysiology and therapeutics. *Neuropsychopharmacology.* 2006;31:1249–1263.

45. Kirschbaum C, Hellhammer DH. Salivary cortisol. In: Fink G (ed.) *Encyclopedia of Stress*. San Diego, CA: Academic; 2000.

46. Levine A, Zagoory-Sharon O, Feldman R, Lewis JG, Weller A. Measuring cortisol in human psychobiological studies. *Physiol Behav.* 2007;90(1):43–53.

47. Callies F, Arlt W, Siekmann L, Hubler D, Bidlingmaier F, Allolio B. Influence of oral dehydroepiandrosterone (DHEA) on urinary steroid metabolites in males and females. *Steroids.* 2000;65(2):98–102.

48. Poor V, Juricskay S, Gati A, Osvath P, Tenyi T. Urinary steroid metabolites and 11beta-hydroxysteroid dehydrogenase activity in patients with unipolar recurrent major depression. *J Affect Disord.* 2004;81(1):55–59.

49. Kirschbaum C, Hellhammer DH. Salivary cortisol in psychobiological research: an overview. *Neuropsychobiology.* 1989;22(3):150–169.

50. Kirschbaum C, Hellhammer DH. Salivary cortisol in psychoneuroendocrine research: recent developments and applications. *Psychoneuroendocrinology.* 1994;19(4):313–333.

51. Lac G. Saliva assays in clinical and research biology. *Pathol Biol (Paris).* 2001;49(8):660–667.

52. Vining RF, McGinley RA, Symons RG. Hormones in saliva: mode of entry and consequent implications for clinical interpretation. *Clin Chem.* 1983;29(10):1752–1756.

53. Granger DA, Schwartz EB, Booth A, Curran M, Zakaria D. Assessing dehydroepiandrosterone in saliva: a simple radioimmunoassay for use in studies of children, adolescents and adults. *Psychoneuroendocrinology.* 1999;24(5):567–579.

54. Shirtcliff EA, Granger DA, Schwartz E, Curran MJ. Use of salivary biomarkers in biobehavioral research: cotton-based sample collection methods can interfere with salivary immunoassay results. *Psychoneuroendocrinology.* 2001;26(2):165–173.

55. Gallagher P, Leitch MM, Massey AE, McAllister-Williams RH, Young AH. Assessment of cortisol and dehydroepiandrosterone (DHEA) in saliva: effects of collection methods. *J Psychopharmacol (Oxf).* 2006;20(5):643–649.

56. Guazzo EP, Kirkpatrick PJ, Goodyer IM, Shiers HM, Herbert J. Cortisol, dehydroepiandrosterone (DHEA), and DHEA sulfate in the cerebrospinal fluid of man: relation to blood levels and the effects of age. *J Clin Endocrinol Metab.* 1996;81(11):3951–3960.

57. Board F, Persky H, Hamburg DA. Psychological stress and endocrine functions; blood levels of adrenocortical and thyroid hormones in acutely disturbed patients. *Psychosom Med.* 1956;18:324–333.

58. Board F, Wadeson R, Persky H. Depressive affect and endocrine functions; blood levels of adrenal cortex and thyroid hormones in patients suffering from depressive reactions. *AMA Arch Neurol Psychiatry.* 1957 78(6):612–620.

59. Gibbons JL, McHugh PR. Plasma cortisol in depressive illness. *J Psychiatr Res.* 1962;2(1):162–171.

60. Gibbons JL. Cortisol secretion rate in depressive illness. *Arch Gen Psychiatry.* 1964;10:572–575.

61. Gibbons JL. The secretion rate of corticosterone in depressive illness. *J Psychosom Res.* 1966;10(3):263–266.

62. Walker E, Mittal V, Tessner K. Stress and the hypothalamic pituitary adrenal axis in the developmental course of schizophrenia. *Annu Rev Clin Psychol.* 2008;4:189–216.

63. Carroll BJ, Curtis GC, Davies BM, Mendels J, Sugerman AA. Urinary free cortisol excretion in depression. *Psychol Med.* 1976 6(1):43–50.

64. Kathol RG, Anton R, Noyes R, Gehris T. Direct comparison of urinary free cortisol excretion in patients with depression and panic disorder. *Biol Psychiatry.* 1989;25(7):873–878.

65. Scott LV, Dinan TG. Urinary free cortisol excretion in chronic fatigue syndrome, major depression and in healthy volunteers. *J Affect Disord.* 1998;47(1–3):49–54.

66. Maes M, Lin A, Bonaccorso S, van Hunsel F, Van Gastel A, Delmeire L, Biondi M, Bosmans E, Kenis G, Scharpé S. Increased 24-hour urinary cortisol excretion in patients with post-traumatic stress disorder and patients with major depression, but not in patients with fibromyalgia. *Acta Psychiatr Scand.* 1998 98(4):328–335.

67. Kathol RG. Persistent elevation of urinary free cortisol and loss of circannual periodicity in recovered depressive patients. *J Affect Disord.* 1985 8(2):137–145.

68. Oldehinkel AJ, van den Berg MD, Flentge F, Bouhuys AL, ter Horst GJ, Ormel J. Urinary free cortisol excretion in elderly persons with minor and major depression. *Psychiatry Res.* 2001;104(1):39–47.

69. Tollefson GD, Haus E, Garvey MJ, Evans M, Tuason VB. 24 hour urinary dehydroepiandrosterone sulfate in unipolar depression treated with cognitive and/or pharmacotherapy. *Ann Clin Psychiatry.* 1990;2:39–45.

70. Wedekind D, Preiss B, Cohrs S, Ruether E, Huether G, Adler L. Relationship between nocturnal urinary cortisol excretion and symptom severity in subgroups of patients with depressive episodes. *Neuropsychobiology.* 2007;56(2–3):119–122.

71. Raven PW, Taylor NF. 11Beta-HSD and 17beta-HSD as biological markers of depression: sex differences and correlation with symptom severity. *Endocr Res.* 1998;24(3–4):659–662.

72. Diebold K, Kick H, Schmidt G. Urinary free cortisol excretion in endogenously depressed and schizophrenic patients. *Psychiatr Clin (Basel).* 1981;14(1):43–48.

73. Yehuda R, Boisoneau D, Mason JW, Giller EL. Glucocorticoid receptor number and cortisol excretion in mood, anxiety, and psychotic disorders. *Biol Psychiatry.* 1993;34(1–2):18–25.

74. Federenko I, Wust S, Hellhammer DH, Dechoux R, Kumsta R, Kirschbaum C. Free cortisol awakening responses are

influenced by awakening time. *Psychoneuroendocrinology.* 2004;29(2):174–184.

75. Bhagwagar Z, Hafizi S, Cowen PJ. Increased salivary cortisol after waking in depression. *Psychopharmacology (Berl).* 2005;182(1):54–57.

76. Bhagwagar Z, Hafizi S, Cowen PJ. Increase in concentration of waking salivary cortisol in recovered patients with depression. *Am J Psychiatry.* 2003;160(10):1890–1891.

77. Deshauer D, Duffy A, Alda M, Grof E, Albuquerque J, Grof P. The cortisol awakening response in bipolar illness: a pilot study. *Can J Psychiatry.* 2003;48(7):462–466.

78. Mannie ZN, Harmer CJ, Cowen PJ. Increased waking salivary cortisol levels in young people at familial risk of depression. *Am J Psychiatry.* 2007;164(4):617–621.

79. Kaneko M, Yokoyama F, Hoshino Y, Takahagi K, Murata S, Watanabe M, Kumashiro H. Hypothalamic-pituitary-adrenal axis function in chronic schizophrenia: association with clinical features. *Neuropsychobiology.* 1992;25(1):1–7.

80. Jansen LMC, Gispen-de Wied CC, Gademan PJ, De Jonge RCJ, van der Linden JA, Kahn RS. Blunted cortisol response to a psychosocial stressor in schizophrenia. *Schizophr Res.* 1998;33(1–2):87–94.

81. Jansen LM, Gispen-de Wied CC, Kahn RS. Selective impairments in the stress response in schizophrenic patients. *Psychopharmacol Bull.* 2000;149(3):319–325.

82. Wong ML, Kling MA, Munson PJ, Listwak S, Licinio J, Prolo P, Karp B, McCutcheon IE, Geracioti TD, DeBellis MD, Rice KC, Goldstein DS, Veldhuis JD, Chrousos GP, Oldfield EH, McCann SM, Gold PW. Pronounced and sustained central hypernoradrenergic function in major depression with melancholic features: relation to hypercortisolism and corticotropin-releasing hormone. *Proc Natl Acad Sci USA.* 2000;97(1):325–330.

83. Carroll BJ, Curtis GC, Mendels J. Cerebrospinal fluid and plasma free cortisol concentrations in depression. *Psychol Med.* 1976 6(2):235–244.

84. Ryan MC, Sharifi N, Condren R, Thakore JH. Evidence of basal pituitary-adrenal overactivity in first episode, drug naive patients with schizophrenia. *Psychoneuroendocrinology.* 2004;29(8):1065–1070.

85. van Broekhoven F, Verkes RJ. Neurosteroids in depression: a review. *Psychopharmacology (Berl).* 2003;165:97–110.

86. Hechter O, Grossman A, Chatterton RT. Relationship of dehydroepiandrosterone and cortisol in disease. *Med Hypotheses.* 1997;49(1):85–91.

87. Parker LN, Levin ER, Lifrak ET. Evidence for adrenocortical adaptation to severe illness. *J Clin Endocrinol Metab.* 1985;60(5):947–952.

88. Oberbeck R, Benschop RJ, Jacobs R, Hosch W, Jetschmann JU, Schurmeyer TH, Schmidt RE, Schedlowski M. Endocrine mechanisms of stress-induced DHEA-secretion. *J Endocrinol Invest.* 1998;21(3):148–153.

89. Kalimi M, Shafagoj Y, Loria R, Padgett D, Regelson W. Anti-glucocorticoid effects of dehydroepiandrosterone (DHEA). *Mol Cell Biochem.* 1994;131(2):99–104.

90. Muller C, Hennebert O, Morfin R. The native anti-glucocorticoid paradigm. *The J Steroid Biochem Mol Biol.* 2006;100(1–3):95–105.

91. Cardounel A, Regelson W, Kalimi M. Dehydroepiandrosterone protects hippocampal neurons against neuro-toxin-induced cell death: mechanism of action. *Proc Soc Exp Biol Med.* 1999;222(2):145–149.

92. Browne ES, Wright BE, Porter JR, Svec F. Dehydroepiandrosterone: antiglucocorticoid action in mice. *Am J Med Sci.* 1992;303:366–371.

93. Wolf OT, Köster B, Kirschbaum C, Pietrowsky R, Kern W, Hellhammer DH, Born J, Fehm HL. A single administration of dehydroepiandrosterone does not enhance memory performance in young healthy adults, but immediately reduces cortisol levels. *Biol Psychiatry.* 1997;42(9):845–848.

94. Kroboth PD, Amico JA, Stone RA, Folan M, Frye RF, Kroboth FJ, Bigos KL, Fabian TJ, Linares AM, Pollock BG, Hakala C. Influence of DHEA administration on 24-hour cortisol concentrations. *J Clin Psychopharmacol.* 2003;23(1):96–99.

95. Goodyer IM, Herbert J, Altham PM. Adrenal steroid secretion and major depression in 8- to 16-year-olds, III. Influence of cortisol/DHEA ratio at presentation on subsequent rates of disappointing life events and persistent major depression. *Psychol Med.* 1998;28(2):265–273.

96. Wolkowitz OM, Epel ES, Reus VI. Stress hormone-related psychopathology: pathophysiological and treatment implications. *World J Biol Psychiatry.* 2001;2(3):115–143.

97. Gallagher P, Young A. Cortisol/DHEA ratios in depression. *Neuropsychopharmacology.* 2002;26(3):410.

98. Goodyer IM, Herbert J, Altham PM, Pearson J, Secher SM, Shiers HM. Adrenal secretion during major depression in 8- to 16-year-olds, I. Altered diurnal rhythms in salivary cortisol and dehydroepiandrosterone (DHEA) at presentation. *Psychol Med.* 1996;26(2):245–256.

99. Herbert J, Goodyer IM, Altham PM, Pearson J, Secher SM, Shiers HM. Adrenal secretion and major depression in 8- to 16-year-olds, II. Influence of co-morbidity at presentation. *Psychol Med.* 1996;26(2):257–263.

100. Goodyer IM, Herbert J, Tamplin A, Altham PM. First-episode major depression in adolescents. Affective, cognitive and endocrine characteristics of risk status and predictors of onset. *Br J Psychiatry.* 2000;176:142–149.

101. Goodyer IM, Herbert J, Tamplin A, Altham PM. Recent life events, cortisol, dehydroepiandrosterone and the onset of major depression in high-risk adolescents. *Br J Psychiatry.* 2000;177:499–504.

102. Goodyer IM, Park RJ, Herbert J. Psychosocial and endocrine features of chronic first-episode major depression in 8–16 year olds. *Biol Psychiatry.* 2001;50(5):351–357.

103. Goodyer IM, Park RJ, Netherton CM, Herbert J. Possible role of cortisol and dehydroepiandrosterone in human development and psychopathology. *Br J Psychiatry.* 2001;179:243–249.

104. Goodyer IM, Herbert J, Tamplin A. Psychoendocrine antecedents of persistent first-episode major depression in adolescents: a community-based longitudinal enquiry.[see comment]. *Psychol Med.* 2003;33(4):601–610.

105. Young AH, Gallagher P, Porter RJ. Elevation of the cortisol-dehydroepiandrosterone ratio in drug-free depressed patients. *Am J Psychiatry.* 2002;159(7):1237–1239.

106. Michael A, Jenaway A, Paykel ES, Herbert J. Altered salivary dehydroepiandrosterone levels in major depression in adults. *Biol Psychiatry.* 2000;48(10):989–995.

107. Heuser I, Deuschle M, Luppa P, Schweiger U, Standhardt H, Weber B. Increased diurnal plasma concentrations of

dehydroepiandrosterone in depressed patients. *J Clin Endocrinol Metab.* 1998;83(9):3130–3133.

108. Weber B, Lewicka S, Deuschle M, Colla M, Heuser I. Testosterone, androstenedione and dihydrotestosterone concentrations are elevated in female patients with major depression. *Psychoneuroendocrinology.* 2000;25(8):765–771.

109. Osran H, Reist C, Chen CC, Lifrak ET, Chicz-DeMet A, Parker LN. Adrenal androgens and cortisol in major depression. *Am J Psychiatry.* 1993;150(5):806–809.

110. Kahl KG, Bens S, Ziegler K, Rudolf S, Dibbelt L, Kordon A, Schweiger U. Cortisol, the cortisol-dehydroepiandrosterone ratio, and pro-inflammatory cytokines in patients with current major depressive disorder comorbid with borderline personality disorder. *Biol Psychiatry.* 2006;59(7):667–671.

111. Ferrari E, Mirani M, Barili L, Falvo F, Solerte SB, Cravello L, Pini L, Magri F. Cognitive and affective disorders in the elderly: a neuroendocrine study. *Arch Gerontol Geriatr.* 2004;(9):171–182.

112. Ferrari E, Casarotti D, Muzzoni B, Albertelli N, Cravello L, Fioravanti M, Solerte SB, Magri F. Age-related changes of the adrenal secretory pattern: possible role in pathological brain aging. *Brain Res Brain Res Rev.* 2001;37(1–3):294–300.

113. Ferrari E, Cravello L, Muzzoni B, Casarotti D, Paltro M, Solerte SB, Fioravanti M, Cuzzoni G, Pontiggia B, Magri F. Age-related changes of the hypothalamic-pituitary-adrenal axis: pathophysiological correlates. *Eur J Endocrinol.* 2001;144(4):319–329.

114. Scott LV, Salahuddin F, Cooney J, Svec F, Dinan TG. Differences in adrenal steroid profile in chronic fatigue syndrome, in depression and in health. *J Affect Disord.* 1999;54(1–2):129–137.

115. Jozuka H, Jozuka E, Takeuchi S, Nishikaze O. Comparison of immunological and endocrinological markers associated with major depression. *J Int Med Res.* 2003;31(1):36–41.

116. Assies J, Visser I, Nicolson NA, Eggelte TA, Wekking EM, Huyser J, Lieverse R, Schene AH. Elevated salivary dehydroepiandrosterone-sulfate but normal cortisol levels in medicated depressed patients: preliminary findings. *Psychiatry Res.* 2004;128(2):117–122.

117. Takebayashi M, Kagaya A, Uchitomi Y, Kugaya A, Muraoka M, Yokota N, Horiguchi J, Yamawaki S. Plasma dehydroepiandrosterone sulfate in unipolar major depression. *J Neural Transm.* 1998;105(4–5):537–542.

118. Fabian TJ, Dew MA, Pollock BG, Reynolds CF, Mulsant BH, Butters MA, Zmuda MD, Linares AM, Trottini M, Kroboth PD. Endogenous concentrations of DHEA and DHEA-S Decrease with remission of depression in older adults. *Biol Psychiatry.* 2001;50(10):767–774.

119. Gallagher P, Watson S, Smith MS, Young AH, Ferrier IN. Plasma cortisol-dehydroepiandrosterone (DHEA) ratios in schizophrenia and bipolar disorder. *Schizophr Res.* 2007;90(1–3):258–265.

120. Tourney G, Erb JL. Temporal variations in androgens and stress hormones in control and schizophrenic subjects. *Biol Psychiatry.* 1979;14(2):395–404.

121. Erb JL, Kadane JB, Tourney G, Mickelsen R, Trader D, Szabo R, Davis V. Discrimination between schizophrenic and control subjects by means of plasma dehydroepiandrosterone measurements. *J Clin Endocrinol Metab.* 1981;52(2):181–186.

122. Brophy MH, Rush AJ, Crowley G. Cortisol, estradiol, and androgens in acutely ill paranoid schizophrenics. *Biol Psychiatry.* 1983;18(5):583–590.

123. di Michele F, Caltagirone C, Bonaviri G, Romeo E, Spalletta G. Plasma dehydroepiandrosterone levels are strongly increased in schizophrenia. *J Psychiatr Res.* 2005;39(3):267–273.

124. Ritsner M, Gibel A, Maayan R, Ratner Y, Ram E, Modai I, Weizman A. State and trait related predictors of serum cortisol to DHEA(S) molar ratios and hormone concentrations in schizophrenia patients. *Eur Neuropsychopharmacol.* 2007;17(4):257–264.

125. Ritsner M, Gibel A, Ram E, Maayan R, Weizman A. Alterations in DHEA metabolism in schizophrenia: Two-month case-control study. *Eur Neuropsychopharmacol.* 2006;16(2):137–146.

126. Ritsner M, Maayan R, Gibel A, Weizman A. Differences in blood pregnenolone and dehydroepiandrosterone levels between schizophrenia patients and healthy subjects. *Eur Neuropsychopharmacol.* 2007;17(5):358–365.

127. Ritsner MS, Gibel A, Ratner Y, Tsinovoy G, Strous RD. Improvement of sustained attention and visual and movement skills, but not clinical symptoms, after dehydroepiandrosterone augmentation in schizophrenia: a randomized, double-blind, placebo-controlled, crossover trial. *J Clin Psychopharmacol.* 2006;26:495–499.

128. Strous RD, Maayan R, Lapidus R, Goredetsky L, Zeldich E, Kotler M, Weizman A. Increased circulatory dehydroepiandrosterone and dehydroepiandrosterone-sulphate in first-episode schizophrenia: relationship to gender, aggression and symptomatology. *Schizophr Res.* 2004;71(2–3):427–434.

129. Yilmaz N, Herken H, Cicek HK, Celik A, Yurekli M, Akyol O. Increased levels of nitric oxide, cortisol and adrenomedullin in patients with chronic schizophrenia. *Med Princ Pract.* 2007;16(2):137–141.

130. Ritsner M, Maayan R, Gibel A, Strous RD, Modai I, Weizman A. Elevation of the cortisol/dehydroepiandrosterone ratio in schizophrenia patients. *Eur Neuropsychopharmacol.* 2004;14(4):267–273.

131. Maayan R, Shaltiel G, Poyurovsky M, Ramadan E, Morad O, Nechmad A, Weizman A, Agam G. Chronic lithium treatment affects rat brain and serum dehydroepiandrosterone (DHEA) and DHEA-sulphate (DHEA-S) levels. *Int J Neuropsychopharmacol.* 2004;7:71–75.

132. Nechmad A, Maayan R, Ramadan E, Morad O, Poyurovsky M, Weizman A. Clozapine decreases rat brain dehydroepiandrosterone and dehydroepiandrosterone sulfate levels. *Eur Neuropsychopharmacol.* 2003;13(1):29–31.

133. Oseko F, Morikawa K, Nakano A, Note S, Endo J, Taniguchi A, Kono T, Imura H. Effect of chronic hyperprolactinemia induced by sulpiride on plasma dehydroepiandrosterone (DHA) in normal men. *Andrologia.* 1986;18(5):523–528.

134. Methodology WCCfDS. *Guidelines for ATC Classification and DDD assignment*, 3rd ed. Oslo: WHO; 2000.

135. Silver H, Knoll G, Isakov V, Goodman C, Finkelstein Y. Blood DHEAS concentrations correlate with cognitive function in chronic schizophrenia patients. A pilot study. *J Psychiatr Res.* 2005;39(6):569–575.

136. Maayan R, Yagorowski Y, Grupper D, Weiss M, Shtaif B, Abou Kaoud M, Weizman A. Basal plasma dehydroepiandrosterone sulfate level: a possible predictor for response

to electroconvulsive therapy in depressed psychotic inpatients. *Biol Psychiatry*. 2000;48(7):693–701.

137. Ritsner M, Gibel A, Maayan R, Ratner Y, Ram E, Biadsy H, Modai I, Weizman A. Cortisol/Dehydroepiandrosterone ratio and responses to antipsychotic treatment in schizophrenia. *Neuropsychopharmacology*. 2005;30(10): 1913–1922.

138. Goyal RO, Sagar R, Ammini AC, Khurana ML, Alias AG. Negative correlation between negative symptoms of schizophrenia and testosterone levels. *Ann N Y Acad Sci*. 2004;1032:291–294.

139. Shirayama Y, Hashimoto K, Suzuki Y, Higuchi T. Correlation of plasma neurosteroid levels to the severity of negative symptoms in male patients with schizophrenia. *Schizophr Res*. 2002;58(1):69–74.

140. Harris DS, Wolkowitz OM, Reus VI. Movement disorder, memory, psychiatric symptoms and serum DHEA levels in schizophrenic and schizoaffective patients. *World J Biol Psychiatry*. 2001;2(2):99–102.

141. Morales AJ, Haubrich RH, Hwang JY, Asakura H, Yen SS. The effect of six months treatment with a 100 mg daily dose of dehydroepiandrosterone (DHEA) on circulating sex steroids, body composition and muscle strength in age-advanced men and women. *Clin Endocrinol (Oxf)*. 1998;49(4):421–432.

142. Strous RD, Maayan R, Lapidus R, Stryjer R, Lustig M, Kotler M, Weizman A. Dehydroepiandrosterone augmentation in the management of negative, depressive, and anxiety symptoms in schizophrenia. *Arch Gen Psychiatry*. 2003;60:133–141.

143. Strous RD, Stryjer R, Maayan R, Gal G, Viglin D, Katz E, Eisner D, Weizman A. Analysis of clinical symptomatology, extrapyramidal symptoms and neurocognitive dysfunction following dehydroepiandrosterone (DHEA) administration in olanzapine treated schizophrenia patients: a randomized, double-blind placebo controlled trial. *Psychoneuroendocrinology*. 2007;32(2):96–105.

144. Nachshoni T, Ebert T, Abramovitch Y, Assael-Amir M, Kotler M, Maayan R, Weizman A, Strous RD. Improvement of extrapyramidal symptoms following dehydroepiandrosterone (DHEA) administration in antipsychotic treated schizophrenia patients: a randomized, double-blind placebo controlled trial. *Schizophr Res*. 2005;79(2–3):251–256.

145. Strous RD, Gibel A, Weizman A, Maayan R, Ritsner MS. Hormonal response on dehydroepiandrosterone administration in schizophrenia: findings from a randomized, double-blind, placebo-controlled, crossover study. *J Clin Psychopharmacol*. 2008; 28(4):456–459.

146. Linkowski P, Mendlewicz J, Leclercq R, Brasseur M, Hubain P, Golstein J, Copinschi G, Van Cauter E. The 24-hour profile of adrenocorticotropin and cortisol in major depressive illness. *J Clin Endocrinol Metab*. 1985;61(3):429–438.

147. Souetre E, Salvati E, Belugou JL, Pringuey D, Candito M, Krebs B, Ardisson JL, Darcourt G. Circadian rhythms in depression and recovery: evidence for blunted amplitude as the main chronobiological abnormality. *Psychiatry Res*. 1989;28(3):263–278.

148. Pfohl B, Sherman B, Schlechte J, Winokur G. Differences in plasma ACTH and cortisol between depressed patients and normal controls. *Biol Psychiatry*. 1985;20(10):1055–1072.

149. Deuschle M, Schweiger U, Weber B, Gotthardt U, Korner A, Schmider J, Standhardt H, Lammers CH, Heuser I. Diurnal activity and pulsatility of the hypothalamus-pituitary-adrenal system in male depressed patients and healthy controls. *J Clin Endocrinol Metab*. 1997;82(1):234–238.

150. Mortola JF, Liu JH, Gillin JC, Rasmussen DD, Yen SS. Pulsatile rhythms of adrenocorticotropin (ACTH) and cortisol in women with endogenous depression: evidence for increased ACTH pulse frequency. *J Clin Endocrinol Metab*. 1987;65(5):962–968.

151. Halbreich U, Asnis GM, Shindledecker R, Zumoff B, Nathan RS. Cortisol secretion in endogenous depression. II. Time-related functions. *Arch Gen Psychiatry*. 1985;42(9):909–914.

152. Young EA, Carlson NE, Brown MB. Twenty-four-hour ACTH and cortisol pulsatility in depressed women. *Neuropsychopharmacology*. 2001;25(2):267–276.

153. Swade C, Metcalfe M, Coppen A, Mendlewicz J, Linkowski P. Seasonal variations in the dexamethasone suppression test. *J Affect Disord*. 1987;13(1):9–11.

154. Monteleone P, Piccolo A, Martino M, Maj M. Seasonal variation in the Dexamethasone Suppression Test: a longitudinal study in chronic schizophrenics and in healthy subjects. *Neuropsychobiology*. 1994;30:61–65.

155. Iancu I, Tchernihovsky E, Maayan R, Poreh A, Dannon P, Kotler M, Weizman A, Strous RD. Circulatory neurosteroid levels in smoking and non-smoking chronic schizophrenia patients. *Eur Neuropsychopharmacol*. 2007;17(8):541–545.

156. Read J, Perry BD, Moskowitz A, Connolly J. The contribution of early traumatic events to schizophrenia in some patients: a traumagenic neurodevelopmental model. *Psychiatry*. 2001;64(4).

157. Watson S, Owen BM, Gallagher P, Hearn AJ, Young AH, Ferrier IN. Family history, early adversity and the hypothalamic-pituitary-adrenal (HPA) axis: Mediation of the vulnerability to mood disorders. *Neuropsychiatr Dis Treat*. 2007;3(5):647–653.

158. BDWG. Biomarkers and surrogate endpoints: preferred definitions and conceptual framework. *Clin Pharmacol Ther*. 2001;69(3):89–95.

159. Nierenberg AA, Feinstein AR. How to evaluate a diagnostic marker test. Lessons from the rise and fall of dexamethasone suppression test. *JAMA*. 1988;259(11):1699–1702.

160. Lowy MT, Meltzer HY. Dexamethasone bioavailability: implications for DST research. *Biol Psychiatry*. 1987; 22(3):373–385.

161. Cassidy F, Ritchie JC, Verghese K, Carroll BJ. Dexamethasone metabolism in dexamethasone suppression test suppressors and nonsuppressors. *Biol Psychiatry*. 2000;47(7):677–680.

162. Heuser I, Yassouridis A, Holsboer F. The combined dexamethasone/CRH test: a refined laboratory test for psychiatric disorders. *J Psychiatr Res*. 1994;28(4):341–356.

163. Rybakowski JK, Twardowska K. The dexamethasone/corticotropin-releasing hormone test in depression in

bipolar and unipolar affective illness. *J Psychiatr Res.* 1999;33(5):363–370.

164. Watson S, Gallagher P, Ritchie JC, Ferrier IN, Young AH. Hypothalamic-pituitary-adrenal axis function in patients with bipolar disorder. *Br J Psychiatry.* 2004; 184(6):496–502.

165. Watson S, Gallagher P, Smith MS, Ferrier IN, Young AH. The dex/CRH test – is it better than the DST? *Psychoneuroendocrinology.* 2006;31:889–894.

166. Watson S, Gallagher P, Smith MS, Nicol Ferrier I, Young AH. The dex/CRH test and the DST in patients with bipolar disorder and major depressive disorder. *Psychoneuroendocrinology.* 2007;32(1):92–94.

167. Ribeiro SC, Tandon R, Grunhaus L, Greden JF. The DST as a predictor of outcome in depression: a meta-analysis. *Am J Psychiatry.* 1993;150(11):1618–1629.

168. Ising M, Kunzel HE, Binder EB, Nickel T, Modell S, Holsboer F. The combined dexamethasone/CRH test as a potential surrogate marker in depression. *Prog Neuropsychopharmacol Biol Psychiatry.* 2005;29(6):1085–1093.

169. Ising M, Horstmann S, Kloiber S, Lucae S, Binder EB, Kern N, Kunzel HE, Pfennig A, Uhr M, Holsboer F. Combined dexamethasone/Corticotropin Releasing Hormone test predicts treatment response in major depression-a potential biomarker? *Biol Psychiatry.* 2007;62(1):47–54.

170. Wolkowitz OM, Reus VI, Keebler A, Nelson N, Friedland M, Brizendine L, Roberts E. Double-blind treatment of major depression with dehydroepiandrosterone. *Am J Psychiatry.* 1999;156(4):646–649.

171. Wolkowitz OM, Reus VI, Roberts E, Manfredi F, Chan T, Ormiston S, Johnson R, Canick J, Brizendine L, Weingartner H. Antidepressant and cognition-enhancing effects of DHEA in major depression. *Ann N Y Acad Sci.* 1995;774:337–339.

172. Gallagher P, Malik N, Newham J, Young AH, Ferrier IN, Mackin P. Antiglucocorticoid treatments for mood disorders. *Cochrane Database Syst Rev.* 2008(1):CD005168.

173. Morrow AL. Recent developments in the significance and therapeutic relevance of neuroactive steroids – Introduction to the special issue. *Pharmacol Ther.* 2007;116 1–6.

174. Hosie AM, Wilkins ME, Smart TG. Neurosteroid binding sites on GABAA receptors. *Pharmacol Ther.* 2007;116:7–19.

175. Watson S, Thompson JM, Malik N, Nicol Ferrier I, Young AH. Temporal stability of the dex/CRH test in patients with rapid-cycling bipolar I disorder: a pilot study. *Aust N Z J Psychiatry.* 2005;39(4):244–248.

176. Halbreich U, Zumoff B, Kream J, Fukushima DK. The mean 1300–1600 h plasma cortisol concentration as a diagnostic test for hypercortisolism. *J Clin Endocrinol Metab.* 1982;54(6):1262–1264.

177. Sachar EJ, Puig-Antich J, Ryan ND, Asnis GM, Rabinovich H, Davies M, Halpern FS. Three tests of cortisol secretion in adult endogenous depressives. *Acta Psychiatr Scand.* 1985;71(1):1–8.

Chapter 31
Neuroactive Steroid Biomarkers of Alcohol Sensitivity and Alcoholism Risk

A. Leslie Morrow and Patrizia Porcu

Abstract Neuroactive steroids are endogenous modulators of neuronal excitability. The GABAergic neuroactive steroids contribute to regulation of synaptic and extrasynaptic inhibitory transmission across brain, hypothalamic–pituitary–adrenal (HPA) axis function, as well as inflammatory processes and myelin formation. These actions are being translated to new treatments for neurologic and psychiatric conditions. Recent data in animal models suggests a potential therapeutic role for neuroactive steroids in the treatment of alcoholism, depression and premenstrual dysphoric disorders. In addition, neuroactive steroid responses to physiological and/or pharmacological challenge may represent useful biomarkers of alcoholism risk. In particular, the lack of dexamethasone suppression of plasma deoxycorticosterone in ethanol-naïve monkeys predicts the subsequent development of heavy voluntary alcohol consumption. These studies point to the opportunity for biomarker development related to heavy drinking and possibly alcoholism risk. In addition, neuroactive steroid responses to activation of the HPA axis may predict ethanol sensitivity and this factor has been shown to predict alcoholism risk in sons and daughters of alcoholics. Hence, neuroactive steroid responses to pharmacological challenges have potential utility as biomarkers of alcoholism risk.

Keywords GABAergic neuroactive steroid · alcoholism

A. L. Morrow
Departments of Psychiatry and Pharmacology, Bowles Center for Alcohol Studies, University of North Carolina School of Medicine, Chapel Hill, USA

P. Porcu
Department of Psychiatry, Bowles Center for Alcohol Studies, University of North Carolina School of Medicine, Chapel Hill, USA

Abbreviations GABA: γ-Aminobutyric acid; DHEA: Dihydroepiandrosterone; $3\alpha,5\alpha$-THP: 3α-Hydroxy-5α-pregnan-20-one; $3\alpha,5\alpha$-THDOC: $3\alpha,21$-Dihydroxy-5α-pregnan-20-one; HPA: Hypothalamic–pituitary–adrenal; CNS: Central nervous system; CRF: Corticotropin releasing factor; ACTH: Adrenocorticotropic hormone; PMDD: Premenstrual dysphoric disorder

Physiological Relevance of Neuroactive Steroids

Neuroactive steroids are endogenous neuromodulators that can be synthesized de novo in the brain as well as adrenal glands, ovaries and testes (for review, see Biggio and Purdy[1]). The biosynthetic pathway for these steroids is shown in Fig. 31.1. Among these compounds, the $3\alpha,5\alpha$- and $3\alpha,5\beta$-reduced metabolites of progesterone,[2,3] deoxycorticosterone,[2,3] dihydroepiandrosterone (DHEA)[4–6] and testosterone[6,7] enhance GABAergic neurotransmission and produce inhibitory neurobehavioral effects such as anxiolytic, anticonvulsant and sedative actions. The excitatory neuroactive steroids include the sulfated derivatives of pregnenolone and DHEA[8] as well as the $3\alpha,5\alpha$- and $3\alpha,5\beta$-reduced metabolites of cortisol.[9] All of these steroids are weak inhibitors of γ-aminobutyric acid type A ($GABA_A$) receptors while both pregnenolone sulfate and DHEA sulfate also act as weak N-methyl-D-aspartate (NMDA) receptor agonists (micromolar potency). Pregnenolone sulfate also inhibits GABA release with nanomolar potency,[10] however it is controversial whether it is present in brain under normal physiological conditions.[11,12]

The physiological significance of these endogenous modulators continues to emerge. The GABAergic neuroactive steroids clearly regulate synaptic and extrasynaptic inhibitory transmission across brain,[13] hypothalamic–pituitary–adrenal (HPA) axis function,[14] inflammatory processes[15] and myelin formation.[16] Recently, the binding sites for neuroactive steroids on GABA$_A$ receptors have been discovered[17] and clinical trials for epilepsy[18] and traumatic brain injury[19,20] have been successful. Furthermore, data in animal models points to the potential value of neuroactive steroids in other central nervous system (CNS) disorders including depression,[21,22] schizophrenia,[23] alcoholism,[24] multiple sclerosis[16] and other neurodegenerative conditions.[25,26] Most of these studies have focused on the progesterone and deoxycorticosterone derivatives, 3α-hydroxy-5α-pregnan-20-one (3α,5α-THP) and 3α,21-dihydroxy-

5α-pregnan-20-one (3α,5α-THDOC), although the presence of GABAergic neuroactive derivatives of DHEA and testosterone as well as the 3α,5β derivatives of these steroids may also be relevant in specific situations. Considering the abundance of precursors and the common metabolic enzymes, it is likely that the GABAergic metabolites of progesterone, deoxycorticosterone, DHEA and testosterone are both singularly and coordinately significant physiological regulators of CNS excitability.

Systemic administration of 3α,5α-THP or 3α,5α-THDOC induces anxiolytic, anticonvulsant and sedative-hypnotic effects, similar to other GABA$_A$ receptor positive modulators and ethanol (for review see Morrow et al.[27]). Neuroactive steroids interact with GABA$_A$ receptors via specific binding sites on α subunits[28] that allosterically modulate binding to GABA

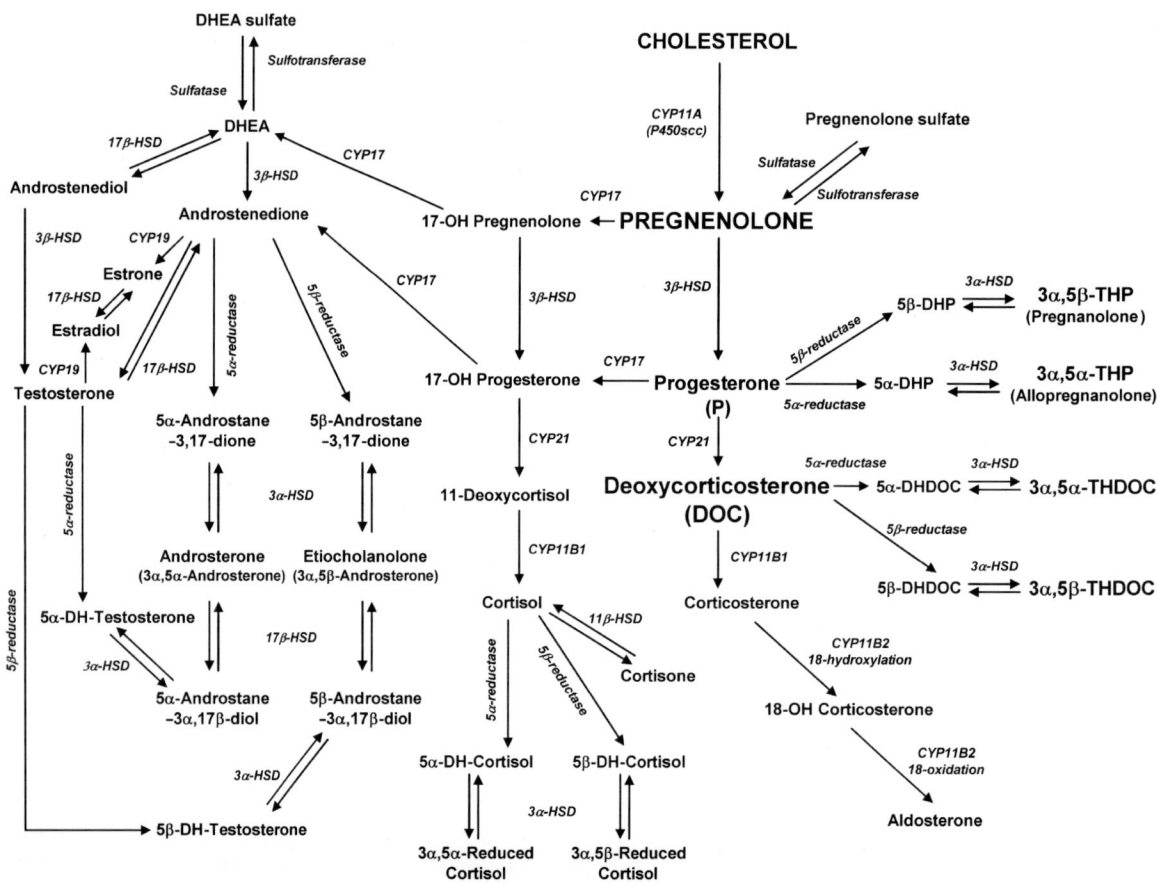

Fig. 31.1 Schematic representation of the neuroactive steroids biosynthetic pathway

and benzodiazepine recognition sites.[29] These steroids alter $GABA_A$ receptor function by enhancing GABA-mediated Cl^- conductance and directly stimulating Cl^- conductance in voltage clamp studies and [$^{36}Cl^-$] flux studies.[2,3,30] Neuroactive steroids appear to interact with multiple neurosteroid recognition sites[31,32] and these sites may differentiate direct gating of Cl^- versus allosteric modulation of GABA-mediated conductance[32] or represent different properties of recognition sites on distinct $GABA_A$ receptor subtypes.[33,34] Neuroactive steroids modulate both synaptic and extra-synaptic $GABA_A$ receptors with lower potency at synaptic receptors that contain $\gamma2$ subunits. Recent studies have highlighted the importance of tonic inhibition that is mediated by extrasynaptic receptors.[35,36] Studies with $\delta^{-/-}$ mice demonstrate that the receptors mediating tonic conductance contain a δ subunit.[37] Furthermore, the enhancement of the tonic conductance by $3\alpha,5\alpha$-THDOC is blunted in neurons derived from $\delta^{-/-}$ mice.[37] Recombinant $\alpha1\beta\delta$, $\alpha4\beta\delta$ and $\alpha6\beta\delta$ receptor subtypes are also more sensitive to neuroactive steroid enhancement of GABA-evoked responses than those mediated by receptors containing the γ subunit.[38,39]

Studies of the structural requirements for neuroactive steroid activity at $GABA_A$ receptors include 3α reduction and $5\alpha/5\beta$ reduction of the A ring as well as hydroxylation of C21.[40] The 5β-reduced metabolites of progesterone and deoxycorticosterone, $3\alpha,5\beta$-THP and $3\alpha,5\beta$-THDOC, are equipotent modulators of GABAergic transmission.[31,41,42] Humans synthesize these 5β-reduced neuroactive steroids and the concentrations of $3\alpha,5\beta$-THP are physiologically relevant and comparable to those of $3\alpha,5\alpha$-THP in human plasma and cerebrospinal fluid.[21,43] In addition, $3\alpha,5\alpha$- and $3\alpha,5\beta$-reduced cortisol have antagonist properties at both GABA and neurosteroid recognition sites of $GABA_A$ receptors and these compounds are the most abundant metabolites of cortisol in human urine.[9]

Role of Neuroactive Steroids in HPA Axis Function

Stress increases plasma and brain levels of GABAergic neuroactive steroids in rodents.[44,45] In rat brain, the increase in $3\alpha,5\alpha$-THP reaches pharmaco-logically significant concentrations in brain between 50 and 100 nM that is sufficient to enhance $GABA_A$ receptor activity and produce behavioral effects. Similarly, stress or corticotropin releasing factor (CRF) infusion elevates $3\alpha,5\alpha$-THP levels in human plasma.[46-48] The levels detected in human plasma are lower than rodent plasma and brain. However, $3\alpha,5\alpha$-THP levels in post-mortem human brain are similar to rat brain and sufficient to have GABAergic activity.[49]

The increase in neuroactive steroid levels elicited by stressful stimuli appears to be mediated by activation of the HPA axis, since it is no longer apparent in adrenalectomized animals.[44,50] Adrenalectomized animals exhibit no circulating concentrations of $3\alpha,5\alpha$-THP and $3\alpha,5\alpha$-THDOC, but brain levels are still detectable,[44] suggesting that brain synthesis plays an important role in neuroactive steroid actions.

The activation of the HPA axis in response to acute stress increases the release of CRF from the hypothalamus that stimulates the release of adrenocorticotropic hormone (ACTH) from the pituitary, which, in turn, stimulates the adrenal cortex to release glucocorticoids, neuroactive steroid precursors and GABAergic neuroactive steroids. Glucocorticoids, mainly cortisol in humans and non-human primates, and corticosterone in rodents, provide negative feedback upon the hypothalamus and pituitary. Likewise, GABAergic neuroactive steroids inhibit CRF production, release, ACTH release and subsequent corticosterone levels in rodents.[51-53] The ability of neuroactive steroids to reduce HPA axis activation may play an important role in returning the animal to homeostasis following stressful events.

This physiological coping response mediated by the HPA axis appears to be critical for mental health since it is dysregulated in various mood disorders including depression, post-traumatic stress disorder and premenstrual dysphoric disorder (PMDD).[54] Restoration of hypothalamic and pituitary homeostasis may be critical for the maintenance of normal responses to subsequent stress challenges. Hence, the blunting of the HPA axis in psychiatric disorders such as depression, alcoholism and PMDD may be indicative of the loss of neuroactive steroid responses to stress as well as insensitivity to glucocorticoid feedback.

Alcohol Mimics Effects of Acute and Chronic Stress

Like stress, systemic administration of moderate ethanol doses (1–2.5 g/kg) increases both brain and plasma levels of 3α,5α-THP and 3α,5α-THDOC in rodents.[55–58] The ethanol-induced increase in neuroactive steroids is mediated by the HPA axis, since it is no longer observed in adrenalectomized animals.[59,60] Indeed, brain synthesis of 3α,5α-THP can be increased by ethanol in adrenalectomized immature animals allowed sufficient time for adaptation,[61] suggesting that brain synthesis of neuroactive steroids may exhibit plasticity in response to physiological challenges. Furthermore, ethanol can increase neuroactive steroids in the hippocampal slice and this results in an enhancement of GABAergic inhibition that can be blocked by the neuroactive steroid biosynthesis inhibitor finasteride.[62]

It is well known that chronic stress results in adaptation of the HPA axis leading to decreases in stress-induced changes in the levels of corticosterone in rats.[63] In contrast, an exaggerated neuroactive steroid response to stress and acute ethanol administration has been shown in animals reared in social isolation,[64,65] and this is accompanied by heightened HPA axis responsiveness.[66] Repeated exposure to alcohol blunts the response of the HPA axis to a second ethanol challenge.[67] This blunting of the HPA axis is associated with reduction in CRF and ACTH elevations following ethanol challenge.[68] In line with these observations, chronic ethanol consumption in rats results in blunted elevation of cerebral cortical 3α,5α-THP[27] and plasma and brain deoxycorticosterone levels following acute ethanol challenge,[69] compared to pair-fed control rats. These findings suggest that there is tolerance to ethanol or stress-induced increases in neuroactive steroid levels. Since decreases in brain neuroactive steroid levels were concomitant with decreases in plasma neuroactive steroid levels, it is likely that the observed decreases in 3α,5α-THP and deoxycorticosterone levels are dependent on blunted HPA axis activity. Thus, adaptations of the HPA axis may include either sensitization or loss of stress-induced increases in GABAergic neuroactive steroids that contribute to the regulation of homeostasis.

Chronic ethanol administration to rodents and humans produces tolerance to ethanol and cross-tolerance to benzodiazepines and barbiturates. In contrast, ethanol dependent rats are sensitized to the anticonvulsant effects of 3α,5α-THP and 3α,5α-THDOC.[70–72] These studies also show that $GABA_A$ receptor sensitivity to 3α,5α-THP and 3α,5α-THDOC is enhanced in ethanol dependent rats, likely due to the reduction of ethanol-induced levels in these animals described above. Since ethanol dependent rats are sensitized to anticonvulsant actions of neuroactive steroids, this class of compounds may be therapeutic during ethanol withdrawal. Indeed, neuroactive steroid therapy may have advantages over benzodiazepine therapy since benzodiazepines exhibit cross-tolerance with ethanol. Further studies are needed to explore this possibility.

Neuroactive Steroid Adaptations in Human Neuropsychiatric Disease

Depression and Premenstrual Dysphoric Disorder

Neuroactive steroid concentrations are altered in various pathological conditions that involve dysfunction of the HPA axis. The HPA axis plays an important role in the pathology of depression: patients with major depression have elevated cortisol levels, a consequence of hypersecretion of CRF due to lowered feedback mechanisms,[73] which also contributes to a blunted dexamethasone response.[74] Some neuroactive steroid concentrations are decreased in patients with major depression as well as in animal models of depression[21,43,64,75] and administration of antidepressant drugs increases these neuroactive steroids in patients and in rodent brain and plasma.[76–80] This increase might be mediated by the HPA axis via an increased serotonin neurotransmission that stimulates the release of CRF (for review see Pisu and Serra[81]). While acute fluoxetine administration increases brain levels of 3α,5α-THP, chronic administration of fluoxetine decreases 3α,5α-THP and 3α,5α-THDOC in rat brain and plasma,[79] probably as a consequence of a reduced basal HPA axis activity induced by antidepressant treatments.[73] Hence, neuroactive steroids may contribute to the therapeutic efficacy of antidepressant medications by contributing to the modulation of the HPA axis as well as by GABAergic inhibition that promotes anxiolysis.

Neuroactive steroids are also altered in PMDD, although the literature is controversial, reporting either decrease, no change or increase in 3α,5α-THP plasma levels.[47,82–89] Differences in analytic methods, diagnostic criteria, or presence of other co-morbid psychiatric disorders might account for these discrepancies. Furthermore, PMDD patients had a blunted 3α,5α-THP response to stress[47] and to HPA axis challenges.[89] Women with a history of depression, regardless of PMDD symptoms, also had a blunted 3α,5α-THP response to stress.[90] All this experimental evidence emphasizes the important link between HPA axis function and neuroactive steroid levels in the maintenance of homeostasis and healthy brain function.

HPA Axis and Neuroactive Steroid Adaptations in Alcoholism

Alterations in HPA axis responsiveness are also found in alcoholism during drinking and abstinence. ACTH and cortisol secretion are increased during ethanol intoxication and acute alcohol withdrawal.[91–98] In contrast, an attenuated responsiveness of the HPA axis has been found in both drinking and abstinent alcohol-dependent patients. Alcohol-dependent patients have low cortisol and 11-deoxycortisol basal levels, show a greater suppression in cortisol and ACTH concentrations following dexamethasone test, and have a reduced cortisol response to exogenous ACTH administered after dexamethasone.[99] Moreover, they have attenuated ACTH and cortisol responses after pituitary stimulation by ovine or human CRF[100–103] and an altered ACTH response to naloxone.[104] An altered cortisol and ACTH response to ovine CRF and naloxone have also been found in sons of alcoholics.[105–107] These data are consistent with the idea that HPA axis dysregulation may contribute to altered neuroactive steroid responses in human alcoholism.

Basal levels of the GABAergic neuroactive steroid 3α,5α-THP are reduced during ethanol withdrawal in humans,[108] when circulating cortisol levels are elevated.[91,92,94] Furthermore, Porcu et al.[109] recently showed that abstinent alcoholic patients show a blunted pregnenolone sulfate response to adrenal stimulation and a delayed deoxycorticosterone response to ovine CRF challenge, suggesting that blunting of the HPA axis also impacts the GABAergic neuroactive steroid responses to stress or ethanol.

Further studies of the GABAergic neuroactive steroids in alcoholics are needed.

Rationale for Neuroactive Steroid Biomarkers of Disease Risk

Potential Development of Neuroactive Steroid Biomarkers of Risk for Alcoholism

The risk of developing alcoholism is influenced by a myriad of factors, both genetic and environmental. Two factors that increase alcoholism risk are low alcohol sensitivity[110] and heavy/binge drinking.[111] In studies over the past 30 years, Schuckit's group has demonstrated that reduced sensitivity to alcohol challenge, including a reduced cortisol response, in sons and daughters of alcoholics is predictive of the eventual development of alcoholism.[110,112] The relationship between low alcohol sensitivity and high alcohol consumption is also observed in many rodent models of alcoholism. Data from preclinical and clinical studies presented below will suggest that elevations in GABAergic neuroactive steroids may contribute to the sensitivity to ethanol challenge and therefore be protective against the development of alcoholism. Heavy drinking is a requisite for the development of alcoholism, therefore a biomarker for risk of heavy drinking may be useful, particularly in those subjects with genetic vulnerability. Dexamethasone suppression of deoxycorticosterone in ethanol naïve cynomolgus monkeys has been demonstrated to predict subsequent heavy drinking[113] and might therefore represent a biomarker of risk. Therefore, neuroactive steroid responses may have potential utility as biomarkers in alcoholism.

Neuroactive Steroids Mediate Ethanol Sensitivity

Systemic administration of moderate ethanol doses (1–2.5 g/kg) increases both plasma and brain levels of 3α,5α-THP and 3α,5α-THDOC in rodents.[55–58,60] Ethanol-induced elevations in neuroactive steroids reach physiologically relevant concentrations that are capable of enhancing GABAergic transmission. A large body of evidence from multiple laboratories suggests that ethanol-induced elevations of GABAergic neuroactive steroids contribute to many behavioral effects of ethanol in rodents. Neuroactive steroids have been shown to modulate ethanol's anticonvulsant effects,[58] sedation,[59] impairment of spatial memory,[27,114] anxiolytic-like[115] and antidepressant-like[116] actions. Each of these behavioral responses is prevented by pretreatment with the neuroactive steroid biosynthesis inhibitor finasteride and/or by prior adrenalectomy. The hypnotic effect of ethanol is partially blocked by adrenalectomy. Importantly, administration of 5α-dihydroprogesterone, the immediate precursor of 3α,5α-THP, to adrenalectomized rats restores effects of ethanol, showing that brain synthesis of neuroactive steroids modulates effects of ethanol.[59] However, neuroactive steroids do not appear to influence the motor incoordinating effects of ethanol, since neither finasteride administration or adrenalectomy diminish these actions.[117] Taken together, these studies suggest that elevations in neuroactive steroids influence many of the GABAergic effects of ethanol *in vivo* and contribute to sensitivity to behavioral effects of ethanol.

We have suggested that ethanol-induced elevations of GABAergic neuroactive steroids contribute to ethanol sensitivity and protect against the risk for ethanol dependence.[24] Diminished elevations of GABAergic neuroactive steroids following ethanol exposure would result in reduced sensitivity to the anxiolytic, sedative, anticonvulsant, cognitive-impairing and discriminative stimulus properties of ethanol. Reduced sensitivity to ethanol is associated with greater risk for the development of alcoholism in individuals with genetic vulnerability to alcoholism.[110,118] Moreover, individuals with the GABA$_A$ receptor α2 subunit polymorphism associated with alcohol dependence are markedly less sensitive to the subjective effects of ethanol compared to individuals that lack this polymorphism.[119] Likewise, rats and mice with low sensitivity to various behavioral effects of alcohol tend to self-administer greater amounts of ethanol in laboratory settings. Taken together, these observations suggest that ethanol-induced elevations of GABAergic neuroactive steroids in brain may underlie important aspects of ethanol sensitivity that may serve to prevent excessive alcohol consumption. The loss of these responses may promote excessive alcohol consumption to achieve the desired effects of ethanol. The lack of neuroactive steroid elevations in response to ethanol could underlie innate ethanol tolerance or ethanol

tolerance induced by long-term ethanol use. Indeed, the observation that finasteride diminished the subjective effects of ethanol in subjects lacking the GABA$_A$ receptor α2 subunit polymorphism, but had no effect in subjects that harbor the alcoholism-associated polymorphism[119] is consistent with the idea that GABAergic neuroactive steroid responses contribute to ethanol sensitivity in humans and may be a factor in the risk for alcoholism. The use of neuroactive steroids as a biomarker of ethanol response is not likely in humans since circulating levels of these compounds are not elevated following challenge with safe doses of ethanol.[120,121] However, neuroactive steroid elevations in response to naloxone or CRF challenge may be useful to evaluate healthy stress responses in vulnerable individuals.

Dexamethasone Suppression of Deoxycorticosterone Predicts Heavy Drinking in Cynomolgus Monkeys

While stimulation of the HPA axis increases GABAergic neuroactive steroids and their precursors in rodent brain and plasma, little data are available for non-human primates. We have recently

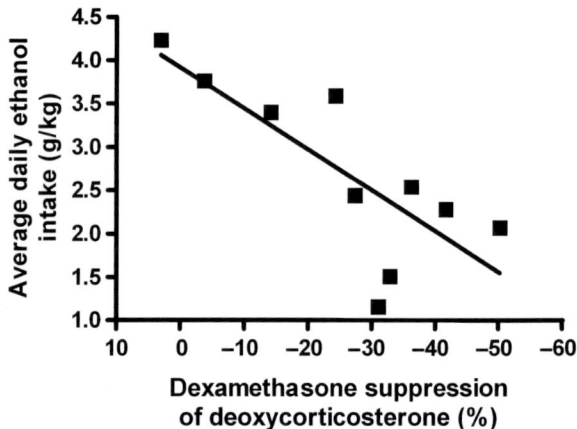

Fig. 31.2 Dexamethasone-induced changes in plasma deoxycorticosterone levels are correlated with subsequent average daily ethanol intake (g/kg) over 12 months of alcohol self-administration in cynomolgus monkeys. Pearson correlation: r = −0.78, p = 0.006. All data are obtained from ten monkeys (Adapted from Porcu et al.[113])

demonstrated that plasma deoxycorticosterone and pregnenolone levels in ethanol naïve cynomolgus monkeys are differentially regulated by various challenges to the HPA axis.[113,122] Plasma deoxycorticosterone levels were sensitive to hypothalamic and pituitary activation of the axis and to negative feedback mechanisms assessed by the dexamethasone test. In contrast, administration of ACTH (10 ng/kg) 4–6 h after 0.5 mg/kg dexamethasone had no effect on plasma deoxycorticosterone levels, suggesting that deoxycorticosterone synthesis is independent of ACTH stimulation of the adrenals.[113] Pregnenolone levels were also elevated by naloxone administration, but not altered by CRF or dexamethasone, while ACTH, 4–6 h following 0.5 mg/kg dexamethasone, decreased plasma pregnenolone levels by 43%. Although pregnenolone is a precursor of deoxycorticosterone and both steroids were measured over 120 min following the same challenges, circulating levels were clearly subject to complex regulation involving factors other than direct HPA axis modulation. The mechanisms that underlie the regulation of these neuroactive steroids are therefore not well understood and not predicted by classical theories of HPA axis activation[24].

Despite the surprising regulation of deoxycorticosterone by challenges of the HPA axis, studies of dexamethasone suppression in alcohol naïve monkeys were linked to subsequent drinking levels following induction of drinking and voluntary consumption for over 18 months. The degree of dexamethasone suppression of deoxycorticosterone, but not cortisol, was predictive of subsequent alcohol drinking in these monkeys. That is, the highest alcohol drinking was found in the monkeys that showed the lowest suppression of deoxycorticosterone levels in response to dexamethasone[113] (Fig. 31.2). In contrast, no other deoxycorticosterone responses to HPA axis stimulation in ethanol naïve monkeys were predictive of subsequent voluntary drinking. The effect of dexamethasone on plasma deoxycorticosterone levels in monkeys appears to be a trait marker of risk for high alcohol consumption. Since deoxycorticosterone is a precursor of GABAergic neuroactive epimers of THDOC formed by 3α,5α and 3α,5β reduction, and these steroids can be altered by acute and chronic ethanol exposure,[108,56] the levels of these steroids may also be involved in the association of drinking risk.

Conclusions and Future Directions

The elevation of GABAergic neuroactive steroids in response to various stress challenges may be indicative of physiological coping, normal ethanol sensitivity and reduced vulnerability to depression or alcoholism. Blunted GABAergic neuroactive steroid stress responses might be useful as a biomarker of risk or vulnerability to these disorders. Indeed, this factor may contribute to the frequent comorbidity of depression and alcoholism and therefore this biomarker would not be expected to differentiate between diagnoses. In contrast, absent or weak dexamethasone suppression of deoxycorticosterone, but not cortisol, holds promise as a biomarker of risk for alcoholism. This biomarker was not accompanied by blunting of deoxycorticosterone responses to naloxone, CRF or ACTH challenges of the HPA axis. Further studies are needed to determine whether this trait will translate to humans and show predictive validity for alcoholism risk.

Acknowledgements This work was supported by NIH grants R37-AA10564, UO1-AA016672 (ALM).

References

1. Biggio G and Purdy RH. Neurosteroids and brain function. In: Biggio G and Purdy RH (ed) International Review of Neurobiology, Vol 46. Academic Press, New York; 2001
2. Majewska MD, Harrison NL, Schwartz RD, et al. Steroid hormone metabolites are barbiturate-like modulators of the GABA receptor. Science 1986;232:1004–1007
3. Morrow AL, Suzdak PD and Paul SM. Steroid hormone metabolites potentiate GABA receptor-mediated chloride ion flux with nanomolar potency. Eur J Pharmacol 1987;142:483–485
4. Frye CA, Van Keuren KR, and Erskine MS. Behavioral effects of 3α-androstanediol 1 Modulation of sexual receptivity and promotion of GABA-stimulated chloride flux. Behav Brain Res 1996;79:109–118
5. Park-Chung M, Malayev A, Purdy RH, et al. Sulfated and unsulfated steroids modulate gamma-aminobutyric acid$_A$ receptor function through distinct sites. Brain Res 1999;830:72–87
6. Kaminski RM, Marini H, Kim W and Rogawski MA. Anticonvulsant activity of androsterone and etiocholanolone. Epilepsia 2005;46:819–827
7. Kaminski RM, Marini H, Ortinski PI, et al. The pheromone androstenol (5 alpha-androst-16-en-3 alpha-ol) is a neurosteroid positive modulator of GABA$_A$ receptors. J Pharmacol Exp Ther 2006;317:694–703
8. Farb DH and Gibbs TT, Steroids as modulators of amino acid receptor function. New York: CRC Press; 1996: 23–36

9. Penland S and Morrow AL. 3α,5β-Reduced cortisol exhibits antagonist properties on cerebral cortical GABA$_A$ receptors. Eur J Pharmacol 2004;506:129–132

10. Mtchedlishvili Z and Kapur J. A presynaptic action of the neurosteroid pregnenolone sulfate on GABAergic synaptic transmission. Mol Pharmacol 2003;64:857–864

11. Liu S, Griffiths WJ and Sjovall J. Capillary liquid chromatography/electrospray mass spectrometry for analysis of steroid sulfates in biological samples. Anal Chem 2003;75:791–797

12. Liere P, Pianos A, Eychenne B, et al. Novel lipoidal derivatives of pregnenolone and dehydroepiandrosterone and absence of their sulfated counterparts in rodent brain. J Lipid Res 2004;45:2287–2302

13. Herd MB, Belelli D and Lambert JJ. Neurosteroid modulation of synaptic and extrasynaptic GABA$_A$ receptors. Pharmacol Ther 2007;116:20–34

14. Morrow AL. Recent developments in the significance and therapeutic relevance of neuroactive steroids – Introduction to the special issue. Pharmacol Ther 2007;116(1):1–6

15. He J, Evans CO, Hoffman SW, et al. Progesterone and allopregnanolone reduce inflammatory cytokines after traumatic brain injury. Exp Neurol 2004;189:404–412

16. Schumacher M, Guennoun R, Stein DG and De Nicola AF. Progesterone: therapeutic opportunities for neuroprotection and myelin repair. Pharmacol Ther 2007;116:77–106

17. Hosie AM, Wilkins ME and Smart TG. Neurosteroid binding sites on GABA$_A$ receptors. Pharmacol Ther 2007;116:7–19

18. Pieribone VA, Tsai J, Soufflet C, et al. Clinical evaluation of ganaxolone in pediatric and adolescent patients with refractory epilepsy. Epilepsia 2007;48:1870–1874

19. Wright DW, Kellermann AL, Hertzberg VS, et al. ProTECT: a randomized clinical trial of progesterone for acute traumatic brain injury. Ann Emerg Med 2007;49:391–402

20. Stein DG. Progesterone exerts neuroprotective effects after brain injury. Brain Res Rev 2008;57:386–397

21. Uzunova V, Sheline Y, Davis JM, et al. Increase in the cerebrospinal fluid content of neurosteroids in patients with unipolar major depression who are receiving fluoxetine or fluvoxamine. Proc Natl Acad Sci USA 1998;95:3239–3244

22. Uzunova V, Sampson L and Uzunov DP. Relevance of endogenous 3α-reduced neurosteroids to depression and antidepressant action. Psychopharmacology 2006;186:351–361

23. Marx CE, Shampine LJ, Khisti RT, et al. Olanzapine and fluoxetine administration and coadministration increase rat hippocampal pregnenolone, allopregnanolone and peripheral deoxycorticosterone: implications for therapeutic actions. Pharmacol Biochem Behav 2006;84:609–617

24. Morrow AL, Porcu P, Boyd KN and Grant KA. Hypothalamic-pituitary-adrenal axis modulation of GABAergic neuroactive steroids influences ethanol sensitivity and drinking behavior. Dialogues Clin Neurosci 2006;8:463–477

25. Griffin LD, Gong W, Verot L and Mellon SH. Niemann-Pick type C disease involves disrupted neurosteroidogenesis and responds to allopregnanolone. Nat Med 2004;10:704–711

26. Mellon SH. Neurosteroid regulation of central nervous system development. Pharmacol Ther 2007;116:107–124

27. Morrow AL, VanDoren MJ, Penland SN and Matthews DB. The role of GABAergic neuroactive steroids in ethanol action, tolerance and dependence. Brain Res Brain Res Rev 2001;37:98–109

28. Hosie AM, Wilkins ME, da Silva HM and Smart TG. Endogenous neurosteroids regulate GABA$_A$ receptors through two discrete transmembrane sites. Nature 2006;444:486–489

29. Gee KW, Brinton RE, Chang WC and McEwen BS. Gamma-aminobutyric acid-dependent modulation of the chloride ionophore by steroids in rat brain. Eur J Pharmacol 1987;136:419–423

30. Harrison NL, Majewska MD, Harrington JW and Barker JL. Structure-activity relationships for steroid interaction with the gamma-aminobutyric acid-A receptor complex. J Pharmacol Exp Ther 1987;241:346–353

31. Morrow AL, Pace JR, Purdy RH and Paul SM. Characterization of steroid interactions with γ-aminobutyric acid receptor-gated chloride ion channels: evidence for multiple steroid recognition sites. Mol Pharmacol 1990;37:263–270

32. Puia G, Santi M, Vicini S, et al. Neurosteroids act on recombinant human GABA$_A$ receptors. Neuron 1990;4:759–765

33. Puia G, Vicini S, Seeburg PH and Costa E. Influence of recombinant GABA$_A$ receptor subunit composition on the action of allosteric modulators of GABA-gated Cl$^-$ currents. Mol Pharmacol 1991;39:691–696

34. Mihalek RM, Banerjee PK, Korpi ER, et al. Attenuated sensitivity to neuroactive steroids in gamma-aminobutyrate type A receptor delta subunit knockout mice. Proc Natl Acad Sci USA 1999;96:12905–12910

35. Semyanov A, Walker MC, Kullmann DM and Silver RA. Tonically active GABA A receptors: modulating gain and maintaining the tone. Trends Neurosci 2004;27:262–269

36. Farrant M and Nusser Z. Variations on an inhibitory theme: phasic and tonic activation of GABA$_A$ receptors. Nat Rev Neurosci 2005;6:215–229

37. Stell BM, Brickley SG, Tang CY, Farrant M and Mody I. Neuroactive steroids reduce neuronal excitability by selectively enhancing tonic inhibition mediated by delta subunit-containing GABA$_A$ receptors. Proc Natl Acad Sci USA 2003;100:14439–14444

38. Vicini S. Pharmacologic significance of the structural heterogeneity of the GABA$_A$ receptor-chloride ion channel complex. Neuropsychopharmacology 1991;4:9–15

39. Belelli D, Casula A, Ling A and Lambert JJ. The influence of subunit composition on the interaction of neurosteroids with GABA$_A$ receptors. Neuropharmacology 2002;43:651–661

40. Paul SM and Purdy RH. Neuroactive steroids. FASEB J 1992;6:2311–2322

41. Callachan H, Cottrell GA, Hather NY, et al. Modulation of the GABA$_A$ receptor by progesterone metabolites. Proc R Soc Lond B Biol Sci 1987;231:359–369

42. Xue BG, Whittemore ER, Park CH, et al. Partial agonism by 3α,21-dihydroxy-5β-pregnan-20-one at the gamma-aminobutyric acid, receptor neurosteroid site. J Pharmacol Exp Ther 1997;281:1095–1101

43. Romeo E, Ströhle A, Spalletta G, et al. Effects of antidepressant treatment on neuroactive steroids in major depression. Am J Psychiatry 1998;155:910–913

44. Purdy RH, Morrow AL, Moore PH, Jr. and Paul SM. Stress-induced elevations of gamma-aminobutyric acid type A receptor-active steroids in the rat brain. Proc Natl Acad Sci USA 1991;88:4553–4557

45. Barbaccia ML, Roscetti G, Trabucchi M, et al. Time-dependent changes in rat brain neuroactive steroid concentrations and

GABA_A receptor function after acute stress. Neuroendocrinology 1996;63:166–172

46. Genazzani AR, Petraglia F, Bernardi F, et al. Circulating levels of allopregnanolone in humans: gender, age, and endocrine influences. J Clin Endocrinol Metab 1998;83(6):2099–2103

47. Girdler SS, Straneva PA, Light KC, Pedersen CA and Morrow AL. Allopregnanolone levels and reactivity to mental stress in premenstrual dysphoric disorder. Biol Psychiatry 2001;49:788–797

48. Droogleever Fortuyn HA, van Broekhoven F, Span PN, et al. Effects of PhD examination stress on allopregnanolone and cortisol plasma levels and peripheral benzodiazepine receptor density. Psychoneuroendocrinology 2004;29:1341–1344

49. Marx CE, Stevens RD, Shampine LJ, et al. Neuroactive steroids are altered in schizophrenia and bipolar disorder: relevance to pathophysiology and therapeutics. Neuropsychopharmacology 2006;31:1249–1263

50. Barbaccia ML, Roscetti G, Trabucchi M, et al. The effects of inhibitors of GABAergic transmission and stress on brain and plasma allopregnanolone concentrations. Brit J Pharmacol 1997;120:1582–1588

51. Owens MJ, Ritchie JC and Nemeroff CB. 5α-Pregnane-3α,21-diol-20-one (THDOC) attenuates mild stress-induced increases in plasma corticosterone via a non-glucocorticoid mechanism: comparison with alprazolam. Brain Res 1992;573:353–355

52. Patchev VK, Shoaib M, Holsboer F and Almeida OFX. The neurosteroid tetrahydroprogesterone counteracts corticotropin-releasing hormone-induced anxiety and alters the release and gene expression of corticotropin-releasing hormone in the rat hypothalamus. Neuroscience 1994;62:265–271

53. Patchev VK, Hassan AHS, Holsboer F and Almeida OFX. The neurosteroid tetrahydroprogesterone attenuates the endocrine response to stress and exerts glucocorticoid-like effects on vasopressin gene transcription in the rat hypothalamus. Neuropsychopharmacology 1996;15:533–540

54. Girdler SS and Klatzkin R. Neurosteroids in the context of stress: implications for depressive disorders. Pharmacol Ther 2007;116:125–139

55. Morrow AL, VanDoren MJ and Devaud LL. Effects of progesterone or neuroactive steroid? Nature 1998;395:652–653

56. Barbaccia ML, Affricano D, Trabucchi M, Purdy RH, Colombo G, Agabio R and Gessa GL. Ethanol markedly increases "GABAergic" neurosteroids in alcohol-preferring rats. Eur J Pharmacol 1999;384:R1–R2

57. Morrow AL, Janis GC, VanDoren MJ, et al. Neurosteroids mediate pharmacological effects of ethanol: A new mechanism of ethanol action? Alcohol Clin Exp Res 1999;23:1933–1940

58. VanDoren MJ, Matthews DB, Janis GC, Grobin AC, Devaud LL and Morrow AL. Neuroactive steroid 3α-hydroxy-5α-pregnan-20-one modulates electrophysiological and behavioral actions of ethanol. J Neurosci 2000;20:1982–1989

59. Khisti RT, VanDoren MJ, O'Buckley TK and Morrow AL. Neuroactive steroid 3α-hydroxy-5α-pregnan-20-one modulates ethanol-induced loss of righting reflex in rats. Brain Res 2003;980:255–265

60. O'Dell LE, Alomary AA, Vallee M, et al. Ethanol-induced increases in neuroactive steroids in the rat brain and plasma are absent in adrenalectomized and gonadectomized rats. Eur J Pharmacol 2004;484:241–247

61. Follesa P, Biggio F, Talani G, et al. Neurosteroids, GABA_A receptors, and ethanol dependence. Psychopharmacology 2006;186:267–280

62. Sanna E, Talani G, Busonero F, Pisu MG, Purdy RH, Serra M and Biggio G. Brain steroidogenesis mediates ethanol modulation of GABA_A receptor activity in rat hippocampus. J Neurosci 2004;24:6521–6530

63. Spencer RL and McEwen BS. Adaptation of the hypothalamic-pituitary-adrenal axis to chronic ethanol stress. Neuroendocrinology 1990;52:481–489

64. Serra M, Pisu MG, Littera M, et al. Social isolation-induced decreases in both the abundance of neuroactive steroids and GABA_A receptor function in rat brain. J Neurochem 2000;75:732–740

65. Serra M, Pisu MG, Floris I, et al. Social isolation-induced increase in the sensitivity of rats to the steroidogenic effect of ethanol. J Neurochem 2003;85:257–263

66. Serra M, Pisu MG, Floris I and Biggio G. Social isolation-induced changes in the hypothalamic-pituitary-adrenal axis in the rat. Stress 2005;8:259–264

67. Lee S and Rivier C. Altered ACTH and corticosterone responses to interleukin-1α in male rats exposed to an alcohol diet: possible role of vasopressin and testosterone. Alcohol Clin Exp Res 1995;19:200–208

68. Lee S, Schmidt ED, Tilders FJ, and Rivier C. Effect of repeated exposure to alcohol on the response of the hypothalamic-pituitary-adrenal axis of the rat: I Role of changes in hypothalamic neuronal activity. Alcohol Clin Exp Res 2001;25:98–105

69. Khisti RT, Boyd KN, Kumar S and Morrow AL. Systemic ethanol administration elevates deoxycorticosterone levels and chronic ethanol exposure attenuates this response. Brain Res 2005;1049:104–111

70. Devaud LL, Purdy RH, Finn DA and Morrow AL. Sensitization of γ-aminobutyric acid_A receptors to neuroactive steroids in rats during ethanol withdrawal. J Pharmacol Exp Ther 1996;278:510–517

71. Devaud LL, Fritschy J-M and Morrow AL. Influence of gender on chronic ethanol-induced alternations of GABA_A receptors in rats. Brain Res 1998;796:222–230

72. Cagetti E, Pinna G, Guidotti A, Baicy K and Olsen RW. Chronic intermittent ethanol (CIE) administration in rats decreases levels of neurosteroids in hippocampus, accompanied by altered behavioral responses to neurosteroids and memory function. Neuropharmacology 2004;46:570–579

73. Holsboer F. Stress, hypercortisolism and corticosteroid receptors in depression: implications for therapy. J Affect Disord 2001;62:77–91

74. Carroll BJ, Feinberg M, Greden JF, et al. A specific laboratory test for the diagnosis of melancholia Standardization, validation, and clinical utility. Arch Gen Psychiatry 1981;38:15–22

75. Matsumoto K, Pinna G, Puia G, et al. Social isolation stress-induced aggression in mice: a model to study the pharmacology of neurosteroidogenesis. Stress 2005;8:85–93

76. Uzunov DP, Cooper TB, Costa E and Guidotti A. Fluoxetine-elicited changes in brain neurosteroid content measured by negative ion mass fragmentography. Proc Natl Acad Sci USA 1996;93:12599–12604

77. Marx CE, Duncan GE, Gilmore JH, Lieberman JA and Morrow AL. Olanzapine increases allopregnanolone in the rat cerebral cortex. Biol Psychiatry 2000;47:1000–1004

78. Barbaccia ML, Affricano D, Purdy RH, et al. Clozapine, but not haloperidol, increases brain concentrations of neuroactive steroids in the rat. Neuropsychopharmacology 2001;25:489–497

79. Serra M, Pisu MG, Muggironi M, et al. Opposite effects of short- versus long-term administration of fluoxetine on the concentrations of neuroactive steroids in rat plasma and brain. Psychopharmacology 2001;158:48–54

80. Marx CE, VanDoren MJ, Duncan GE, et al. Olanzapine and clozapine increase the GABAergic neuroactive steroid allopregnanolone in rodents. Neuropsychopharmacology 2003;28:1–13

81. Pisu MG, Serra M. Neurosteroids and neuroactive drugs in mental disorders. Life Sci 2004;74:3181–3197

82. Schmidt PJ, Purdy RH, Moore PH, et al. Circulating levels of anxiolytic steroids in the luteal phase in women with premenstrual syndrome and in control subjects. J Clin Endocrinol Metab 1994;79:1256–1260

83. Wang M, Seippel L, Purdy RH and Backstrom T. Relationship between symptom severity and steroid variation in women with premenstrual syndrome: study on serum pregnenolone, pregnenolone sulfate, 5α-pregnane-3,20-dione and 3α-hydroxy-5α-pregnan-20-one. J Clin Endocrinol Metab 1996;81:1076–1082

84. Rapkin AJ, Morgan M, Goldman L, et al. Progesterone metabolite allopregnanolone in women with premenstrual syndrome. Obstet Gynecol 1997;90:709–714

85. Bicikova M, Dibbelt L, Hill M, et al. Allopregnanolone in women with premenstrual syndrome. Horm Metab Res 1998;30:227–230

86. Sundstrom I and Backstrom T. Patients with premenstrual syndrome have decreased saccadic eye velocity compared to control subjects. Biol Psychiatry 1998;44:755–764

87. Monteleone P, Luisi S, Tonetti A, et al. Allopregnanolone concentrations and premenstrual syndrome. Eur J Endocrinol 2000;142:269–273

88. Epperson CN, Haga K, Mason GF, et al. Cortical gamma-aminobutyric acid levels across the menstrual cycle in healthy women and those with premenstrual dysphoric disorder: a proton magnetic resonance spectroscopy study. Arch Gen Psychiatry 2002;59:851–858

89. Lombardi I, Luisi S, Quirici B, et al. Adrenal response to adrenocorticotropic hormone stimulation in patients with premenstrual syndrome. Gynecol Endocrinol 2004;18:79–87

90. Klatzkin RR, Morrow AL, Light KC, Pedersen CA and Girdler SS. Histories of depression, allopregnanolone responses to stress, and premenstrual symptoms in women. Biol Psychol 2006;71:2–11

91. Mendelson JH, Ogata M, and Mello NK. Adrenal function and alcoholism I Serum cortisol. Psychosom Med 1971;33:145–157

92. Iranmanesh A, Veldhuis JD, Johnson ML and Lizarralde G. 24-hour pulsatile and circadian patterns of cortisol secretion in alcoholic men. J Androl 1989;10:54–63

93. Adinoff B, Martin PR, Bone GH, et al. Hypothalamic-pituitary-adrenal axis functioning and cerebrospinal fluid corticotropin releasing hormone and corticotropin levels in alcoholics after recent and long-term abstinence. Arch Gen Psychiatry 1990;47:325–330

94. Adinoff B, Risher-Flowers D, De Jong J, et al. Disturbances of hypothalamic-pituitary-adrenal axis functioning during ethanol withdrawal in six men. Am J Psychiatry 1991; 148:1023–1025

95. Waltman C, Blevins LSJ, Boyd G and Wand GS. The effects of mild ethanol intoxication on the hypothalamic-pituitary-adrenal axis in non-alcoholic men. J Clin Endocrinol Metab 1993;77:518–522

96. Inder WJ, Joyce PR, Wells JE, et al. The acute effects of oral ethanol on the hypothalamic-pituitary-adrenal axis in normal human subjects. Clin Endocrinol (Oxf) 1995;42:65–71

97. Hundt W, Zimmermann U, Pottig M, Spring K and Holsboer F. The combined dexamethasone-suppression/CRH-stimulation test in alcoholics during and after acute withdrawal. Alcohol Clin Exp Res 2001;25:687–691.

98. Adinoff B, Ruether K, Krebaum S, et al. Increased salivary cortisol concentrations during chronic alcohol intoxication in a naturalistic clinical sample of men. Alcohol Clin Exp Res 2003;27:1420–1427

99. Adinoff B, Krebaum SR, Chandler PA, et al. Dissection of hypothalamic-pituitary-adrenal axis pathology in 1-month-abstinent alcohol-dependent men, part 1: adrenocortical and pituitary glucocorticoid responsiveness. Alcohol Clin Exp Res 2005;29:517–527

100. Wand GS and Dobs AS. Alterations in the hypothalamic-pituitary-adrenal axis in actively drinking alcoholics. J Clin Endocrinol Metab 1991;72:1290–1295

101. Costa A, Bono G, Martignoni E, Merlo P, Sances G and Nappi G. An assessment of hypothalamo-pituitary-adrenal axis functioning in non-depressed, early abstinent alcoholics. Psychoneuroendocrinology 1996;21:263–275

102. Ehrenreich H, Schuck J, Stender N, etal. Endocrine and hemodynamic effects of stress versus systemic CRF in alcoholics during early and medium term abstinence. Alcohol Clin Exp Res 1997;21:1285–1293

103. Adinoff B, Krebaum SR, Chandler PA, Ye W, Brown MB and Williams MJ. Dissection of hypothalamic-pituitary-adrenal axis pathology in 1-month-abstinent alcohol-dependent men, part 2: response to ovine corticotropin-releasing factor and naloxone. Alcohol Clin Exp Res 2005;29:528–537

104. Inder WJ, Joyce PR, Ellis MJ, et al. The effects of alcoholism on the hypothalamic-pituitary-adrenal axis: interaction with endogenous opioid peptides. Clin Endocrinol (Oxf) 1995;43:283–290

105. Waltman C, McCaul ME and Wand GS. Adrenocorticotropin responses following administration of ethanol and ovine corticotropin-releasing hormone in the sons of alcoholics and control subjects. Alcohol Clin Exp Res 1994;18:826–830

106. Wand GS, Mangold D, El Deiry S, McCaul ME and Hoover D. Family history of alcoholism and hypothalamic opioidergic activity. Arch Gen Psychiatry 1998;55:1114–1119

107. Hernandez-Avila CA, Oncken C, Van Kirk J, Wand GS and Kranzler HR. Adrenocorticotropin and cortisol responses to a naloxone challenge and risk for alcoholism. Biol Psychiatry 2002;51:652–658

108. Romeo E, Brancati A, De Lorenzo A, et al. Marked decrease of plasma neuroactive steroids during alcohol withdrawal. J Pharmacol Exp Ther 1996;247:309–322

109. Porcu P, O'Buckley TK, Morrow AL and Adinoff B. Differential hypothalamic-pituitary-adrenal activation of the neuroactive steroids pregnenolone sulfate and deoxycorticosterone in healthy controls and alcohol-dependent subjects. Psychoneuroendocrinology 2008;33:214–226

110. Schuckit MA. Low level of response to alcohol as a predictor of future alcoholism. Am J Psychiatry 1994;151:184–189

111. Dawson DA, Grant BF and Li TK. Quantifying the risks associated with exceeding recommended drinking limits. Alcohol Clin Exp Res 2005;29:902–908

112. Schuckit MA and Smith TL. An evaluation of the level of response to alcohol, externalizing symptoms, and depressive symptoms as predictors of alcoholism. J Stud Alcohol 2006;67:215–227

113. Porcu P, Grant KA, Green HL, Rogers LS and Morrow AL. Hypothalamic-pituitary-adrenal axis and ethanol modulation of deoxycorticosterone levels in cynomolgus monkeys. Psychopharmacology 2006;186:293–301

114. Matthews DB, Morrow AL, Tokunaga S and McDaniel JR. Acute ethanol administration and acute allopregnanolone administration impair spatial memory in the Morris water task. Alcohol Clin Exp Res 2002;26:1747–1751

115. Hirani K, Sharma AN, Jain NS, Ugale RR and Chopde CT. Evaluation of GABAergic neuroactive steroid 3alpha-hydroxy-5alpha-pregnane-20-one as a neurobiological substrate for the anti-anxiety effect of ethanol in rats. Psychopharmacology 2005;180:267–278

116. Hirani K, Khisti RT and Chopde CT. Behavioral action of ethanol in Porsolt's forced swim test: modulation by 3α-hydroxy-5α-pregnan-20-one. Neuropharmacology 2002; 43:1339–1350

117. Khisti RT, VanDoren MJ, Matthews DB and Morrow AL. Ethanol-induced elevation of 3alpha-hydroxy-5alpha-pregnan-20-one does not modulate motor incoordination in rats. Alcohol Clin Exp Res 2004;28: 1249–1256

118. Schuckit MA and Smith TL. An 8-year follow-up of 450 sons of alcoholic and control subjects. Arch Gen Psychiatry 1996;53:202–210

119. Pierucci-Lagha A, Covault J, Feinn R, et al. GABRA2 alleles moderate the subjective effects of alcohol, which are attenuated by finasteride. Neuropsychopharmacology 2005;30:1193–1203

120. Holdstock L, Penland SN, Morrow AL and De Wit H. Moderate doses of ethanol fail to increase plasma levels of neurosteroid 3α-hydroxy-5α-pregnan-20-one-like immunoreactivity in healthy men and women. Psychopharmacology 2006;186:442–450

121. Pierucci-Lagha A, Covault J, Feinn R, et al. Subjective effects and changes in steroid hormone concentrations in humans following acute consumption of alcohol. Psychopharmacology 2006;186:451–461

122. Porcu P, Rogers LSM, Morrow AL and Grant KA. Plasma pregnenolone levels in cynomolgus monkeys following pharmacological challenges of the hypothalamic-pituitary-adrenal axis. Pharmacol Biochem Behav 2006;84: 618–627

Chapter 32
Neuroendocrine Markers of Psychopathy

Andrea L. Glenn

Abstract Psychopathy is a distinct subtype of antisocial personality disorder that is associated with a pronounced lack of emotion – psychopathic individuals
are described as lacking fear, shame, remorse, and
empathy toward others. Psychopathy has been associated with severe antisocial behavior and high rates of
criminal recidivism, making it a particularly important
area of study. Recent research has begun to uncover
several biological markers that may be important in the
etiology and maintenance of the disorder. One such
example is hormones. Hormones are biological markers that may be of particular importance because they
can affect brain functioning as well as be an indicator
of brain functioning. Hormones are also relatively easy
to measure and could be potential targets for treatment.
Hormones such as cortisol and testosterone have been
associated with several features that are observed in
psychopathy, including blunted stress reactivity, fearlessness, aggression, and stimulation seeking. In this
chapter, the research on hormones in psychopathy will
be discussed, including how hormones such as cortisol
and testosterone might impact the development and
maintenance of psychopathic features, and how hormones contribute to current neurobiological theories
of psychopathy. In addition, the salivary enzyme alpha-
amylase, a biomarker for the neurotransmitter norepinephrine, will be discussed as a prospect for future
research. Finally, the implications of future hormone
research in psychopathy for intervention and treatment
will be discussed.

Keywords Psychopathy • antisocial • hormone • cortisol • estosterone

A. L. Glenn
Department of Psychology, University of Pennsylvania, USA

Abbreviations ACTH: Adrenocorticotrophic hormone;
CRF: Corticotrophin releasing factor; DSM-IV:
Diagnostic and Statistical Manual of Mental Disorders;
DTI: Diffusion tensor imaging; EEG: ectroencepha-
lography; HPA: Hypothalamic-pituitary-adrenal; HPG:
Hypothalamic-pituitary-gonadal; NE: Norepinephrine;
PCL-R: Psychopathy Checklist-Revised;

Introduction

Psychopathy is a personality type that was first
described in depth by Hervey Cleckley the 1941 book
"The Mask of Sanity." Cleckley's description has
become the most influential modern-day description of
psychopathy. Psychopaths are described as individuals
who demonstrate a pronounced lack of guilt, remorse,
and empathic concern for others. They appear to lack
emotional distress and are impervious to distress in
others. In addition, psychopaths are described as being
superficially charming, manipulative, egocentric, and
grandiose.[1] They have been shown to display severe
aggression and high rates of criminal recidivism.[2]
Psychopathy is present in approximately 15–20% of
criminal offenders,[3] making the study of psychopathy
an especially important issue for the criminal justice
system.

Psychopathy in some ways overlaps with the clinical diagnosis of antisocial personality disorder in the
Diagnostic and Statistical Manual of Mental Disorders
(DSM-IV), however many features are unique to psychopathy and are only observed in a subset of individuals with antisocial personality disorder. The gold
standard for the assessment of psychopathy is the

Psychopathy Checklist-Revised (PCL-R),[3] which consists of a structured interview to assess 20 items, supplemented by official records of criminal history and other documents of the individual case history. The 20 items are listed in Table 32.1.

Some of the earliest work on the biological features of psychopathy showed that psychopaths are autonomically underaroused.[4] Since then, studies have continued to find indicators of autonomic system impairments in psychopathy. At rest, psychopaths have been found to have lower levels of electrodermal activity (skin conductance). They also show reduced electrodermal responses in anticipation of aversive stimuli,[5,6] and reduced startle responses to unpleasant pictures and cues of punishment.[7] This suggests they may not generate a fear response. Indeed, psychopaths have been shown to have deficits in fear conditioning.[8,9] This means that psychopaths may lack the ability to predict impending harm from signals of threat. Their lack of fear may result in a failure to avoid negative outcomes such as legal consequences or negative social evaluation.

Peripheral measures of the psychopath's hyporesponsivity to cues of fear and punishment are likely indicators of central nervous system dysfunction. In the brain, the limbic system is involved in generating a response to stressful or threatening stimuli. Recently, brain imaging studies have begun to identify impairments in key regions of the limbic system in psychopaths. Studies have revealed structural and functional differences in regions such as the amygdala[8,10] and orbitofrontal cortex[11,12]; findings have also suggested hypofunctioning of in the general paralimbic circuitry, which includes the orbitofrontal cortex and amygdala.[13,14] Impaired functioning in these brain regions is thought to underlie a wide range findings in psychopathy, including deficits in stress reactivity, sensitivity to punishment, autonomic functioning, fear conditioning, and decision-making.[15] To date, the underlying causes of impaired brain functioning in psychopathy remain unknown.[13] Abnormalities in hormone levels are a potential source of these brain impairments.

Cortisol

Cortisol is a stress hormone that may be important to understanding the biological bases of psychopathy in different ways. First, it is released during stress from a system that is closely linked to the limbic system and therefore may be an important biomarker for deficits in the body's stress response system. Secondly, cortisol acts in the brain to *potentiate* the state of fear and therefore may affect the way that the brain functions, making it a potential target for treatment research.

Cortisol is a glucocorticoid hormone that is part of the hypothalamic-pituitary-adrenal (HPA) axis, which is a stress reactivity network connecting the central nervous system to the endocrine system. When a stressor occurs, limbic, cortical, and other afferent inputs trigger the hypothalamus to release corticotrophin releasing factor (CRF) into the blood. CRF stimulates the secretion of adrenocorticotrophic hormone (ACTH) from the anterior pituitary gland. ACTH then triggers the adrenal gland to release glucocorticoid stress hormones such as cortisol. The role of cortisol is to mobilize the body's resources and to provide energy in times of stress.[16] Cortisol in the blood travels to the brain where it binds to receptors on neurons in the amygdala, hippocampus, and prefrontal cortex. Cortisol, through its actions particularly in the amygdala, is thought to be involved in potentiating the state of fear,[17] generating a sensitivity to punishment, and inducing withdrawal behavior.[18] Differential levels of circulating cortisol may be an *indicator* of differential functioning in structures such as the amygdala that are

Table 32.1 The 20 characteristics of psychopathy listed on the psychopathy checklist-revised (PCL-R)[3]

Glibness/superficial charm
Grandiose sense of self-worth
Pathological lying
Conning/manipulative
Lack of remorse or guilt
Shallow affect
Callous/lack of empathy
Failure to accept responsibility
Need for stimulation
Lack of realistic, long-term goals
Impulsivity
Irresponsibility
Poor behavioral controls
Promiscuous sexual behavior
Many marital relationships
Juvenile delinquency
Revocation of conditional release
Criminal versatility

involved in the generating a response to fear and stress. Alternatively, if cortisol is reduced, it may *affect* the functioning of the amygdala during stress.[19]

In accordance with this hypothesis, low resting cortisol levels have been associated with impaired fear reactivity in young children,[20] increased sensation seeking in men,[21] and poor performance on the Iowa gambling task.[22] In the latter study, it was suggested that low cortisol levels may decrease punishment sensitivity and increase reward dependency.

Psychopaths have been found to have reduced amygdala functioning, are fearless, sensation seeking, and perform poorly on the Iowa Gambling Task. These factors lead to the hypothesis that psychopathy would be associated with reduced cortisol levels. Although the measurement of cortisol levels has been generally overlooked in the psychopathy literature,[23] a few recent studies have found relationships between cortisol and psychopathy. Holi et al.[24] measured serum cortisol levels in young adult male psychopathic offenders with a history of violence and found a negative correlation with psychopathy. Cima et al.[25] showed that psychopathic offenders showed lower cortisol levels than non-psychopathic offenders. O'Leary et al.[26] measured cortisol *reactivity*, or changes in cortisol in response to an environmental stressor or threat. Results showed that undergraduate males scoring higher on a self-report psychopathy measure lacked cortisol responsivity to a social stress test when compared with low-scoring males. Studies examining cortisol reactivity may be especially key in understanding the abnormal stress reactivity in antisocial groups. Together the studies of cortisol in psychopathy suggest some initial evidence for an association, but more studies with larger samples sizes are needed to establish the relationship. Furthermore, it is unclear whether low cortisol levels are a factor that leads to the development of psychopathy, or whether they are simply an indicator of reduced functioning in certain brain regions.

Although further studies need to be conducted to clarify the relationship between cortisol and psychopathy specifically, several studies have demonstrated evidence for reduced cortisol in antisocial groups, particularly in children and adolescents. For example, low cortisol levels have been observed in aggressive children,[27] adolescents with conduct disorder[28] and in violent adults.[29] Low salivary cortisol levels were also observed in adolescents with callous-unemotional traits, which are thought to be similar to psychopathic

traits in adults.[23] Reduced cortisol responsivity has been observed in boys with externalizing behaviors and low anxiety,[30] adolescent males with conduct problems,[31] and in children with symptoms of conduct disorder.[32] In a 5-year longitudinal study, Shoal et al.[33] found that low cortisol in preadolescent boys (age 10–12 years) was associated with low harm-avoidance, low self-control, and more aggressive behavior 5 years later during adolescence (age 15–17 years). These studies suggest that, at least in antisocial groups, low cortisol is present at early ages, and therefore it might be hypothesized that it is involved in the development of antisocial behavior.

A few studies have found *increased* cortisol in antisocial groups. van Bokhoven et al.[34] found increased cortisol levels in adolescents with conduct disorder. Within that group, they found that those with an aggressive form of conduct disorder showed particularly high cortisol levels. They suggest that this finding, which is contradictory to previous literature, may be due to differences in the sample. Whereas previous studies of conduct disorder involved clinically referred samples, which may represent an extreme subset of individuals, this study examined a large, population-based sample. Additional studies will need to be conducted to further clarify the relationship. In an adult sample, Soderstrom et al.[35] found higher serum and CSF concentrations of cortisol in violent offenders, although they suggest that this effect may be due to the fact that the participants were involved in a particularly stressful situation, as they were awaiting trial for serious crimes. Although there are a few contradictory findings, research generally supports the hypothesis of low cortisol in antisocial groups.

Suggestions for Future Research

Additional research is necessary to further elucidate the role of cortisol in psychopathy. Some key areas for future research include,

- Clarifying the basic associations between cortisol and psychopathy – is psychopathy primarily associated with low baseline cortisol levels, reduced cortisol reactivity in response to a stressor, or both?
- Determining whether reduced cortisol is present throughout the lifespan, or if it develops at a certain age. Longitudinal studies involving periodic cortisol

assessments beginning at a very young age and following through to adulthood may help determine whether individuals who became psychopathic in adulthood have reduced cortisol levels as children, or whether some type of "burn-out" effect occurs in which cortisol levels are normal in childhood but decrease due to other biological or environmental factors.

• Identifying causal factors that may lead to low cortisol levels. By measuring a variety of additional biological and social factors in a very large sample, it may be possible to determine specific factors that contribute to low cortisol.

Testosterone

Another hormone that may prove to be important in our understanding of the biological bases of crime is testosterone. Testosterone is a sex hormone that is part of the hypothalamic-pituitary-gonadal (HPG) axis. It is primarily released by the testes in males and the ovaries in females. Males have several times the amount of testosterone as females. Because there are large sex differences in antisocial behavior, with the male-to-female ratio being about 4:1 for antisocial personality disorder and as large as 10:1 for violent crimes,[36] it has been hypothesized that testosterone may be involved in aggressive behavior.

Studies of testosterone have produced observations that are strikingly similar to those found in psychopathy (Table 32.2).[37–57] Many studies have found that individuals with high testosterone levels have characteristics that have been observed in psychopathy. For example, individuals with elevated testosterone levels have been described as impulsive,[37] sensation seeking,[39] and more forward than individuals with low testosterone.[43] They have more sexual partners,[39] marital instability,[46] and are more likely to engage in violent crime.[45] Interestingly, these are all characteristics used to describe psychopaths.[3]

Other parallels come from studies that have examined behavioral changes following injections of testosterone. For example, injections of testosterone have been shown to reduce fear potentiated startle,[53] reduce skin conductance responding to affective stimuli,[55] and impair performance on the Iowa Gambling task,[56]

Table 32.2 Parallels between testosterone and psychopathy findings

High testosterone levels associated with:	Ref.	Psychopathy associated with:	Ref.
Impulsivity	[37, 38]	Impulsivity	[3; PCL-R item]
Sensation seeking	[39]	Stimulation seeking	[40]
Sociability	[39]	Sociability	[41]
Disinhibition	[39]	Poor behavioral controls PCL-R item	[3]
Showing off	[42]	Grandiose sense of self-worth PCL-R item	[3]
Forwardness	[43]	Glibness/superficial charm PCL-R item	[3]
Increased number of sexual partners	[39]	Promiscuous sexual behavior; increased number of sexual partners PCL-R item	[3, 44]
More criminal violence	[45]	More criminal violence	[2]
Not marrying, marital instability, poor marital quality	[46]	Many short term marital relationships	[3; PCL-R item]
Law breaking	[47]	Law breaking	[3]
Reckless driving	[48]	Poor behavioral controls, risk-taking behavior	[3; PCL-R item]
Failure to plan ahead	[48]	Impulsivity/failure to plan ahead	[3; PCL-R item]
Judged more harshly by parole board/violate prison rules more often	[45]	Revocation of conditional release	[3; PCL-R item]
Juvenile delinquency	[49]	Juvenile delinquency	[3; PCL-R item]
Injections of testosterone associated with:		**Psychopathy associated with:**	
Reduced emotional response to fearful faces	[50]	Reduced brain activity to fearful faces	[51]
Reduced empathetic behavior (facial mimicry)	[52]	Callous/lack of empathy	[3; PCL-R item]
Reduced fear-potentiated startle	[53]	Deficient fear-potentiated startle	[54]
Reduced skin conductance response to affective stimuli	[55]	Reduced skin conductance responding	[5]
Poor decision-making on the Iowa Gambling task	[56]	Poor performance on the Iowa Gambling task	[57]

which likely occurs due to reduced sensitivity to punishment and increased sensitivity to reward. Psychopaths have reduced fear potentiated startle,[7] reduced skin conductance responses,[5] and exhibit poor performance on the Iowa Gambling task.[57] The parallels between these studies lead to the strong hypothesis that psychopathy is associated with high testosterone levels.

As can be observed in Table 32.2, testosterone has been associated with a variety of traits and behaviors; these have led to a variety of hypotheses regarding the function of testosterone. A recent theory has attempted to unify hypotheses regarding the role of testosterone. Archer[58] suggests that individual testosterone levels may reflect a particular type of life history strategy. A life history strategy is the behaviors and physiological traits an organism uses across the lifespan to maximize reproductive fitness. For example, age of sexual maturation, number of matings, and time invested in caring for offspring are all components of an individual's life history strategy. Archer[58] argues that individuals with higher testosterone may preferentially pursue mating effort and use their energy to try to mate with as many individuals as possible, while individuals with lower testosterone tend to pursue parental effort, spending less time mating and more time caring for offspring. Indeed, individuals with high testosterone levels have an increased number of sexual partners[39] and are more likely to not marry, exhibit marital instability, and have increased extramarital sex.[46] In addition, characteristics associated with high testosterone levels such as aggression, antisocial behavior, sensation seeking, and impulsivity, which have been associated with high testosterone levels, may be byproducts of behaviors necessary to facilitate such a life history strategy and increase mating opportunities.

Similarly, it has been theorized that psychopathic traits such as fearlessness, risk taking, aggression, and blunted emotion may be indicators that psychopaths pursue a life history strategy involving many short-term mating opportunities as opposed to long-term cooperative relationships involving significant parental care.[59,60] Studies have found psychopathy to be associated with the pursuit of short-term, uncommitted relationships with multiple sexual partners and a lack of parental investment.[2,59] Psychopathy has been associated with an increased number of sexual partners,[44,61] early sexual behavior,[59] an uncommitted approach to mating, increased mating effort, sexual coercion,[44]

many short marital relationships, sexual promiscuity,[3] and poor performance as parents.[62] The similarities in the hypotheses regarding psychopaths and individuals with high testosterone levels lead to the prediction that testosterone may be a key biological component of psychopathy.

The association between testosterone and aggression/antisocial behavior has been the focus of much research, but to the author's knowledge, only one study thus far has examined testosterone levels in psychopaths. Stalenheim et al.[63] found that testosterone levels were positively related to scores on Factor 2 of the PCL-R, although it is possible the results may be confounded by comorbid substance abuse and other psychiatric disorders; Factor 2 of psychopathy represents the antisocial behavior and lifestyle features. Several studies have examined testosterone levels in antisocial groups. Loney et al.[23] examined testosterone levels in boys with callous-unemotional traits, which are thought to be analogous to psychopathic traits in adulthood, but found no effects. In other antisocial and aggressive groups, higher testosterone levels have been found in girls with conduct disorder,[64] adolescent boys with externalizing behaviors,[49] young criminals,[45,65,66] and in criminal women.[67] Furthermore, testosterone has been associated with a variety of antisocial behaviors including difficulties on the job, nonobservance of the law, marriage failures, drug use, alcohol abuse, and violent behaviors.[47] This evidence suggests that testosterone may also be increased in psychopathic individuals.

Suggestions for Future Research

Given the strong evidence supporting the hypothesis that psychopathy is associated with increased testosterone, surprisingly little research has been done to test the hypothesis. Suggestions for future studies include

- Testing whether testosterone is in fact involved in psychopathy.
- Exploring whether testosterone is associated with the personality features (e.g. being conning, manipulative, lacking guilt and remorse), or if it is primarily associated with the antisocial behavior features, as was found by Stalenheim et al.[63] and as has been observed in other antisocial groups.

- Conducting longitudinal studies to determining whether increased testosterone is present in youth with psychopathic traits, or whether the association develops later, perhaps during or after puberty.

Interactions

The hormones cortisol and testosterone do not act in isolation, but can interact with each other to produce effects. Testosterone and cortisol have been shown to have mutually antagonistic properties. Cortisol suppresses the activity of the HPG axis on all levels, diminishing the production and inhibiting the effects of testosterone.[68] In turn, testosterone inhibits activity of the HPA axis.[69] Van Honk et al. have found that injections of testosterone reduce fearfulness,[50] promote responding to angry faces,[70] and shift the balance from punishment to reward sensitivity[56]; each of these effects is opposite to the effects of cortisol. In addition to the future research needed on cortisol and testosterone, it may also be important to test for interactions between the two hormones.

In a recent study of hormones in delinquent boys, Popma et al.[71] found an interaction between cortisol and testosterone; at low levels of cortisol a positive association was found between testosterone and overt aggression, but at high levels of cortisol no relationship was observed. This study raises the possibility that if an individual has high cortisol levels, it may act as a protective factor against the adverse effects of high testosterone. Thus, there would be no relationship between testosterone and criminal behavior when cortisol is sufficiently high.

Suggestions for Future Research

- Clarifying whether the interaction observed by Popma et al.[71] is also observed in psychopaths.
- Determining whether abnormalities in one hormone (cortisol or testosterone) may cause an abnormality in the other hormone. Cortisol and testosterone systems have mutually antagonistic properties, so it is possible that when the levels of one hormone become abnormal, it leads to abnormalities in the

other hormone. Two types of studies could be conducted to test this:

1. A longitudinal study could be done to see if abnormalities in one hormone appear before abnormalities in the other.
2. A study could manipulate either cortisol or testosterone to see how it affected the other hormone. For example, in the van Honk et al.[50] study mentioned above, it would be interesting to see whether fearfulness was actually reduced because of testosterone, or whether the testosterone lowered cortisol levels, which in turn altered fearfulness.

Connectivity

In addition to possible interactions, the development of psychopathy may involve the combination of the two hormones cortisol and testosterone acting in concert. A new, unifying theory in psychopathy suggests that the underlying source of the emotional and behavioral deficits observed in psychopaths may result from an imbalance in *both* cortisol and testosterone,[72] specifically through decreased cortisol *and* increased testosterone. This imbalance is argued to affect amygdala functioning, which has been widely implicated in psychopathy,[15,73] as well as the connectivity between subcortical structures such as the amygdala, and the prefrontal cortex. The amygdala is a major binding site for cortisol and testosterone, where these hormones can affect neuropeptide gene expression and thus alter amygdala responsivity to fearful or threatening stimuli.[18] Cortisol is involved in potentiating the state of fear and increasing sensitivity to punishment,[19] while testosterone is associated with reward sensitivity, fear reduction, and increased approach behavior, including aggression.[74] Therefore, an imbalance involving decreased cortisol (low fear) and increased testosterone levels (high approach/reward seeking behavior) can change the responsivity of the amygdala to reduce sensitivity to punishment cues or fearful stimuli, and increase sensitivity to reward. Psychopaths are described as fearless, insensitive to punishment, impaired in fear conditioning,[7] and reward dependent.[75] Several studies have demonstrated reduced functioning of the amygdala in psychopaths.[10,14] van Honk and Schutter[72] predict that decreased cortisol and increased testosterone may be an underlying cause for the impaired amygdala functioning and related behavioral deficits observed in psychopathy.

In addition to effects on the amygdala, the hormonal imbalance involving decreased cortisol and increased testosterone has also been shown to disrupt the connectivity between subcortical and cortical structures. Studies have used electroencephalography (EEG) to demonstrate relative increases or decreases in subcortico-cortical "cross-talk"; injections of testosterone have been shown to reduce subcortico-cortical communication[76] and increased cortisol has been shown to strengthen it.[77] Neuroimaging data in psychopaths suggests that connectivity between the amygdala and prefrontal regions is compromised.[10,14] The result is that cortical regions involved in cognitive evaluation and decision-making no longer receive emotional input from areas such as the amygdala. van Honk and Schutter[72] predict that low cortisol and high testosterone contributes to this disruption in connectivity.

One method that could be used to test this theory is diffusion tensor imaging (DTI). DTI is a neuroimaging technique used to measure the white matter tracts between brain regions. It would be interesting to see if cortisol and testosterone imbalances are associated with reduced volume of the white matter tracts that connect subcortical to cortical structures, and also if the connectivity is associated with psychopathy.

A proper hormonal balance appears to be essential to maintain homeostatic regulation of emotions; imbalances in cortisol and testosterone have been shown to be highly significant in other disorders that involve emotion.[56] For example, in anxiety disorders, patients frequently demonstrate *increased* cortisol relative to testosterone levels, which results in *hyper*active amygdala functioning.[18] The increase in amygdala activity results in exaggerated fearfulness and hypersensitivity to punishment. Furthermore, the hormone imbalance may cause a "hypercoupling" of the amygdala to prefrontal regions such as the orbitofrontal cortex, resulting in cognition that is overly emotion-based. Thus, when these hormones become imbalanced in either direction, brain functioning can be affected, resulting in disorders of emotion.

van Honk and Schutter[72] provide theoretical support for their hypothesis, but to date it has not been empirically tested. As mentioned, a variety of studies could be conducted to test this hypothesis, as well as individual hypotheses about cortisol, testosterone, and their interactions. In addition to exploring cortisol and testosterone levels in psychopathy, future studies may also want to examine the distributions and sensitivity of different receptors. For example, depression has been associated with an increased ratio between two different types of cortisol receptors, and also with decreased sensitivity in one type of receptor.[78] Thus, it may be important to explore different aspects of neuroendocrine functioning in psychopathy.

Alpha-amylase

In addition to cortisol and testosterone, another biomarker that has promise for research in psychopathy is alpha-amylase. Alpha-amylase is an enzyme found in saliva and is a marker for the release of the neurotransmitter norepinephrine (NE) into the blood.[79] NE is theoretically important to the study of psychopathy because it is an essential neurotransmitter for emotional processing in the amygdala.[80] In addition, it has been found to interact with cortisol.[81] The administration of a beta-blocker, which blocks the effects of NE, has been shown to inhibit functioning of the amygdala when individuals view highly emotional pictures such as of mutilation or accidents; this suggests that amygdala activity in response to emotional stimuli is NE-dependent.[82] Since stress responses of the amygdala appear to be impaired in psychopaths, it is possible that NE transmission may be disrupted.

Neurotransmitters such as norepinephrine interact with cortisol to play a key role in maintaining homeostasis and normal responding to stress.[83] NE is released immediately in response to stress as part of the fast-acting "fight-or-flight" response. The release of cortisol is part of a second, slower-acting response. The interaction of these systems is thought to occur in the amygdala,[84] a region thought to be important in the development of psychopathy.[73] In a recently published study, van Stegeren et al.[85] found that when NE is present, cortisol moderates amygdala responsivity to emotional stimuli (i.e., lower cortisol is associated with less amygdala activation). However, if NE is blocked using an adrenergic blocker, amygdala functioning is impaired regardless of cortisol levels. This suggests that NE in the amygdala is critical for cortisol to moderate amygdala functioning. At low levels of NE, it may be the case that cortisol does not have an effect on amygdala functioning. Therefore it may be critical to assess both cortisol and norepinephrine levels to gain a complete picture of the biology underlying the functioning of the amygdala.

The novel advantage of measuring alpha-amylase is that it allows for a parallel investigation of two stress response systems (endocrine and neurotransmitter) through saliva samples, and is less invasive and stress-inducing than taking blood samples to measure NE, as salivary measures of NE have been found to reflect NE levels in the blood.[86] Salivary alpha-amylase levels are predictive of NE levels under a variety of stressors, including exercise, exposure to heat and cold, and written examinations.[87] A recent placebo-controlled study showed that stress-related increases in alpha-amylase can be inhibited by administering an adrenergic blocker, which blocks NE.[88] Thus, one hypothesis would be that if NE release during stressors is disrupted in psychopathy, alpha-amylase will be reduced.

Alpha-amylase as a marker of NE transmission has not been studied in psychopathic or antisocial personality disordered populations, and thus has great potential for contributing to our understanding of the factors that may affect brain functioning in psychopathy. One published study has examined alpha-amylase reactivity in relation to aggressive behavior. While no direct effects were found between alpha-amylase reactivity and aggression, Gordis et al.[89] found that alpha-amylase reactivity affected the relationship between cortisol reactivity and aggression. At high levels of alpha-amylase (i.e., high NE levels), cortisol was not related to aggression, yet at low alpha-amylase levels, low cortisol was associated with increased aggression. This could be another example of a protective factor – it suggests that if NE is sufficiently high, it may act as a buffer against the development of aggressive behavior, even in individuals with low cortisol.

The studies by Gordis et al.[89] and van Stegeren et al.,[85] discussed above, are difficult to compare, as one measures brain activity in a particular region and the other measures a behavior; however, when considered together they may suggest the following relationships: When NE is absent, amygdala functioning is impaired regardless of cortisol levels. At normal levels of NE, cortisol moderates amygdala functioning and thus is associated with related behaviors such as aggression. At higher levels of NE, the effects of

cortisol on amygdala functioning are masked, so cortisol no longer predicts behaviors such as aggression.

These studies demonstrate that interactions between cortisol and NE (as measured by alpha-amylase) affect both brain functioning and behavior. It remains to be seen whether such interactions may be associated with adult psychopathy, but it seems crucial to consider the two systems together in order to gain a complete understanding of the biology. Studies examining the role of cortisol *and* alpha-amylase in psychopathy could prove beneficial in furthering our understanding of psychopathy.

Treatment

If future research begins to uncover the role of hormones in psychopathy, it may be possible to use hormone therapies to increase the functioning of key brain regions such as the amygdala that have been found to be impaired in psychopathy. If studies were to establish a definitive relationship between cortisol levels or reactivity and psychopathy, additional studies could seek to determine which factors might be able to change cortisol levels, how much cortisol levels can be changed, and whether changing cortisol levels can change the functioning of regions such as the amygdala. The same could be true for testosterone levels. With respect to the interaction between the two hormones, pharmacological therapies that would restore the homeostatic balance between cortisol and testosterone could potentially help to sensitize a psychopath's emotional responsiveness so that behavioral therapies that previously failed may gain efficacy.

One hormone treatment study in antisocial children has provided some insight into how treatment might work. Brotman, et al.[90] conducted a 22-week family-based intervention in preschool children at risk for antisocial behavior. They measured salivary cortisol levels before an after a social challenge involving entry into an unfamiliar peer group. Results showed that children who had received the intervention had increased cortisol levels in anticipation of the social challenge. This study suggests that, at least in the early years of life, the cortisol response is malleable and can be changed by experience. It remains unknown whether similar treatments could be effective in adolescents and adults and whether they would be effective in psychopaths, a group that has been found to be particularly treatment resistant. It is also unclear how long these changes may last. By intervening at key points in development, it may be possible to lessen some of the biological risk factors for antisocial behavior.

Finally, while psychopathic individuals tend to be resistant to treatment relative to other disorders, researchers in the field may be able to gain insight from understanding the neurobiology and biological treatments of other disorders. For example, depression has been linked to most of the same structures that have been implicated in psychopathy, but in the opposite direction. Depression has been associated with *hyper*activation in areas such as the amygdala, hippocampus, ventromedial prefrontal cortex, and anterior cingulate. As in psychopathy, it has been proposed that the connectivity between limbic and cortical areas may be disrupted, compromising the cross-talk between regions. Hyperactivity of the limbic system leads to stimulation of the hypothalamus, resulting in imbalances in the endocrine system, including increased cortisol levels.[91] Future research in psychopathy may benefit from paying attention to ongoing research on seemingly unrelated psychopathology. An exploration of the factors that may cause some individuals to develop *hyper*activity in certain brain regions while others experience *hypo*activity may provide essential clues to the development of psychopathy and other disorders. In addition, by examining the effects of various pharmacologic and behavioral treatments for disorders such as depression, we may be able to form new hypotheses about possible treatments for psychopathy.

Conclusion and Future Directions

Evidence from several neuroendocrine studies has begun to suggest that hormones may be an important biomarker for psychopathy in the future. With additional research, a particular hormonal profile may become a very useful biomarker for psychopathy for several reasons:

- It represents pathology in biological systems that have an effect on behavior by inducing chemical changes in specific brain regions and is specific to the core features of psychopathy, including blunted stress reactivity, fearlessness, aggression, and mating versus parental effort.
- With additional research a specific hormone profile can be narrowed down that is consistent with the interpersonal and affective features that are unique to psychopathy compared to other disorders involving antisocial behavior.
- Because they modulate neural pathways, hormones represent a specific pattern of brain functioning that results in psychopathy, rather than representing specific symptoms.
- Hormones can be measured repeatedly over time and are reproducible under the same conditions (e.g. time of day).
- Hormones can be easily measured through saliva samples.
- The acquisition of saliva samples is noninvasive and not harmful to the individual being tested.
- When following recommended procedures for saliva sample collection, storage, and assaying, hormones are reliable in various testing environments.
- Hormones are relatively inexpensive to collect and test.

Thus, future research to explore the potential of hormones as a biomarker for psychopathy is an important step in furthering our ability to identify, prevent, and treat such a devastating disorder to society.

The involvement of hormones in the development and maintenance of psychopathy is significant because it may help to explain numerous findings in the field. Decreased cortisol levels and increased testosterone levels may help to explain poor decision-making, blunted stress reactivity, fearlessness, poor conditioning, and increased instrumental aggression, all of which have been observed in psychopathy. In addition, examining the role of hormones may also lead to a deeper understanding of neurobiological findings in psychopathy. Brain imaging studies have highlighted several key brain regions that appear to be hypofunctioning in psychopathy, but have thus far have not been able to explain the source of this hypofunctioning. The consistent findings of reduced amygdala and orbitofronal cortex activity may be a result of an imbalance in cortisol and testosterone levels. In addition, alpha-amylase may be an important moderator that may buffer or potentiate the effects of cortisol or testosterone. Thus, hormone studies have the potential to take psychopathy research to the next level of understanding, moving from the systems neuroscience level to explore some of the underlying molecular mechanisms.

Most importantly, hormone studies may have significant implications for treatment. If it is found that psychopathy is associated with an imbalance between cortisol and testosterone, it is possible that the balance could be restored through bottom-up endocrinological manipulations targeting the activity of the HPA and/or HPG axis, or targeting cortisol and testosterone levels. This could be a key neurobiological first step in sensitizing a psychopath's emotional system so that previously failed attempts at behavioral therapies may begin to have efficacy. Several studies were suggested that may help to further elucidate the relationships between psychopathy and hormones, which could provide important information about how to develop therapies that may help in treatment and prevention.

References

1. Cleckley H. The Mask of Sanity. St. Louis, MO: Mosby; 1941
2. Hare RD. Manual for the Hare Psychopathy Checklist-Revised. Toronto: Multi-Health Systems; 1991
3. Hare RD. Hare Psychopathy Checklist-Revised (PCL-R), 2nd edition. Toronto: Multi-Health Systems; 2003
4. Hare RD. Psychopathy, fear arousal and anticipated pain. Psychol Rep 1965;16:499–502
5. Hare RD. Electrodermal and cardiovascular correlates of psychopathy. In: Hare RD, Schalling D, eds. Psychopathic Behavior: Approaches to Research. New York:Wiley; 1978:107–144
6. Hare RD. Psychopathy, affect and behavior. In: Cooke DJ, Forth AE, Hare RD, eds. Psychopathy: Theory, Research and Implications for Society. Dordrecht, The Netherlands: Kluwer; 1998:105–137
7. Patrick CJ. Emotion and psychopathy: Startling new insights. Psychophysiology 1994;31:319–330
8. Birbaumer N, Viet R, Lotze M et al. Deficient fear conditioning in psychopathy: a functional magnetic resonance imaging study. Arch Gen Psychiat 2005;62:799–805
9. Flor H, Birbaumer N, Hermann C et al. Aversive Pavlovian conditioning in psychopaths: Peripheral and central correlates. Psychophysiology 2002;39:505–518
10. Kiehl KA, Smith AM, Hare RD et al. Limbic abnormalities in affective processing by criminal psychopaths as revealed by functional magnetic resonance imaging. Biol Psychiat 2001;50:677–684
11. Yang Y, Raine A, Lencz T et al. Volume reduction in prefrontal gray matter in unsuccessful criminal psychopaths. Biol Psychiat 2005;15:1103–1108
12. Rilling JK, Glenn AL, Jairam MR et al. Neural correlates of social cooperation and non-cooperation as a function of psychopathy. Biol Psychiat 2007;61:1260–1271
13. Kiehl KA. A cognitive neuroscience perspective on psychopathy: Evidence for paralimbic system dysfunction. Psychiat Res 2006;142:107–128
14. Viet R, Flor H, Erb M et al. Brain circuits involved in emotional learning in antisocial behavior and social phobia in humans. Neurosci Lett 2002;328:233–236
15. Blair RJ. The amygdala and ventromedial prefrontal cortex in morality and psychopathy. Trends Cogn Sci 2007;11: 387–392
16. Kudielka BM, Kirschbaum C. Sex differences in HPA axis responses to stress: a review. Biol Psychiat 2005;69:1 13–132
17. Schulkin J, Gold PW, McEwen BS. Induction of corticotropin-releasing hormone gene expression by glucocorticoids: implication for understanding the states of fear and anxiety and allostatic load. Psychoneuroendocrinology 1998;23: 219–243
18. Schulkin J. Allostasis: A neural behavioral perspective. Horm Behav 2003;43:21–27
19. Rosen JB, Schulkin J. From normal fear to pathological anxiety. Psychol Rev 1998;105:325–350
20. Kagan J, Reznick JS, Snidman N. Biological bases of childhood shyness. Science 1988;240:167–171
21. Rosenblitt JC, Soler H, Johnson SE et al. Sensation seeking and hormones in men and women: exploring the link. Horm Behav 2002;40:396–402
22. van Honk J, Schutter DJLG, Hermans EJ et al. Low cortisol levels and the balance between punishment sensitivity and reward dependency. Neuroreport 2003;14:1993–1996
23. Loney BR, Butler MA, Lima EN et al. The relation between salivary cortisol, callous-unemotional traits, and conduct problems in an adolescent non-referred sample. J Child Psychol Psychiat 2006;47:30–36
24. Holi M, Auvinen-Lintunen L, Lindberg N et al. Inverse correlation between severity of psychopathic traits and serum cortisol levels in young adult violent male offenders. Psychopathology 2006;39:102–104
25. Cima M, Smeets T, Jelicic M. Self-reported trauma, cortisol levels, and aggression in psychopathic and non-psychoathic prison inmates. Biol Psychiat 2008;78:75–86
26. O'Leary MM, Loney BR, Eckel LA. Gender differences in the association between psychopathic personality traits and cortisol response to induced stress. Psychoneuroendocrinology 2006;32:183–191
27. McBurnett K, Lahey BB, Rathouz PJ et al. Low salivary cortisol and persistent aggression in boys referred for disruptive behavior. Arch Gen Psychiat 2000;57:38–43
28. Pajer K, Gardner W, Rubin RT et al. Decreased cortisol levels in adolescent girls with conduct disorder. Arch Gen Psychiat 2001;58:297–302
29. Virkkunen M. Urinary free cortisol secretion in habitually violent offenders. Acta Psychiat Scand 1985;72:40–44
30. van Goozen SHM, Matthys W, Cohen-Hettenis PT et al. Salivary cortisol and cardiovascular activity during stress in oppositional defiant disorder boys and normal controls. Biol Psychiat 1998;43:531–539
31. McBurnett K, Raine A, Stouthamer-Loeber M et al. Mood and hormone responses to psychological challenge in adolescent males with conduct problems. Biol Psychiat 2005;57:1109–1116

32. Oosterlaan J, Geurts HM, Sergeant JA. Low basal salivary cortisol is associated with teacher-reported symptoms of conduct disorder. Psychiat Res 2005;134:1–10

33. Shoal GD, Giancola PR, Kilrillova GP. Salivary cortisol, personality, and aggressive behavior in adolescent boys: a 5-year longitudinal study. Child Adol Psychiat Ment Health 2003;42:1101–1107

34. van Bokhoven I, van Goozen SHM, van Engeland H et al. Salivary cortisol and aggression in a population-based longitudinal study of adolescent males. J Neural Transm 2005;112:1083–1096

35. Soderstrom H, Soderstrom K, Blennow K et al. A controlled study of tryptophan and cortisol in violent offenders. J Neural Transm 2004;111:1605–1610

36. van Honk J, Schutter DJLG. Testosterone reduces conscious detection of signals serving social correction: Implications for antisocial behavior. Psychol Sci 2007;18:663–667

37. Bjork JM, Moeller FG, Dougherty DM et al. Endogenous plasma testosterone levels and commission errors in women: A preliminary report. Physiol Behav 2001;73:217–221

38. Baucom DH, Besch PK, Callahan S. Relation between testosterone concentration, sex role identity and personality among females. J Pers Soc Psychol 1985;48:1218–1226

39. Daitzman R, Zuckerman M. Disinhibitory sensation seeking, personality and gonadal hormones. Pers Individ Differ 1980;1:103–110

40. Blackburn R. Cortical and autonomic response arousal in primary and secondary psychopaths. Psychophysiology 1979;16:143–150

41. Benning SD, Patrick CJ, Iacono WG. Estimating facets of psychopathy from normal personality traits: A step toward community-epidemiological investigations. Assessment 2005;12:3–18

42. Udry JR, Talbert LM. Sex hormone effects on personality at puberty. J Pers Soc Psychol 1988;54:291–295

43. Dabbs JM, Bernieri FJ, Strong RK et al. Going on stage: testosterone in greetings and meetings. J Res Pers 2001;35:27–40

44. Lalumiere ML, Quinsey VL. Sexual deviance, antisociality, mating effort, and the use of sexually coercive behaviors. Pers Individ Differ 1996;21:33–48

45. Dabbs JM, Jurkovic GJ, Frady RL. Salivary testosterone and cortisol among late adolescent male offenders. J Abnorm Child Psychol 1991;19:469–478

46. Booth A, Dabbs J, Testosterone and men's marriages. Soc Forces 1993;72:463–477

47. Mazur A, Booth A. Testosterone and dominance in men. Behav Brain Sci 1998;21:353–397

48. Aromaki AS, Lindman RE, Eriksson CJP. Testosterone, aggressiveness, and antisocial personality. Aggress Behav 1999;25:113–123

49. Maras A, Laucht M, Gerdes D et al. Association of testosterone and dihydrotestosterone with externalizing behavior in adolescent boys and girls. Psychoneuroendocrinology 2003; 28:932–940

50. van Honk J, Peper JS, Schutter DJLG. Testosterone reduces unconscious fear but not consciously experienced anxiety: implications for the disorders of fear and anxiety. Biol Psychiat 2005;58:218–225

51. Deeley Q, Surguladze S, Tunstall N et al. Facial emotion processing in criminal psychopathy. Preliminary functional magnetic resonance imaging study. Br J Psychiat 2006; 189:533–539

52. Hermans EJ, Putman P, van Honk J. Testosterone administration reduces empathic behavior: A facial mimicry study. Psychoneuroendocrinology 2006;31:859–866

53. Hermans EJ, Putman P, Baas JM et al. A single administration of testosterone reduces fear-potentiated startle in humans. Biol Psychiat 2006;59:872–874

54. Benning SD, Patrick CJ, Iacono WG. Psychopathy, startle blink modulation, and electrodermal reactivity in twin men. Psychophysiology 2005;42:753–762

55. Hermans EJ, Putman P, Baas JM et al. Exogenous testosterone attenuates the integrated central stress response in healthy young women. Psychoneuroendocrinology 2007; 32:1052–1061

56. van Honk J, Schutter DJLG, Hermans EJ et al. Testosterone shifts the balance between sensitivity for punishment and reward in healthy young women. Psychoneuroendocrinology 2004;29:937–943

57. Mitchell DGV, Colledge E, Leonard A et al. Risky decisions and response reversal: is there evidence of orbitofrontal cortex dysfunction in psychopathic individuals? Neuropsychologica 2002;40:2013–2022

58. Archer J. Testosterone and human aggression: an evaluation of the challenge hypothesis. Neurosci Biobehav Rev 2006;30:319–345

59. Harris GT, Rice ME, Hilton NZ et al. Coercive and precocious sexuality as a fundamental aspect of psychopathy. J Pers Disord 2007;21:1–27

60. Mealey L. The sociobiology of sociopathy: an integrated evolutionary model. Behav Brain Sci 1995;18:523–599

61. Halpern CT, Campbell B, Agnew CR et al. Associations between stress reactivity and sexual and nonsexual risk taking in young adult human males. Horm Behav 2002;42:387–398

62. Cleckley H. The Mask of Sanity, 5th edition. St. Louis, MO: Mosby; 1976

63. Stalenheim EG, Eriksson E, von Knorring L et al. Testosterone as a biological marker in psychopathy and alcoholism. Psychiatry Res 1998;77:79–88

64. Pajer K, Tabbah R, Gardner W et al. Adrenal androgen and gonadal hormone levels in adolescent girls with conduct disorder. Psychoneuroendocrinology 2006;31:1245–1256

65. Kreuz LE, Rose RM. Assessment of aggressive behavior and plasma testosterone in a young criminal population. Psychosom Med 1972;34:321–332

66. Dabbs JM, Frady RL, Carr TS. Saliva testosterone and criminal violence in young adult prison inmates. Psychosom Med 1987;49:174–182

67. Banks T, Dabbs, J.M. Salivary testosterone and cortisol in a delinquent and violent urban subculture. J Soc Psychol 1996;136:49–56

68. Tilbrook AJ, Turner AI, Clark IJ. Effects of stress on reproduction in non-rodent mammals: The role of glucocorticoids and sex differences. Rev Reprod 2000;5:105–113

69. Viau V. Functional cross-talk between the hypothalamic-pituitary-gonadal and adrenal axes. J Neuroendocrinol 2002;14:506–513

70. van Honk J, Tuiten A, Hermans EJ et al. A single administration of testosterone induces cardiac accelerative responses

to angry faces in healthy young women. Behav Neurosci 2001;115:238–242

71. Popma A, Vermeiren R, Geluk CAML et al. Cortisol moderates the relationship between testosterone and aggression in delinquent male adolescents. Biol Psychiat 2007;61:405–411

72. van Honk J, Schutter DJLG. Unmasking feigned sanity: a neurobiological model of emotion processing in primary psychopathy. Cognit Neuropsyhiat 2006;11:285–306

73. Blair RJ. Subcortical brain systems in psychopathy. In: Patrick CJ, ed. Handbook of Psychopathy. New York: Guilford; 2006:296–312

74. Boissy A, Bouissou MF. Effects of androgen treatment on behavioral and physiological responses to heifers to fear-eliciting situations. Horm Behav 1994;28:66–83

75. Newman JP, Kosson DS, Patterson CM. Delay of gratification in psychopathic and nonpychopathic offenders. J Abnorm Psychol 1992;101:630–636

76. Schutter DJLG, van Honk J. Decoupling of midfrontal delta-beta oscillations after testosterone administration. Int J Psychophysiol 2004;53:71–73

77. Schutter DJLG, van Honk J. Salivary cortisol levels and the coupling of midfrontal delta-beta oscillations. Int J Psychophysiol 2005;55:127–129

78. Young EA, Lopez JF, Murphy-Weinbert V et al. Mineralocorticoid receptor function in major depression. Arch Gen Psychiat 2003;60:24–28

79. Chrousous GP, Gold PW. The concepts of stress and stress system disorders: overview of physical and behavioral homeostasis. JAMA 1992;267:1244–1252

80. McGaugh JL. Memory – a century of consolidation. Science 2000;287:248–251

81. Granger DA, Kivlighan KT, El-Sheikh M et al. Salivary alpha-amylase in biobehavioral research: Recent developments and applications. Ann NY Acad Sci 2007;1098:122–144

82. van Stegeren A, Goekoop R, Everaerd W et al. Noradrenaline mediates amygdala activation in men and women during encoding of emotional material. Neuroimage 2005;24:898–909

83. de Kloet ER, Joels M, Holsboer F. Stress and the brain: from adaptation to disease. Nat Rev Neurosci 2005;6:463–475

84. Roozendaal B, Okuda S, Van der Zee EA et al. Glucocorticoid enhancement of memory requires arousal-induced noradrenergic activation in the basolateral amygdala. Proc Natl Acad Sci USA 2006;103:6741–6746

85. van Stegeren A, Wolf OT, Everaerd W et al. Endogenous cortisol level interacts with noradrenergic activation in the human amygdala. Neurobiol Learn Mem 2007;87:57–66

86. Schwab KO, Heubel G, Bartels H. Free epinephrine, norepinephrine and dopamine in saliva and plasma of healthy adults. Eur J Clin Chem Clin Biochem 1992;30:541–544

87. Chatterton RT, Vogelsong KM, Lu Y et al. Salivary alpha-amylase as a measure of endogenous adrenergic activity. Clin Physiol 1996;16:433–448

88. van Stegeren A, Rohleder N, Everaerd W et al. Salivary alpha amylase as marker for adrenergic activity during stress: effect of betablockade. Psychoneuroendocrinology 2006;31:137–141

89. Gordis EB, Granger DA, Susman EJ et al. Asymmetry between salivary cortisol and alpha-amylase reactivity to stress: Relation to agressive behavior in adolescents. Psychoneuroendocrinology 2006;31:976–987

90. Wolf M, van Doorn GS, Leimar O et al. Life-history trade-offs favour the evolution of animal personalities. Nature 2007;447:581–584

91. Maletic V, Robinson M, Oakes T et al. Neurobiology of depression: an integrated view of key findings. Int J Clin Pract 2007;61:2030–2040

Chapter 33
Mitochondrial Complex I as a Possible Novel Peripheral Biomarker for Schizophrenia

Dorit Ben-Shachar

Abstract Schizophrenia is heritable and is believed to be caused by variations in multiple genes, each contributing a subtle effect, which combine with each other and with environmental stimuli. At present schizophrenia clinical heterogeneity as well as difficulties relating brain's emergent properties to its physiological substrates hinders the identification of a single anatomical, physiological, molecular or genetic abnormality. However, the goal of neuroscience research in schizophrenia is the discovery of biomarkers that may later be used to develop diagnostics and drug treatments for the disease. In the present chapter we suggest mitochondrial first complex (complex I) of the oxidative phosphorylation system as an attractive potential biomarker for schizophrenia by demonstrating its role in the pathological processes of the disorder, as well as its disease specific alterations in blood cells. Mitochondrial complex I plays a major role in controlling oxidative phosphorylation and therefore its dysfunction can cause alterations in ATP production, cytoplasmatic calcium concentrations, as well as increase in ROS and NO production. All of the latter processes have been well established as leading to altered neuronal activity and thereby to abnormal neuronal circuitry. Recent evidence supports the impairment of mitochondria in schizophrenia, including mitochondrial hypoplasia, dysfunction of the oxidative phosphorylation system and altered mitochondrial related gene expression including those encoding for complex I subunits. The CNS abnormalities in complex I are reflected in blood cells, as its activity demonstrates schizophrenia as well as disease-state specific alterations. In addition, the transcriptoms as well as the proteome of three of complex I subunits are abnormal in blood cells in schizophrenia. This chapter suggests that the biological complexity of schizophrenia, its overlap of symptoms with other mental disorders as well as co-morbidity with other common disorders, leads to the inevitable conclusion that any biomarker would have to be an array of interrelated factors or even a set of several such arrays in which mitochondrial complex I will be a major constituent.

Keywords Schizophrenia • biomarker • mitochondria • complex I • postmortem brains • blood cells

Abbreviations ATP: Adenosine triphosphate; DA: Dopamine; ^{1}H-MRS: Hydrogen magnetic resonance spectroscopy; Pi: Inorganic phosphate; mt DNA: Mitochondrial DNA; NAA: N-acetylaspartate; nDNA: Nuclear DNA; (OXPHOS): Oxidative phosphorylation system; PCr: Phosphocreatine; PMEs: Phosphomonoesters; ^{31}P-MRS: Phosphorous or hydrogen magnetic resonance spectroscopy.

Introduction

Schizophrenia is a major psychotic disorder, which afflicts 1% of the world population. The disorder is characterized by various abnormal cognitive, affective and motor behavioral features. Its main symptoms involve multiple psychophysiological processes, such as perception (hallucinations), ideation, reality testing (delusions), thought processes (loose associations), feeling (flatness, inappropriate affect), behavior (catatonia, disorganization), attention, concentration, motivation (avolition, impaired intention and planning) and judgment. These psychological and behavioral

D. Ben-Shachar
Laboratory of Psychobiology, Department of Psychiatry Rambam Medical Center and B. Rappaport Faculty of Medicine Technion, Haifa Israel

characteristics are associated with a variety of impairments in occupational and social functioning. No single symptom is pathognomonic of schizophrenia, consequently the disorder is noted for its great heterogeneity across individuals and for its variability within individuals over time.[1–3] Despite intensive research, the etiology and the pathophysiology of the disease have remained elusive. A variety of hypotheses have been raised regarding the etiology of schizophrenia, one of the first and still the most substantiated is the 'dopamine hypothesis', which postulates a malfunction of the dopaminergic system as the pathophysiological basis of the disease.[4] Additional neurotransmitters, such as glutamate and GABA have also been implicated in schizophrenia.[5–9] Another hypothesis that research has converged to, is the 'neurodevelopmental hypothesis', which posits prenatal or early postnatal cerebral aberration as the causative factor of schizophrenia at adulthood.[10] One main feature of schizophrenia is abnormal brain circuits' activity observed by imaging studies. While this observation can be consistent with the neurodevelopmental component of the disease, it also points to the possibility of prepuberty and/or postpuberty events, secondary to the *in-utero* genetic and epigenetic processes, taking place in the developing as well as in the mature brain. A conceivable secondary mechanism is abnormal intrinsic- or extrinsic-dependent modulation of neuronal activity-dependent pathological mechanisms.

Neuronal activity, or neuronal firing rate, impinges on neurotransmitter transmission, neuronal net communication and the consequent intracellular processes, and vise versa. It is the core process by which the nervous system adapts to changes in the environment. Neuronal firing governs the interaction between synaptic strength and efficacy, gene expression and protein activity, which leads to changes in neurogenesis, cell migration as well as slow spatiotemporal dynamic morphogenesis and neuronal plasticity. These processes are associated with learning and memory, as well as with adaptive changes in emotional, cognitive and sensorimotor function, all abnormal in schizophrenia.

Mitochondria are the energy source for driving the biochemical processes involved in various cell functions and take part in intracellular Ca^{2+} homeostasis, both fundamental for neuronal activity. This chapter suggests that the pathophysiology of schizophrenia involves abnormalities in neuronal activity in which mitochondria play a pivotal role. The accumulating data for abnormal brain activity and energy metabolism, dysfunction of mitochondria and specifically the aberrations in the first complex (complex I) of its oxidative phosphorylation system (OXPHOS), in schizophrenia will be reviewed. In the second part, experimental evidence suggesting mitochondrial complex I as a possible biomarker for schizophrenia will be discussed.

Brain Energy Metabolism in Schizophrenia

Brain imaging studies in schizophrenia largely reveal decreased metabolism in the prefrontal cortex.[11–21] Although less consistent, alterations in brain metabolic rates in other brain regions, including the temporal and parietal cortices, thalamus, basal ganglia and cerebellum, were also observed.[14,22,23] It is currently suggested that aberrations in brain metabolism in schizophrenia vary with disease state. Patients with predominantly negative symptoms demonstrate lower metabolic rates in areas of the right frontal, temporal and parietal cortices, and higher metabolic rates in the cerebellar cortex and in the lower deep cerebellar nuclei during cognitive task performance, compared to subjects with predominantly positive symptoms.[17,19,24] Alteration in brain energy metabolism strongly suggests the involvement of mitochondria, the key players in cellular energy production.

Imaging studies using phosphorous or hydrogen magnetic resonance spectroscopy (^{31}P-MRS, and ^{1}H-MRS, respectively) provide more direct evidence for mitochondrial dysfunction in schizophrenia. Studies using ^{31}P-MRS demonstrated reduced mitochondrial originated high energy phosphates, such as ATP and phosphocreatine (PCr), in the frontal lobe, the caudate nucleus and the left temporal lobe of schizophrenic patients.[25–28] Deficits have been described in additional cellular factors, such as phospholipids, whose metabolism is strongly suggested to be linked to mitochondrial ATP production.[29,30] Decreased levels of phosphomonoesters (PMEs) and inorganic phosphate (Pi) have been described, in the dorsal prefrontal and the temporal cortices of neuroleptic-naive, first-episode and chronic schizophrenia subjects.[31,32] ^{1}H-MRS studies in schizophrenia have shown regional brain decreases in N-acetylaspartate (NAA), which is suggested to play an integral role in mitochondrial energy production,[33] particularly in the prefrontal cortex, the temporal cortex and the hippocampus.[34,35]

Although less consistent, abnormalities in mitochondria-related cellular compounds were also observed in brains of patients with affective disorders, mainly in bipolar disorder. Thus, in bipolar patients, mostly in medicated patients, ^{31}P- and ^{1}H-MRS imaging studies identified changes in different cellular factors tightly connected with mitochondrial function, including PCr, PMEs, intracellular pH, lactate and NAA in the prefrontal cortex, temporal cortex and in the hippocampus.[30,36–38] In major depression, the current literature on MRS studies is sparse with large diversity between studies, some reporting no change, while others a decrease or even an increase in Cr and NAA in prefrontal, medial temporal, orbitofrontal, and anterior cingulate cortices as well as in the basal ganglia of adult patients.[39–41] In addition, a decrease in ATP and an increase in PME and in pH in the frontal cortex were reported.[26,42]

The similarity between all three mental disorders regarding brain energy metabolism and mitochondrial function analyzed by different imaging techniques, raises the question of disease-specificity of this pathology. One possible factor affecting brain activity and blunting the differences between disorders is medication, as most studies have been conducted in non-drug naïve patients. Indeed, the putative contribution of medication to altered brain metabolism was studied. For example, drug-naïve schizophrenic patients, both at first episode and when chronically ill, demonstrated reduced blood flow in the prefrontal, associative frontal, parietal and temporal gyri, and increased perfusion in the thalamus, cingulate cortex and cerebellum, suggesting that these abnormalities were neither progressive nor a consequence of medication.[43–45] Findings in the basal ganglia of unmedicated schizophrenic patients are inconsistent. Several studies reported no significant difference between patients and control subjects,[46] whereas others found relatively higher or lower striatal metabolism or blood flow in patients.[24,47,48] In contrast, most studies reported an increase in basal ganglia metabolism or blood flow in patients receiving antipsychotic medication as compared to non-medicated patients and controls.[49,50] Variations in blood flow in medicated patients were also reported in other brain areas such as thalamus and cortex,[12] suggesting a drug effect that is mediated through the striato-thalamo-cortical neuronal circuitry, rather than a direct effect on each of these brain areas.[16] Thus, at least in some brain areas, such as the basal ganglia, medication seems to affect brain energy metabolism.

An additional explanation for limited specificity of the imaging studies may be the overlapping of symptoms between the disorders. Regardless, the similarities in the abnormal brain activation between the mental disorders may be attributed to methodology, as imaging usually samples a relatively broad brain area in which glia cells outnumber the neurons, and can not exclude circuit-dependent inductions. These limitations of imaging studies calls for the need to unravel a more functionally and/or neuroanatomically specific biological process, which may be able to differentiate between disorders or symptoms' categories.

Mitochondria and Neuronal Activity

The role of mitochondria in neuronal activity and thereby long-term structural and functional changes, which modulate synaptic connectivity associated with adaptive changes in emotional and cognitive function, was first inferred from histochemical evidence demonstrating mitochondria recruitment to location of high activity zones in the neuron. Mitochondria are highly mobile organelles and their movement along microtubules can be adjusted according to changes in neuronal activity and local energy demands. Indeed, mitochondria have been shown to move to active zones during neurotransmitter release in response to an increase in synaptic activity.[51] In addition, it has been shown that mitochondria play a key role in the establishment of neuronal polarity by concentrating at the site of axogenesis.[52] This may be relevant to neurite sprouting, elongation and growth cone motility of axons as well as dendrites. More recent studies have shown that loss of mitochondria from axon terminal in *Drosophila* results in defective synaptic transmission.[53–55] Moreover, a role for axonal mitochondria (pre-synaptic) in short-term facilitation[56] as well as for dendritic mitochondria (post-synaptic) in morphogenesis and plasticity of spines and synapses[57] was demonstrated in mice hippocampal slice cultures. Substantial evidence for mitochondrial role in neuronal activity and plasticity is depicted in studies of the visual system in cats and rats, which serves as a paradigm for neuronal plasticity, in as much as anatomical and physiological development can be altered by visual experience.[58] Thus, in rats exposed to light, as opposed to those deprived of light, optic synapses in the suprachiasmatic nucleus had larger boutons with larger mitochondria, as well as more and larger mitochondria in the postsynaptic dendrites.[59] Others have shown that in rats exposed to complex environments, synaptogenesis and an increase in the volume of the visual cortex is associated with infiltration of new mitochondria and capillaries.[60] In addition, it has been reported that in the visual system several genes encoding mitochondrial OXPHOS complexes (several subunits of cytochrome oxidase, NADH dehydrogenase and of ATPase as well as cytochrome b) are regulated by neuronal activity, and their expression is correlated with the extent of plasticity in the visual cortex.[61–63] Mitochondrial genes up-regulation, primarily that of 12 S rRNA, which was indicative of a generalized elevation in mitochondrial transcription, has been also observed in the hippocampus following synaptic activity, both of sufficient strength to induce LTP and of reduced strength that may be related to the induction of a metaplastic state.[64] Additional evidence supporting the role of mitochondria in mediating synaptic activity is ensued from the findings that inhibition of succinate dehydrogenase, the second complex of the OXPHOS, induces a long-term potentiation of the NMDA-mediated synaptic excitation, which depends on dopamine acting via its D_2 receptors.[65] Finally, mitochondrial permeability transition pores and their constituents, the porin proteins, which have a significant role in diverse cellular processes, including regulation of mitochondrial ATP and calcium efflux, have been found to have a dynamic functional role in amending neuronal activity, learning and synaptic plasticity.[66,67] In all, these studies render a growing body of evidence for the important contribution of mitochondria to neuronal activity and thereby both to short-term modulation and to long-term phenomena.

Mitochondria in Schizophrenia

Mitochondrial Morphometry

Mitochondrial morphological abnormalities in the frontal cortex, the striatum and the substantia nigra in schizophrenia have been demonstrated by microscopic analysis of autopsy specimens. Ultrastructure of autopsied anterior limbic cortex from schizophrenic patients showed deformation and reduction in the number of mitochondria.[68] An additional small-sample study reported that in the caudate nucleus and the putamen throughout the neuropil, mitochondrial density was significantly reduced in a mixed sample of drug-treated and off-drug cases as compared to control level.[69] Their findings indicate that mitochondrial size in both dendrites and axon terminals was smaller than normal in drug-off cases. Antipsychotic treatment appeared to normalize mitochondrial density and volume in the putamen more than in the caudate. A hypoplasia of mitochondria was also observed in the substantia nigra pars compacta and in the presynaptic terminals of tyrosine hydroxylase immunoreactive neurons. This was associated with deformed terminals and altered connectivity.[70] In oligodendrocytes, the most affected cells in schizophrenia and bipolar disorder, a significant reduction of approximately 33% in both number and volume of their mitochondria was observed in the caudate nucleus and the prefrontal cortex in patients with schizophrenia.[71] Despite the small number of patients and the use of postmortem tissue, which may suffer from inherent artifacts due to the process of tissue collection, these structural findings hold potential importance for the involvement of mitochondria in schizophrenia, and are strengthened by the convergence of additional lines of evidence discussed throughout this chapter.

Mitochondrial Oxidative Phosphorylation System

Mitochondria are intracellular organelles that are composed of four functionally specific compartments including the relatively non-selective outer membrane, the intermembrane space, the highly selective inner membrane and its cristae and the matrix. Mitochondria are the "powerhouse" of the cell as they provide ATP, used as a source of chemical energy. In addition to supplying cellular energy, mitochondria are involved in a range of other processes, such as signaling, cellular differentiation, cell death, as well as the control of the cell cycle and cell growth. In excitatory cells, such as neurons and muscles they are essential for cell activity, firing and contraction, respectively. In addition the mitochondria have their own DNA (mtDNA), contributing to the mitochondrial proteome, which is mostly encoded by the nuclear DNA (nDNA). Cellular energy is primarily generated by mitochondrial oxidative phosphorylation system (OXPHOS), a process requiring a coordinated action of four respiratory enzyme complexes arranged in a specific orientation in the inner mitochondrial membrane, termed also the mitochondrial respiratory chain. Electrons generated from reduced electron carriers NADH and FADH$_2$, which are produced from oxidation of nutrients such as glucose, are ultimately transferred through the respiratory chain to molecular oxygen. This process is coupled to proton translocation across the inner membrane forming an electrochemical gradient, which stores energy that is then used for ATP synthesis by the fifth complex, ATP synthase. Each complex of the OXPHOS system consists of multiple components, or subunits. Apart from complex II subunits, which are exclusively encoded by the nuclear genome, the subunits of the other four complexes are encoded either by the nDNA, approximately 70 genes, which are randomly distributed over the chromosomes with no obvious clustering, or by the mtDNA which encodes 13 genes.

Complex I (NADH: Ubiquinone Oxidoreductase)

Among the enzyme complexes that compose the OXPHOS, complex I has the most complex structure and the least understood mechanism of electron transfer and proton translocation.[72] Complex I plays a major role in controlling oxidative phosphorylation in mitochondria and its abnormal activity can lead to mitochondrial dysfunction.[73] It is therefore not surprising that many human mitochondrial diseases result from complex I deficiencies, including Leber's hereditary optic neuropathy, severe and fetal lactic acidosis, and various neuromuscular myopathies. In our studies we have focused on the 51-kDa (NDUFV1) and the 24-kDa (NDUFV2) subunits, both iron-sulfur flavoproteins having catalytic properties, including the site for transhydrogenation from NADH to NAD$^+$, and the 75 kDa (NDUFS1), the largest iron-sulfur structural protein,[72] all forming one functional subunit of complex I.[74–79]

Accumulating molecular data point to abnormalities in mitochondrial mRNA and protein expression, both in periphery and brain in schizophrenia.[80–87] Studies using transcriptomic, proteomic and metabolomic approaches on human brain tissue, mostly the prefrontal cortex and one study in the hippocampus have demonstrated a specific robust change in gene and protein expression associated with mitochondrial function in schizophrenia.[86–88] The evidence for the alterations in nuclear encoded mitochondrial gene expression obtained by microarrays is not unequivocal as some studies demonstrated that sample pH as well as statistical complications including multiple comparisons may have a strong effect on the results.[89] Focusing on the mitochondrial oxidative phosphorylation system reveals a schizophrenia specific neuroanatomical pattern of alteration in the transcripoms and proteoms of complex I subunits NDUFV1 (51-kDa), NDUFV2 (24-kDa) and the 75-kDa was observed in brain.[84]

Mitochondrial gene expression studies were also performed in patients with affective disorder with some, yet not all, reporting alterations in mitochondrial related genes and proteins. In bipolar disorder a reduction in the expression level of mitochondrial related genes in hippocampal and prefrontal postmortem specimens was observed,[93,94] with one paper demonstrating reductions in genes of the OXPHOS in the prefrontal cortex,[95] while another reported an increase in complex I subunits NDUFV1 and NDUFV2 in the parieto-occipital cortex.[84] In major depression, although most studies did not show cortical modifications in mitochondrial related genes, some reports suggest alterations in the expression of nuclear as well as of mitochondrial DNA encoded genes in the prefrontal cortex.[84,93] In addition, it was demonstrated that muscle mitochondria in depressive patients produced less ATP and that the activity of the oxidative phosphorylation system complexes I + III and II + III was impaired.[96]

Genetic studies also implicate mitochondria abnormalities in schizophrenia. For example, missense variants in mitochondrial DNA (mtDNA), one of which encodes for the ND4 subunit of mitochondrial complex I, are present in schizophrenic patients while absent in healthy controls,[97] and a single nucleotide polymorphism (SNP) of complex I ND3 subunit is associated with bipolar disorder.[98] Additional association studies reported a significant association of a haplotype consisting of two SNPs in a nuclear encoded subunit of complex I, the NDUFV2, with schizophrenia and with bipolar disorder.[99,100] These studies suggest complex I as a risk factor in both disorders. In major depression, an association with 3243A to G mutation in tRNA$^{\text{Leu(UUR)}}$ gene of the mtDNA[101] was observed.

Several groups have studied the enzymatic activity of different complexes of the OXPHOS in schizophrenia post mortem brain specimens. However, functional measurements in postmortem brain specimens have to be taken with reservations, as enzymatic activity is particularly sensitive to postmortem delay,[102,103] and its detection is less sensitive in whole tissue than in isolated mitochondria. Nevertheless, depending on the brain area studied several groups reported an increase or a decrease as well as no change in cytochrome c oxidase (complex IV) activity,[85,90,91] the enzyme which was suggested as an endogenous metabolic marker for neuronal activity.[104] Interestingly, another study demonstrated a strong negative correlation between complex IV activity and emotional and intellectual impairments in schizophrenia, but not motor impairment, solely in the putamen.[105] There are controversial findings regarding the activity of complex I in the brains of schizophrenics. A significant reduction in NADH-cytochrome c reductase (complexes I–III) activity was observed in one study in the frontal cortex and in two others in the temporal cortex and in the basal ganglia.[85,91,92]

The imaging, biochemical, molecular and genetic studies described thus far provide strong evidence for mitochondrial dysregulation as part of the pathology of schizophrenia as well as of mood disorders, consequently raising the question as to whether mitochondrial impairment displays disease-specific characteristics or is rather a general non-distinguishing pathology of these mental disorders. Our recent findings concerning brain mitochondrial complex I,[84] may constitute one possible explanation for this intriguing question. We have shown that complex I subunits NDUFV1, NDUFV2 and NDUFS1, were altered in all three psychiatric disorders, albeit in a disease-specific neuroanatomical pattern. In schizophrenia, but not in affective disorders, a selective reduction in mRNA and protein levels of complex I subunits was observed in the prefrontal cortex and the striatum. However, in both affective disorders reductions were observed specifically in the cerebellum with the depressed group demonstrating more consistent alterations. Bipolar disorder, displayed anatomical overlaps also with schizophrenia, as an increase in the expression of complex I subunits was observed in the parieto-ocipital cortex of both disorders. This is in line with the similarities in clinical symptoms of bipolar disorder and the other two disorders. These data may suggest that the similarities in clinical symptoms between mental disorders are reflected in their pathophysiology, however the pattern or extent of the biological impairment can differentiate between disorders.

Peripheral Abnormalities in Mitochondrial Complex I

The mitochondrial oxidative phosphorylation system was found to be impaired not only in brains of schizophrenic patients but also in their platelets and lymphocytes. It was shown that the activity of complex I was significantly reduced in platelets and lymphocytes of a small sample of schizophrenic patients chronically treated with antipsychotics.[106,107] We have further substantiated the implication of complex I in schizophrenia in several more recent studies. We found that in 113 schizophrenic patients, complex I activity, but not that of complex V, was significantly reduced (53% of controls) in patients with residual schizophrenia, while significantly increased (190% of controls) both in patients in an acute psychotic episode and those in a chronic active state, as compared to 37 healthy control subjects.[83] These results suggest disease state-dependent alterations in complex I activity in platelets of schizophrenic patients. Moreover, in 27 patients with affective disorders, either major depression or bipolar disorder (the depressed type), complex I activity did not differ from that of the control group,[82] suggesting that the alterations in complex I activity may not only be state-dependent but also disease-specific. Further support for the relationship between the clinical state and complex I activity in schizophrenia can be inferred from the highly significant positive correlation ($r = 0.7$, $p < 0.0001$) of complex I activity with the severity of patients' positive symptom scores, as well as a tendency towards a negative correlation with the negative symptom scores, as assessed by the Positive and Negative Symptom Scale (PANSS).[83] Interestingly, in the acute active state, increased complex I activity was observed in both 25 medicated and 25 unmedicated patients,[82] suggesting that at least in this state of the disease medication did not affect complex I activity. However, interpretation of the observed effects of medication on complex I are far from straightforward given the findings that complex I can be inhibited by antipsychotic medications in-vitro[107,108] and will be elaborated below. In this connection it is notable that dopamine (DA), which is a major pathological factor in schizophrenia, also inhibits complex I activity both in mitochondrial preparation and in intact respiring cells. DA inhibition of complex I resulted in mitochondrial dysfunction.[109] In schizophrenia, both in platelets and in EBV transformed lymphocytes, DA displayed a pathological interaction with complex I, as it inhibited complex I activity as well as complex I driven cell respiration twice as much as in cells derived from healthy subjects.[109] These findings further support that abnormal complex I activity in schizophrenia is associated with impaired mitochondria and cellular dysfunction. Concomitant with the abnormal activity of complex I both mRNA and protein levels of the NDUFV1 and NDUFV2 subunits were significantly higher in schizophrenic patients than in controls, with no change in the NDUFS1 subunit.[83] Interestingly, in juvenile neuroleptic-naive schizophrenic patients an increase was observed in mRNA levels of the NDUFS1 subunit.[110]

Mitochondria Complex I as a Possible Biomarker for Schizophrenia

The peripheral findings regarding specific alterations in mitochondrial complex I activity and gene expression in schizophrenia, as well as the evidence for brain mitochondria impairment, in general, and complex I in particular, as part of the pathophysiology of the disorder, render complex I as an attractive potential peripheral biomarker for schizophrenia.

The validity of peripheral markers as proxies for measuring brain function has been questioned, and the attempt to demarcate a peripheral biological marker for schizophrenia may appear ambitious. However, mitochondria are unique in that they are partly independent organelles, contain their own DNA and are highly preserved along evolution and in different tissues. In addition, biochemical and pharmacological similarities exist between blood cells, specifically platelets and 5HT or DA containing neurons of the CNS.[111] Numerous studies have shown that platelets from schizophrenic patients behave differently than those isolated from healthy controls in dopamine uptake, 5-HT content arachidonic acid metabolism, inositol phosphate levels and disturbance of calcium homeostasis.[111–113] Reduction in imipramine binding was found both in platelets and in postmortem brains of deceased depressed patients and patients who committed suicide.[114] More recently it has been shown that CNS neurotransmitters can regulate the immune cells, as

activated T cells can cross the blood brain barrier.[115–117] Such regulation involves changes in the expression of lymphocytes' surface markers including neurotransmitter receptors, among them DA D3 and D2 and GluR3 receptors.[115,118,119] These findings suggest brain pathologies can be reflected in peripheral blood cells. In line with the latter, our recent study shows a pathological correlation between platelets complex I activity and cerebral glucose metabolism in several schizophrenia-relevant brain areas in schizophrenic patients.[120]

Interestingly, the extent of complex I activity in healthy controls is highly stable over time. However, the vast heterogeneity between individuals and over time in schizophrenia as well as medication ability to affect complex I activity, may limit the ability of complex I to discriminate between patients. Indeed, complex I activity, could distinguish between patients with positive symptoms and controls or patients with residual schizophrenia with a high accuracy (90.2% and 96.8%, respectively), but not between the two latter groups. One possible explanation for the lack of a significant difference between patients with residual schizophrenia and healthy controls is the inhibitory effect of antipsychotic medication, which annuls the pathological increase in complex I activity at this disease state. The inhibitory effect of antipsychotic drugs fit also with the increased activity of complex I in unmedicated schizophrenic patients but do not explain its increased activity in medicated patients in an acute or chronic active state. An attractive hypothesis is that the lack of medication-induced reduction in complex I activity at these disease-states could imply the presence of a temporary (in acute state) or long lasting (in chronic active patients) of partial 'non-responsiveness' to antipsychotic treatment. The latter is in line with the inability of medication to control positive symptomology. Alongside antipsychotic drugs, an endogenous factor may exist that is altered in association with disease state or treatment and modulates complex I activity. The massive antipsychotic treatment most chronic schizophrenic patients receive may mask the disease-state dependent alterations in complex I activity. Indeed, in a preliminary 4 months follow-up study of medicated patients (chlorpromazine mg equivalents/day 1,500 ± 568) who entered the study at an acute state and showed a significant improvement of symptom severity and executive functioning from baseline to end-of-study, no significant change was observed in complex I activity. However, the abnormality in complex I activity could still be detected by its significantly higher fluctuations throughout the study period as compared to its activity variations in healthy controls.

The ability of antipsychotic medication to alter complex I activity under certain disease related conditions may suggest that a single measure is unlikely to provide a differential diagnostic tool for all patients and calls for additional complex I-related markers. We have already shown that the expression of three subunits of complex I (NDUFV1, NDUFV2 and NDUFS1), both at the level of mRNA and protein, are altered in brains as well as in the periphery of schizophrenic patients. Moreover, accumulating evidence from postmortem brain, tissue culture and animal model of schizophrenia,[84] suggest that antipsychotic drugs have no effect on complex I subunits' expression. However, there remains the question of specificity of the peripheral changes of complex I subunits mRNA and protein levels in schizophrenia. This is underscored by the findings that the expression of complex I subunits in brain was altered in patients with either schizophrenia or mood disorders. Still there could be schizophrenia specific peripheral changes, as although complex I subunits impaired expression was observed in all three disorders in brain, there was a disease-specific neuroanatomical pattern of distribution of complex I abnormality.[84]

Alterations in gene expression can be attributed to epigenetic factors such as DNA or histone methylation and acetylation or to abnormal regulation by transcription factors. We have recently shown that the transcription factor Sp1, which regulates, among other genes, the expression of complex I subunits, is abnormally expressed both in brain and lymphocytes of schizophrenic patients.[121] Moreover, its specific pattern of alteration paralleled that of complex I subunits both in brain and periphery. These findings support the need for an array of related biomarkers that together may become a better biomarker for schizophrenia.

Does Complex I Meet the Criteria for a Biomarker?

The first chapter of the present book mentions eight criteria that a biomarker has to meet in order to become a useful diagnostic tool. The first criterion demands that the biomarker should reflect some basic pathophysiological process, and detect a fundamental feature of the disease with high sensitivity and specificity. Indeed,

the first part of this chapter describes substantiating evidence, ranging from morphological through imaging to molecular and genetic studies, for mitochondria, in general and complex I, in particular, as part of the pathophysiology of schizophrenia. In addition, complex I activity detected with high sensitivity and specificity positive symptomology when compared to patients at the residual state or healthy subjects. Interestingly, in a preliminary study with 11 non-schizophrenic psychotic patients, complex I activity did not differ from that of the healthy group. The latter, however, needs further verification. The residual state of the disease could not be discriminated from health by complex I activity, despite the significant difference in the average activity values between the two cohorts. We suggest that taking into account mRNA and protein levels of complex I subunits and analyzing them together with complex I activity may improve both sensitivity and selectivity of the marker at the residual state of the disease. Nevertheless, at present the enzyme activity appears to meet the first criteria for the high positive symptom subgroup of schizophrenic patients. The second criterion relates to disease specificity. As far as our current data indicate, complex I activity in blood cells demonstrates disease specific abnormalities in patients with positive symptomology as compared to patients with major depression, bipolar disorder (the depressed type) and patients with Alzheimer's disease. Interestingly, in Parkinson's disease a decrease in complex I activity was observed[122] as opposed to its increase observed in schizophrenia with positive symptomology. Despite these encouraging data, further studies are needed to establish disease specificity of our biomarker, particularly as molecular abnormalities in complex I were also observed in brains of patients with mood disorders. However, the findings that molecular abnormalities of complex I subunits showed a disease specific neuroanatomical pattern of distribution in postmortem brains of major depression, bipolar and schizophrenic patients, suggest that despite sharing pathological resemblance, complex I can discriminate between the three disorders. Complex I also fulfills the other six criteria mentioned in the first chapter, the lack of reflection of symptomology, the reproducibility over time, noninvasiveness, non-harmfulness, reliability, easy-to-perform and inexpensiveness. The ability of complex I to meet the majority of the criteria for a biomarker turns it into an attractive candidate for any future study in the field of biomarkers in schizophrenia.

Conclusion

At present, definitive diagnosis of schizophrenia as well as other psychiatric disorders relies on descriptive behavioral and symptomatic criteria that are inevitably inaccurate. Indeed the validity of several psychiatric diagnoses remains a matter of considerable controversy, even among psychiatrists who generally support the conceptual framework of the Diagnostic and Statistical Manual of Mental Disorders, Fourth Edition (DSM-IV). Accurate biomarkers, along with more reliable and valid disease criteria, will help psychiatry achieve greater objectivity in diagnosis. Even more, promising biomarkers may help diagnosing psychiatric disorders in their earliest stages, potentially enhancing the care of patients. Neuroimaging has raised hopes for an objective and reliable diagnostic tool. Yet, the reality of research in psychiatric diseases is such that peripheral markers, despite their inherent disadvantage, are the only alternative to very costly and still insensitive functional neuroimaging, which is unlikely to be applicable as routine diagnostic measures for a large number of patients in the near future. Although more studies are needed to establish complex I as a biomarker for schizophrenia, our finding on the role of mitochondria and complex I in the pathology of schizophrenia together with its peripheral disease specific alterations, renders complex I as a promising convenient peripheral marker for the disease, or at least to specific sub-states of the disease. There still remains the unavoidable effect of medication on complex I activity. However, the expression of complex I subunits is not affected by medication, thus together with complex I activity they may still differentiate between schizophrenia and other mental disorders in heavily medicated patients. The disadvantage of medication effect can turn into an advantage as it may enable the use of complex I activity as a biomarker for responsiveness to medication, management and follow-up of the disorder as well as a diagnostic tool. Finally, we suggest that the biological complexity of schizophrenia, its overlap of symptoms with other mental disorders as well as co-morbidity with other common disorders, leads to the inevitable conclusion that any biomarker would have to be an array of interrelated factors or even a set of several such arrays in which mitochondrial complex I will be a major constituent.

Acknowledgements The following people contributed significantly to the studies on complex I reviewed in this chapter: Rachel Karry, Ph.D., Natlie Dror, Ph.D., Hanit Brenner-Lavie, Ph.D., Rosa Zuk, M.Sc., Haifa Gazawi, M.Sc., Ehud Klein, M.D., Alon Reshef, M.D., Ala Sheinkman, M.D.

The studies reviewed in this chapter were supported by grants from Chief Scientist of the Ministry of Health, Israel, The Center for Absorption in Science, Ministry of Immigrant Absorption, State of Israel, The National Alliance for Research on Schizophrenia and Affective Disorders (NARSAD), USA and The Stanley Medical Research Foundation, USA

References

1. Dingman CW, McGlashan TH. Discriminating characteristics of suicides. Chestnut Lodge follow-up sample including patients with affective disorder, schizophrenia and schizoaffective disorder. Acta Psychiatr. Scand. 1986; 74(1): 91–97.
2. McGlashan TH. A selective review of recent North American long-term follow-up studies of schizophrenia. Schizophr. Bull. 1988; 14: 515–542.
3. McGlashan TH, Fenton WS. The positive/negative distinction in schizophrenia: review of natural history validators. Arch. Gen. Psychiatry 1992; 49: 63–72.
4. Seeman P. Dopamine receptors and the dopamine hypothesis of schizophrenia. Synapse 1987; 1(2): 133–152.
5. Kim JS, Kornhuber HH, Schmid-Burgk W, et al. Low cerebrospinal fluid glutamate in schizophrenic patients and a new hypothesis on schizophrenia. Neurosci. Lett. 1980; 20: 379–382.
6. Carlsson A, Hansson LO, Waters N, et al. A glutamatergic deficiency model of schizophrenia. Br. J. Psychiatry 1999; 37(Suppl): 2–6.
7. Carlsson A, Waters N, Carlsson ML. Neurotransmitter interactions in schizophrenia – therapeutic implications. Biol. Psychiatry 1999; 46: 1388–1395.
8. Lewis DA, Pierri JN, Volk DW, et al. Altered GABA neurotransmission and prefrontal cortical dysfunction in schizophrenia. Biol. Psychiatry 1999; 46: 616–626.
9. Carlsson A, Waters N, Holm-Waters S, et al. Interactions between monoamines, glutamate, and GABA in schizophrenia: new evidence. Annu. Rev. Pharmacol. Toxicol. 2001; 41: 237–260.
10. Altshuler LL, Conrad A, Kovelman JA, et al. Hippocampal Pyramidal cell orientation in schizophrenia. Arch. Gen. Psychiatry 1987; 44: 1094–1098.
11. Buchsbaum MS. The frontal lobes, basal ganglia and temporal lobes as sites for schizophrenia. Schizophr. Bull. 1990; 16: 377–387.
12. Buchsbaum MS, Hazlett EA. Positron emission tomography studies of abnormal glucose metabolism in schizophrenia. Schizophr. Bull. 1998; 24: 343–346.
13. Carter CS, Perlstein W, Ganguli R, et al. Functional hypofrontality and working memory dysfunction in schizophrenia. Am. J. Psychiatry 1998; 155: 1285–1287.
14. Gur RE, Resnick SM, Alavi A, et al. Regional brain function in schizophrenia II: repeated evaluation with positron

emission tomography. Arch. Gen. Psychiatry 1987; 44: 126–129.

15. Hazlett EA, Buchsbaum MS, Jeu LA, et al. Hypofrontality in unmedicated schizophrenia patients studied with PET during performance of a serial verbal learning tasks. Schizophr. Res. 2000; 43: 33–46.

16. Holcomb HH, Cascella NG, Thaker GK, et al. Functional sites of neuroleptic drug action in the human brain: PET/FDG studies with and without haloperidol. Am. J. Psychiatry 1996; 153: 41–49.

17. Lahti AC, Holcomb HH, Medoff DR, et al. Abnormal patterns of regional cerebral blood flow in schizophrenia with primary negative symptoms during an effortful auditory recognition task. Am. J. Psychiatry 2001; 158: 1797–1808.

18. Manoach DS, Press DZ, Thangaraj V, et al. Schizophrenic subjects activate dorsolateral prefrontal cortex during a working memory task, as measured by fMRI. Biol. Psychiatry 1999; 45: 1128–1137.

19. Potkin SG, Alva G, Fleming K, et al. A PET study of the pathophysiology of negative symptoms in schizophrenia. Positron emission tomography. Am. J. Psychiatry 2002; 157: 227–237.

20. Shenton ME, Dickey CC, Frumin M, et al. A review of MRI findings in schizophrenia. Schizophr. Res. 2001; 49: 1–52.

21. Weinberger D. On the plausibility of the neurodevelopmental hypothesis in schizophrenia. Neuropsychopharmacol 1996; 14: 1 S–11 S.

22. Hazlett EA, Buchsbaum MS, Byne W, et al. Three-dimensional analysis with MRI and PET of the size, shape, and function of the thalamus in the schizophrenia spectrum. Am. J. Psychiatry 1999; 156: 1190–1199.

23. Tamminga CA, Thaker GK, Buchanan R, et al. Limbic system abnormalities identified in schizophrenia using positron emission tomography with fluorodeoxyglucose and neocortical alterations with deficit syndrome. Arch. Gen. Psychiatry 1992; 49: 522–530.

24. Wolkin A, Jaeger J, Brodie JD, et al. Persistence of cerebral metabolic abnormalities in chronic schizophrenia as determined by positron emission tomography. Am. J. Psychiatry 1985; 142: 564–571.

25. Fujimoto T, Nakano T, Takano T, et al. Study of chronic schizophrenics using 31P magnetic resonance chemical shift imaging. Acta Psychiatr. Scand. 1992; 86: 455–462.

26. Volz HR, Riehemann S, Maurer I, et al. Reduced phosphodiesters and high-energy phosphates in the frontal lobe of schizophrenic patients: a 31P chemical shift spectroscopic-imaging study. Biol. Psychiatry 2000; 47: 954–961.

27. Jayakumar PN, Venkatasubramanian G, Keshavan MS, et al. MRI volumetric and 31P MRS metabolic correlates of caudate nucleus in antipsychotic-naive schizophrenia. Acta Psychiatr. Scand. 2006; 114(5): 346–351.

28. Jensen JE, Miller J, Williamson PC, et al. Grey and white matter differences in brain energy metabolism in first episode schizophrenia: 31P-MRS chemical shift imaging at 4 Tesla. Psychiatry Res 2006; 146(2): 127–135.

29. Purdon AD, Rapoport SI. Energy requirements for two aspects of phospholipid metabolism in mammalian brain. Biochem. J. 1998; 335 (Pt 2): 313–318.

30. Stork C, Renshaw PF. Mitochondrial dysfunction in bipolar disorder: evidence from magnetic resonance spectroscopy research. Mol. Psychiatry 2005; 10(10): 900–919.

31. Fukuzako H, Fukuzako T, Hashiguchi T, et al. Changes in levels of phosphorus metabolites in temporal lobes of drug-naive schizophrenic patients. Am. J. Psychiatry 1999; 156(8): 1205–1208.

32. Reddy R, Keshavan MS. Phosphorus magnetic resonance spectroscopy: its utility in examining the membrane hypothesis of schizophrenia. Prostaglandins Leukot. Essent. Fatty Acids 2003; 69(6): 401–405.

33. Madhavarao CN, Chinopoulos C, Chandrasekaran K, et al. Characterization of the N-acetylaspartate biosynthetic enzyme from rat brain. J. Neurochem. 2003; 86(4): 824–835.

34. Deicken RF, Johnson C, Pegues M. Proton magnetic resonance spectroscopy of the human brain in schizophrenia. Rev. Neurosci. 2000; 11(2–3): 147–158.

35. Harrison PJ, Weinberger DR. Schizophrenia genes, gene expression, and neuropathology: on the matter of their convergence. Mol. Psychiatry 2005; 10(1): 40–68; image 5.

36. Bertolino A, Frye M, Callicott JH, et al. Neuronal pathology in the hippocampal area of patients with bipolar disorder: a study with proton magnetic resonance spectroscopic imaging. Biol. Psychiatry 2003; 53(10): 906–913.

37. Kato T. Mitochondrial dysfunction in bipolar disorder: from 31P-magnetic resonance spectroscopic findings to their molecular mechanisms. Int. Rev. Neurobiol. 2005; 63: 21–40.

38. Kato T. Mitochondrial dysfunction as the molecular basis of bipolar disorder: therapeutic implications. CNS Drugs 2007; 21(1): 1–11.

39. Kumar A, Thomas A, Lavretsky H, et al. Frontal white matter biochemical abnormalities in late-life major depression detected with proton magnetic resonance spectroscopy. Am. J. Psychiatry 2002; 159(4): 630–636.

40. Coupland NJ, Ogilvie CJ, Hegadoren KM, et al. Decreased prefrontal Myo-inositol in major depressive disorder. Biol. Psychiatry 2005; 57(12): 1526–1534.

41. Yildiz-Yesiloglu A, Ankerst DP. Review of 1 H magnetic resonance spectroscopy findings in major depressive disorder: a meta-analysis. Psychiatry Res. 2006; 147(1): 1–25.

42. Kato T, Takahashi S, Shioiri T, et al. Brain phosphorous metabolism in depressive disorders detected by phosphorus-31 magnetic resonance spectroscopy. J. Affect Disord. 1992; 26(4): 223–230.

43. Andreasen NC, OwLeary DS, Flaum M, et al. Hypofrontality in schizophrenia: Disturbed dysfunctional circuits in neuroleptic-naive patients. Lancet 1997; 349: 1730–1734.

44. Kishimoto H, Yamada K, Iseki E, et al. Brain imaging of affective disorders and schizophrenia. Psychiatry Clin. Neurosci. 1998; 52(Suppl): S212–S214.

45. Kim JJ, Mohamed S, Andreasen NC, et al. Regional neural dysfunctions in chronic schizophrenia studied with positron emission tomography. Am. J. Psychiatry 2000; 157: 542–548.

46. Sheppard G, Gruzelier J, Manchanda R, et al. 15O positron emission tomographic scanning in predominantly never-treated acute schizophrenic patients. Lancet 1983; 2: 1448–1452.

47. Volkow ND, Brodie JD, Wolf AP, et al. Brain metabolism in patients with schizophrenia before and after acute neuroleptic administration. J. Neurol. Neurosurg. Psychiatry 1986; 49: 1199–1202.

48. Early TS, Reiman EM, Raichle ME, et al. Left globus pallidus abnormality in never-medicated patients with schizophrenia. Proc. Natl. Acad. Sci. USA 1987; 84: 561–563.

49. Buchsbaum MS, Potkin SG, Siegel BV, Jr, et al. Striatal metabolic rate and clinical response to neuroleptics in schizophrenia. Arch. Gen. Psychiatry 1992; 49: 966–974.

50. Corson PW, O'Leary DS, Miller DD, et al. The effects of neuroleptic medications on basal ganglia blood flow in schizophreniform disorders: a comparison between the neuroleptic-naive and medicated states. Biol. Psychiatry 2002; 52: 855–862.

51. Brodin L, Bakeeva L, Shupliakov O. Presynaptic mitochondria and the temporal pattern of neurotransmitter release. Philos. Trans. R. Soc. Lond. B Biol. Sci. 1999; 354(1381): 365–372.

52. Mattson MP. Establishment and plasticity of neuronal polarity. J. Neurosci. 1999; 57: 577–589.

53. Guo X, Macleod GT, Wellington A, et al. The GTPase dMiro is required for axonal transport of mitochondria to Drosophila synapses. Neuron 2005; 47(3): 379–393.

54. Stowers RS, Megeath LJ, Gorska-Andrzejak J, et al. Axonal transport of mitochondria to synapses depends on milton, a novel Drosophila protein. Neuron 2002; 36(6): 1063–1077.

55. Verstreken P, Ly CV, Venken KJ, et al. Synaptic mitochondria are critical for mobilization of reserve pool vesicles at Drosophila neuromuscular junctions. Neuron 2005; 47(3): 365–378.

56. Kang JS, Tian JH, Pan PY, et al. Docking of axonal mitochondria by syntaphilin controls their mobility and affects short-term facilitation. Cell 2008; 132(1): 137–148.

57. Li Z, Okamoto K, Hayashi Y, et al. The importance of dendritic mitochondria in the morphogenesis and plasticity of spines and synapses. Cell 2004; 119(6): 873–887.

58. Sherman SM, Spear PD. Organization of visual pathways in normal and visually deprived cats. Physiol. Rev. 1982; 62: 738–755.

59. Guldner FH, Bahar E, Young CA, et al. Structural plasticity of optic synapses in the rat suprachiasmatic nucleus: adaptation to long-term influence of light and darkness. Cell Tissue Res. 1997; 287(1): 43–60.

60. Black JE, Zelazny AM, Greenough WT. Capillary and mitochondrial support of neural plasticity in adult rat visual cortex. Exp. Neurol. 1991; 111: 204–209.

61. Hevner RF, Wong-Riley M. Neuronal expression of nuclear and mitochondrial genes for cytochrome oxidase (CO) subunits analyzed by in situ hybridization: comparison with CO activity and protein. J. Neurosci. 1991; 11: 1942–1958.

62. Kaminska B, Kaczmarek L, Larocque S, et al. Activity-dependent regulation of cytochrome b gene expression in monkey visual cortex. J. Comp. Neurol. 1997; 379: 271–282.

63. Yang C, Silver B, Ellis SR, et al. Bidirectional regulation of mitochondrial gene expression during developmental neuroplasticity of visual cortex. Biochem. Biophys. Res. Commun. 2001; 287: 1070–1074.

64. Williams JM, Thompson VL, Mason-Parker SE, et al. Synaptic activity-dependent modulation of mitochondrial gene expression in the rat hippocampus. Brain Res. Mol. Brain Res. 1998; 60: 50–56.

65. Calabresi P, Gubellini P, Picconi B, et al. Inhibition of mitochondrial complex II induces a long-term potentiation of NMDA-mediated synaptic excitation in the striatum requiring endogenous dopamine. J. Neurosci. 2001; 20: 5110–5120.

66. Albensi BC, Sullivan PG, Thompson MB, et al. Cyclosporin ameliorates traumatic brain-injury-induced alterations of hippocampal synaptic plasticity. Exp. Neurol. 2000; 162: 385–389.

67. Weeber EJ, Levy M, Sampson MJ, et al. The role of mitochondrial porins and the permeability transition pore in learning and synaptic plasticity. J. Biol. Chem. 2002; 277: 18891–18897.

68. Uranova NA, Aganova EA. Ultrastructure of synapses of the anterior limbic cortex in schizophrenia. Zhurnal Nevropatologii I Psikhiatrii Imeni S-S-Korsakova 1989; 89: 56–59.

69. Kung L, Roberts RC. Mitochondrial pathology in human schizophrenic striatum: a postmortem ultrastructural study. Synapse 1999; 31: 67–75.

70. Kolomeet NS, Uranova NA. Synaptic contacts in schizophrenia: studies using immunocytochemical identification of dopaminergic neurons. Neurosci. Behav. Physiol. 1999; 29: 217–221.

71. Uranova N, Orlovskaya D, Vikhreva O, et al. Electron microscopy of oligodendroglia in severe mental illness. Brain Res. Bull. 2001; 55: 597–610.

72. Hatefi Y. The mitochondrial electron transport and oxidative phosphorylation system. Annu. Rev. Biochem. 1985; 54: 1015–1069.

73. Davey GP, Peuchen S, Clark JB. Energy thresholds in brain mitochondria: potential involvement in neurodegeneration. J. Biol. Chem. 1998; 273: 12753–12757.

74. Fecke W, Sled VD, Ohnishi T, et al. Disruption of the gene encoding the NADH-binding subunit of NADH: ubiquinone oxidoreductase in Neurospora crassa. Formation of a partially assembled enzyme without FMN and the iron-sulphur cluster N-3. Eur. J. Biochem. 1994; 220(2): 551–558.

75. Ragan CI, Galante YM, Hatefi Y. Purification of three iron-sulfur proteins from the iron-protein fragment of mitochondrial NADH-ubiquinone oxidoreductase. Biochemistry 1982; 21(10): 2518–2524.

76. Ohnishi T, Ragan CI, Hatefi Y. EPR studies of iron-sulfur clusters in isolated subunits and subfractions of NADH-ubiquinone oxidoreductase. J. Biol. Chem. 1985; 260(5): 2782–2788.

77. Belogrudov GI, Hatefi Y. Intersubunit interactions in the bovine mitochondrial complex I as revealed by ligand blotting. Biochem. Biophys. Res. Commun. 1996; 227(1): 135–139.

78. Zickermann V, Zwicker K, Tocilescu MA, et al. Characterization of a subcomplex of mitochondrial NADH:ubiquinone oxidoreductase (complex I) lacking the flavoprotein part of the N-module. Biochim. Biophys. Acta 2007; 1767(5): 393–400.

79. Clason T, Zickermann V, Ruiz T, et al. Direct localization of the 51 and 24 kDa subunits of mitochondrial complex I by three-dimensional difference imaging. J. Struct. Biol. 2007; 159(3): 433–442.

80. Altar CA, Jurata LW, Charles V, et al. Deficient hippocampal neuron expression of proteasome, ubiquitin, and mitochondrial genes in multiple schizophrenia cohorts. Biol Psychiatry 2005; 58(2): 85–96.

81. Ben-Shachar D. Mitochondrial dysfunction in schizophrenia: a possible linkage to dopamine. J. Neurochem. 2002; 83: 1241–1251.

82. Ben-Shachar D, Zuk R, Gazawi H, et al. Increased mitochondrial complex I activity in platelets of schizophrenic patients. Int. J. Neuropsychopharmacol. 1999; 2: 245–253.

83. Dror N, Klein E, Karry R, et al. State dependent alterations in mitochondrial complex I activity in platelets: A potential peripheral marker for schizophrenia. Mol. Psychiatry 2002; 7: 995–1001.

84. Karry R, Klein E, Ben Shachar D. Mitochondrial complex I subunits expression is altered in schizophrenia: a postmortem study. Biol. Psychiatry 2004; 55(7): 676–684.

85. Maurer I, Zierz S, Moller H. Evidence for a mitochondrial oxidative phosphorylation defect in brains from patients with schizophrenia. Schizophr. Res. 2001; 48(1): 125–136.

86. Middleton FA, Mirnics K, Pierri JN, et al. Gene expression profiling reveals alterations of specific metabolic pathways in schizophrenia. J. Neurosci. 2002; 22: 2718–2729.

87. Prabakaran S, Swatton JE, Ryan MM, et al. Mitochondrial dysfunction in schizophrenia: evidence for compromised brain metabolism and oxidative stress. Mol. Psychiatry 2004; 9(7): 684–697, 643.

88. Mulcrone J, Whatley SA, Ferrier IN, et al. A study of altered gene expression in frontal cortex from schizophrenic patients using differential screening. Schizophr. Res. 1995; 14: 203–213.

89. Shao L, Martin MV, Watson SJ, et al. Mitochondrial involvement in psychiatric disorders. Ann. Med. 2008; 40(4): 281–295.

90. Cavelier L, Jazin E, Eriksson I, et al. Decreased cytochrome c oxidase activity and lack of age related accumulation of mtDNA in brain of schizophrenics. Genomics 1995; 29: 217–228.

91. Prince JA, Blennow K, Gottfries CG, et al. Mitochondrial function in differentially altered in the basal ganglia of chronic schizophrenics. Neuropsychopharmacology 1999; 21: 372–379.

92. Whatley SA, Curi D, Marchbanks RM. Mitochondrial involvement in schizophrenia and other functional psychoses. Neuroch. Res. 1996; 21: 995–1004.

93. Vawter MP, Tomita H, Meng F, et al. Mitochondrial-related gene expression changes are sensitive to agonal-pH state: implications for brain disorders. Mol. Psychiatry 2006; 11(7): 615, 663–679.

94. Iwamoto K, Bundo M, Kato T. Altered expression of mitochondria-related genes in postmortem brains of patients with bipolar disorder or schizophrenia, as revealed by large-scale DNA microarray analysis. Hum. Mol. Genet. 2005; 14(2): 241–253.

95. Konradi C, Eaton M, MacDonald ML, et al. Molecular evidence for mitochondrial dysfunction in bipolar disorder. Arch. Gen. Psychiatry 2004; 61(3): 300–308.

96. Burnett BB, Gardner A, Boles RG. Mitochondrial inheritance in depression, dysmotility and migraine? J. Affect Disord. 2005; 88(1): 109–116.

97. Martorell L, Segues T, Folch G, et al. New variants in the mitochondrial genomes of schizophrenic patients. Eur. J. Hum. Genet. 2006; 14(5): 520–528.

98. McMahon FJ, Chen YS, Patel S, et al. Mitochondrial DNA sequence diversity in bipolar affective disorder. Am. J. Psychiatry 2000; 157(7): 1058–1064.

99. Kato T, Kunugi H, Nanko S, et al. Mitochondrial DNA polymorphisms in bipolar disorder. J. Affect Disord. 2001; 62(3): 151–164.

100. Washizuka S, Kametani M, Sasaki T, et al. Association of mitochondrial complex I subunit gene NDUFV2 at 18p11 with schizophrenia in the Japanese population. Am. J. Med. Genet. B Neuropsychiatr. Genet. 2006; 141(3): 301–304.

101. Onishi H, Kawanishi C, Iwasawa T, et al. Depressive disorder due to mitochondrial transfer RNALeu(UUR) mutation. Biol. Psychiatry 1997; 41(11): 1137–1139.

102. Mizino Y, Suzuki K, Ohta S. Postmortem changes in mitochondrial respiratory enzymes in brain and a preliminary observation in Parkinson's disease. J. Neurol. Sci. 1990; 96: 49–57.

103. Prince JA, Yassin M, Oreland L. A histochemical demonstration of altered cytochrome c oxidase activity in the rat brain by neuroleptics. Eur. Neuropsychopharmacol. 1998; 8: 1–6.

104. Wong-Riley M. Cytochrome c oxidase: An endogenous metabolic marker for neuronal activity. Trends Neurosci. 1989; 12: 94–101.

105. Prince JA, Harro J, Blennow K, et al. Putamen mitochondrial energy metabolism is highly correlated to emotional and intellectual impairment in schizophrenics. Neuropsychopharmacology 2000; 22: 284–292.

106. Whatley SA, Curi D, Das Gupta F. Superoxide, neuroleptics and the ubiquinone and cytochrome b5 reductases in brain and lymphocytes from normals and schizophrenic patients. Mol. Psychiatry 1998; 3: 227–237.

107. Burkhardt C, Kelly JP, Lim YH, et al. Neuroleptic medications inhibit complex I of the electron transport chain. Ann. Neurol. 1993; 33: 512–517.

108. Balijepalli S, Boyd MR, Ravindranath V. Inhibition of mitochondrial complex I by haloperidol: the role of thiol oxidation. Neuropsychopharmacology 1999; 38: 567–577.

109. Brenner-Lavie H, Klein E, Zuk R, et al. Dopamine modulates mitochondrial function in viable SH-SY5Y cells possibly via its interaction with complex I: relevance to dopamine pathology in schizophrenia. Biochim. Biophys. Acta 2008; 1777(2): 173–185.

110. Mehler-Wex C, Duvigneau JC, Hartl RT, et al. Increased mRNA levels of the mitochondrial complex I 75-kDa subunit: A potential peripheral marker of early onset schizophrenia? Eur. Child Adolesc. Psychiatry 2006; 15(8): 504–507.

111. Da Prada M, Cesura AM, Launany JM, et al. Platelets as a model for neurons? Experientia 1988; 44: 115–126.

112. Yao JK, van Kammen DP, Moss HB, et al. Decreased serotonergic responsivity in platelets of drug-free patients with schizophrenia. Psychiatry Res. 1996; 63(2–3): 123–132.

113. Strunecka A, Ripova D. What can the investigation of phosphoinositide signaling system in platelets of schizophrenic patients tell us? Prostaglandins Leukot. Essent. Fatty Acids 1999; 61(1): 1–5.

114. Wirz-Justce A. Platelet research in psychiatry. Experientia 1988; 44: 152–155.

115. Ilani T, Strous RD, Fuchs S. Dopaminergic regulation of immune cells via D3 dopamine receptor: a pathway mediated by activated T cells. FASEB J. 2004; 18(13): 1600–1602.

116. Hickey WF, Hsu BL, Kimura H. T-lymphocyte entry into the central nervous system. J. Neurosci. Res. 1991; 28(2): 254–260.

117. Owens T, Tran E, Hassan-Zahraee M, et al. Immune cell entry to the CNS – a focus for immunoregulation of EAE. Res. Immunol. 1998; 149(9): 781–789; discussion 844–846, 855–860.

118. Levite M. Nerve-driven immunity. The direct effects of neurotransmitters on T-cell function. Ann. N. Y. Acad. Sci. 2000; 917: 307–321.

119. Levite M. Neurotransmitters activate T-cells and elicit crucial functions via neurotransmitter receptors. Curr. Opin. Pharmacol. 2008; 8(4): 460–471.

120. Ben-Shachar D, Bonne O, Chisin R, et al. Cerebral glucose utilization and platelet mitochondrial complex

I activity in schizophrenia: a FDG-PET study. Prog. Neuropsychopharmacol. Biol. Psychiatry 2007; 31(4): 807–813.

121. Ben-Shachar D, Karry R. Sp1 expression is disrupted in schizophrenia; a possible mechanism for the abnormal expression of mitochondrial complex I genes, NDUFV1 and NDUFV2. PLoS ONE 2007; 2(9): e817.

122. Schapira AH, Cooper JM, Dexter D. Mitochondrial complex I deficiency in Parkinson's disease. J. Neurochem. 1990; 54: 823–827.

Chapter 34
Peripheral Biomarkers of Excitotoxicity in Neurological Diseases

Lucio Tremolizzo, Gessica Sala, and Carlo Ferrarese

Abstract Since the proposal that excessive glutamatergic stimulation, also known as excitotoxicity, could be responsible for neuronal suffering and death, several studies have repeatedly confirmed the key role of this mechanism in the pathogenesis of different neurological diseases. Therefore, it is conceivable that assessing the glutamatergic function directly in patients could be extremely useful for early diagnosis, prognostic evaluation, and optimization of therapeutics interventions. For example, a possibility is offered by measuring glutamate levels directly in plasma, in patients affected by stroke, amyotrophic lateral sclerosis, Alzheimer's disease, and AIDS dementia complex, among other diseases. However, the possibility of directly assessing functional glutamatergic parameters, such as the amino acid reuptake in ex vivo cells following different stimuli, would probably mirror closely the actual damage operative in single patients. In this chapter our findings regarding excitotoxicity biomarkers in peripheral ex vivo cells, such as platelets and fibroblasts, will be described focusing on different neurological diseases, together with a review of the available literature.

Keywords Excitotoxicity • peripheral models • glutamate • neurological diseases • biomarkers

Abbreviations Abeta: β-Amyloid; AD: Alzheimer's disease; AIDS: Acquired immune deficiency syndrome; ALS: Amyotrophic lateral sclerosis; AMPA: α-Amino-3-hydroxy-5-methylisoxazolepropionate; APP: Amyloid precursor protein; ATP: Adenosine triphosphate; CNS: Central nervous system; CSF: Cerebrospinal fluid; EAAC1: Excitatory amino-acid carrier 1; EAAT: Excitatory amino acid transporter; FCC: Carbonylcyanide-p-trifluoromethoxyphenylhydrazone; GDH: Glutamate dehydrogenase; GDS: Global deterioration scale; GEE: Glutathione-ethyl-ester; GLAST: Glutamate-aspartate transporter; GLT1: Glial glutamate transporter-1; GTRAP: Glutamate-transporter-associated protein; HAD: HIV-associated dementia; HIV: Human immunodeficiency virus; HNE: 4-Hydroxy-2-nonenal; HPLC: High performance liquid chromatography; LHON: Leber hereditary optic neuropathy; MAO-B: Monoamine oxidase type B: MELAS: Mitochondrial encephalomyopathy, lactic acidosis, and stroke-like episodes; MMSE: Mini-mental state examination; MRS: Magnetic resonance spectroscopy; mtDNA: Mitochondrial DNA; NMDA: N-methyl-D-aspartic acid; PBMC: Peripheral blood mononuclear cell; PD: Parkinson's disease; PSD: Post-synaptic density; RGC: Retinal ganglion cell; ROS: Reactive oxygen species; SNc: Substantia nigra, pars compacta; SOD1: Superoxide dismutase-1; TBARS: Thiobarbituric-acid-reactive species; TNF-α: Tumor necrosis factor-α; VGLUT: Vesicular glutamate transporter

Excitotoxicity: When Glutamate Is Too Much

In these first sections we will focus on the role of the derangement of the homeostasis of glutamate in the pathogenesis of different neurological disorders. Glutamate is the major excitatory neurotransmitter in the vertebrate CNS and its dysregulation plausibly plays a major contributory role in the pathogenesis of various

L. Tremolizzo, G. Sala, and C. Ferrarese
Department of Neurology, "S.Gerardo" Hospital, Monza and Department of Neuroscience and Biomedical Technologies, University of Milano-Bicocca, Italy

neurological diseases. Since the first reports in the early 1960s,[1,2] glutamate role as synaptic mediator has been elucidated and the different elements of the glutamatergic system have been described and studied. Moreover, the concept that glutamate, under specific circumstances, could lead neurons to death was formulated since those early years,[3] and the hypothesis that a dysfunction of the glutamatergic homeostasis might be responsible for both acute[4] and chronic neurological disorders[5,6] (for an extensive review see ref.[7]) was subsequently called "excitotoxicity."[8,9] However, before treating the role of excitotoxicity in neurological diseases, a brief overview about the glutamatergic system in physiology and its key regulators is required.

The Glutamatergic System: A Brief Overview

Glutamate is a non essential dicarboxylic amino acid ($COOH$-CH_2-CH_2-CH-NH_2-$COOH$) involved in metabolic pathways and excitatory neurotransmission in both CNS,[10] and peripheral tissues.[11] Glutamate biosynthesis proceeds mainly from transamination of the respective Krebs cycle component, α-ketoglutarate, by the enzyme glutamate dehydrogenase, and from the deamination of glutamine by the enzyme glutaminase. Although other pathways have been described, the latter is probably the more relevant, accounting for more than 50% of the synthesized neurotransmitter throughout the CNS. Glutamate, packed in secretory vesicles by apposite vesicular glutamate transporters (VGLUT-1, VGLUT-2, and VGLUT-3, see ref.[12]), is released from pre-synaptic cells in response to depolarization and crosses the synaptic cleft activating specific receptors present in the post-synaptic density (PSD), a specialized region of the post-synaptic neurons. In order to terminate the excitatory signal, glutamate is then removed from the synaptic cleft by means of specific transporter molecules, located both on neurons and astrocytes surrounding the synapses. Astrocytic glutamate is then converted into glutamine, and subsequently shuttled back to neurons where it is converted into the neurotransmitter by glutaminase, closing the so-called glutamate-glutamine cycle.[13,14] A short description of glutamate receptors and transporters follows in light of the crucial role played by these molecules in controlling glutamate homeostasis and therefore, possibly, in mediating neuronal death by excitotoxicity in both acute and chronic neurological diseases.

Glutamate Receptors

Glutamate receptors may be classified as ligand-gated structures coupled to either monovalent and/or divalent ion permeable pores (also know as ionotropic), or to G proteins (aka metabotropic), responsible for the intracellular transduction of the message. About 30 genes for different glutamate receptor subunits have been identified up to now, and accumulating evidences show that it is the differential arrangement of these subunits to determine the functional profile of the receptors. Considering the number of possible permutations, this determines the hypothetical availability of a considerable variety of different receptor molecules, a data perfectly fitting with the evidence of a tightly regulated regional (both inside, and outside the CNS) and ontogenetic specificity of the glutamate signaling machinery.

Ionotropic receptors may be further sub-classified in NMDA and AMPA/Kainate receptors in function of the specific agonistic molecules by which they are activated.

NMDA receptors are sensitive to both glutamate, and the co-agonist glycine, and they display an unusual profile for ligand-gated ion channels.[15] In fact, the ionic pore displays a high permeability to divalent cations and is blocked by a magnesium ion, a mechanism that is strongly voltage-dependent.[16] This propriety is crucial in the so-called "secondary" excitotoxicity, in which monovalent cation influx through AMPA/Kainate receptors determines the removal of the Mg^{2+} block from NMDA receptors, with consequent receptor activation and calcium entry, even in presence of glutamate concentrations within the physiologic range.[17] Other agents can modulate NMDA activity, such as polyamines,[18] zinc,[19] and, interestingly, oxidative stress products.[20] NMDA permeability to calcium ions is crucial to understand the role of these receptors in the pathogenesis of neurological disorders. In fact calcium homoeostasis is central in the cellular dysfunction that eventually kills neurons, starting several calcium-dependent enzymatic cascades leading to neuronal damage.[21]

On the other hand, AMPA/Kainate receptors are generally regarded as permeable to monovalent cations only, such as sodium and potassium. They are often defined as non-NMDA receptors due to the difficulty to obtain a clear functional distinction between them. Contrariwise to NMDA molecules, they display very fast activity profiles, but they may be as well associated to neuronal dysfunction through membrane depolarization and alteration of the energetic balance. Among the different subunits, GluR2 is particularly interesting because with its presence confers to the receptor the impermeability to calcium entry. In fact, AMPA receptors lacking GluR2 subunits have been shown to be highly permeable to calcium, potentially determining a perturbation of the intracellular homeostasis of this ion, activating pathways such as the arachidonic acid cascade,[22] or the nitric oxide synthesis,[23] with important implications for the survival of the cell. Interestingly, the permeability of the receptor molecules to divalent cations is determined by the post-transcriptional editing within the second transmembrane domain of the GluR2 AMPA subunit. Here, in position 586, there is normally a glutamine (Q) residue, coded by a CAG triplet, which may be transformed by the enzyme adenosine deaminase into a CIG triplet, coding for an arginine (R) residue.[24] Other mechanisms able to determine different functional characteristics of the AMPA receptors are a similar adenosine deaminase-mediated transition (R→G) within the exon 13 of the GluR2–4 subunits, and the alternative splicing of a 38 amino acid region within the fourth transmembrane domain, denoted as flip/flop editing.

Glutamate metabotropic receptors are membrane-associated G-protein-coupled proteins that determine the activation of second messenger systems (either hydrolysis of membrane phospholipids, or negative regulation of the adenylate cyclase). They appear to be involved in long-term changes in synaptic strength, modulating both post-synaptic ionotropic response, and pre-synaptic glutamate release, functioning as auto-receptors.

Although several of the mechanisms underlying glutamate receptor function, both in CNS,[25–27] and peripheral tissues,[11,28] have been elucidated, more work is needed to address issues such as the differential regulation of the expression of the receptor molecules, or a better definition of the intracellular transduction cascade systems.

Glutamate Transporters

Glutamate homeostasis strictly depends on the efficient removal of the neurotransmitter from the synaptic cleft, a function performed by a class of transporter molecules known as excitatory amino acid transporters (or EAATs; for a review see ref.[29,30]), which includes five members differently expressed throughout the CNS[31,32]: EAAT1 (also known as GLAST in mouse), EAAT2 (aka GLT1), EAAT3 (aka EAAC1), EAAT4, and EAAT5. Brain plasma membranes express also another transporter, the glutamate-cystine exchanger,[33,34] which may be implicated in glutamate release,[35] transporting cystine into cells by using the glutamate transmembrane gradient.[36] Interestingly, brain cells express intracellular glutamate transporters as well,[37] including mitochondrial transporters[38] and the class of vesicular glutamate transporters (VGLUTs; for a review see ref.[12]), responsible for regulating glutamate packing into secretory structures, in order to release it following depolarization of the pre-synaptic axon terminal.

The role of EAATs consists in tightly regulating the extra-cellular concentration of L-glutamate (they also uptake L- and D-aspartate), in order to maintain it between 1 and 3 μM, both preventing neuronal excitotoxic effects, recycling glutamate to avoid metabolic waste, and securing a high signal to noise ratio.[37] The process of glutamate re-uptake is associated with ion co-transport and is modulated by glutamate itself,[39,40] and other soluble factors such as cytokines (for example, TNF-α),[41] or oxidative stress products (for example, 4-hydroxynonenal).[42,43]

EAAT1 and EAAT2 are expressed mainly by astrocytes throughout the entire CNS, although EAAT2 is expressed by neurons as well,[44,45] surely in early developmental stages,[46] and this expression is retained at least at the level of the retina.[47] EAAT1 is the major transporter in the cerebellum[48] and the retina,[47] while EAAT2 takes over in the majority of the brain and spinal cord. On the other hand, EAAT3 is widely expressed by neurons, especially in hippocampus, cerebellum and basal ganglia,[49] although its role appears to be secondary, since EAAT3 knock-out mice display normal uptake activity,[50] while knocking-out EAAT2 uptake values decrease significantly.[51] EAAT4 and EAAT5 display a very restricted pattern of distribution in adult brain; in fact they are located mainly in the cerebellum[52] and the retina, respectively.

Interestingly, the different EAAT molecules display a specific pattern of subcellular distribution and may undergo to rapid changes in their expression at the plasma membrane level. Four proteins[37] have been reported to interact with EAATs, mediating attachment to cytoskeletal structures: Ajuba which binds EAAT2,[53] GTRAP3–18 which binds EAAT3[54]; GTRAP41, and GTRAP48 which bind EAAT4.[55]

The Role of Excitotoxicity in Neurological Diseases

The extracellular concentrations of the excitatory neurotransmitter glutamate are usually maintained at relatively low levels to ensure an appropriate signal-to-noise ratio, preventing an excessive activation of glutamate receptors that can result in cell death. As already mentioned before, this latter phenomenon is known as "excitotoxicity" and has been associated with a wide range of acute and chronic neurological disorders. At the subcellular and molecular levels, excitotoxicity is a very complex process. In fact, it has been shown that the cascade of events triggered by glutamate receptor activation that ultimately results in cell death includes changes in different cell compartments, including cytosol, mitochondria, endoplasmic reticulum and nucleus.

A large body of evidence implicates in different ways excitotoxicity in the pathogenesis of various neurological diseases. For example, in the setting of acute cerebral ischemia, epilepsy or trauma, the primary mechanism of excitotoxic cell death appears to be necrosis, in the so called context of "strong excitotoxicity." The energetic failure and the necrosis induce the release of glutamate from the storage compartments, leading to excessive extracellular concentrations, with ionotropic receptor activation and increased calcium influx, activating death pathways in the surrounding cells. On the other hand, according to some authors the role of excitotoxicity in neurodegenerative diseases is just a bit more than a simple speculation. In fact, in these diseases there is no evidence for an increase in glutamate concentrations, possibly with the exception of glutamate in plasma/CSF of ALS or HIV-associated dementia patients. Moreover, the increase in glutamate concentration may not be sufficient by itself to cause excitotoxicity.[56] Therefore, since the early 1990s, the concept of slow or "weak" excitotoxicity has been proposed.[5,6] Various possibilities may account for this phenomenon, including receptor abnormalities/modified assembly (see below for the GluR2 AMPA subunit issue in ALS), or the impairment in energy metabolism (as suggested below for MELAS and the defect of glutamate uptake).[57] Moreover, up to know, clinical trials with glutamate receptor antagonists that would logically seem to prevent the effects of excessive receptor activation, have been associated with unexpected side effects or little clinical benefit. Nevertheless, the available data still point out that excitotoxicity and the associated mechanisms, such as oxidative stress and mitochondrial failure, might play a pivotal role in the pathophysiology of various acute and chronic neurological disorders; this specific topic clearly goes beyond the scope of our chapter, and, for an extensive review, we may send the reader to the book edited by Ferrarese and Beal.[7] Conceivably, the possibility of assessing peripheral markers mirroring the dysfunction of the glutamatergic homeostasis might help in diagnosing the disease and monitoring its course, possibly allowing the definition of new avenues for therapeutic interventions.[58] In the next sections we will review some of the more accessible and promising peripheral markers of excitotoxicity, pointing out the available evidence regarding their use in the setting of the most prevalent neurological disorders.

Assessing Peripheral Glutamatergic Markers in Neurological Diseases

Despite the recent optimism about mapping brain functions derived from positron emission tomography, functional magnetic resonance imaging, and magnetic resonance spectroscopy, these techniques do not allow us to make any definite statement about the glutamatergic dysfunction that might be operating in each patient. The imaging signal obtained by the aforementioned techniques is actually a measure of neuroenergetics, in terms of blood flow and energy consumption. Nevertheless, recent results from in vivo magnetic resonance spectroscopy (MRS) have established quantitative relations between cerebral metabolic rates and glutamate neurotransmitter cycling, the latter measured as the isotope label flow from glutamate to glutamine.[59] A number of studies have been performed with carbon or proton MRS in neurological patients. Slow-rate of glutamine-glutamate cycling, decreased glutamine content and increase in glutamate content were shown in epileptic human hippocampus.[60] Furthermore, excess of glutamate and glutamine was reported in the medulla of ten amyotrophic lateral sclerosis patients compared to seven control subject tested with MRS.[61] On the contrary, MRS showed no increased striatal glutamate in 12 Parkinson's disease patients compared to 12 healthy controls.[62] One major limit of these neuroimaging techniques applied to excitatory neurotransmission is that their signal indicates solely the total glutamate level in the studied area, without discriminating between its intra- or extra-cellular localization at the synaptic level (Table 34.1).

On the other hand, other means exist to indirectly assess glutamatergic function. In vivo microdialysis is a quantitative measure of glutamate levels in the extra-cellular fluid in brain tissue. This technique has allowed confirming the pathophysiologic role of glutamate toxicity in cerebral ischemia,[63,64] subarachnoid haemorrhage[65,66] and traumatic brain injury.[67,68] Another apparently simple technique consists in measuring glutamate levels in biological fluids, as we will discuss in the next section. However, new interesting functional markers (ex vivo) of glutamatergic transmission may be assessed in peripheral tissues, where specific glutamate receptors and transporters, identical to those expressed in the CNS, are present. Here we strongly argue that, overall, the accessibility of these peripheral tissues should undoubtedly be regarded as their foremost advantage, significantly compensating for the fact that the sampled cells might not be the primary targets of the disease. Moreover, peripheral markers offer the unquestionable advantage of not being affected by post-mortem secondary changes as it happens within the CNS, and that may be misleading when formulating

Table 34.1 Major peripheral glutamate alterations in neurological diseases

Neurological disease	Plasma	Platelets	Leukocytes	Fibroblasts
Acute ischemic stroke	↑ Levels	↓ Uptake in the early weeks (persisting up to 3 months)	Not investigated	Not investigated
Migraine	↑ Levels (with aura > without aura > healthy controls)	↑ Levels in migraine with aura ↑ Release upon collagen stimulation (with aura > without aura > healthy controls) ↑ Uptake in migraine with aura ↓ Uptake in migraine without aura	Not investigated	Not investigated
Alzheimer's disease	Conflicting results	↓ Uptake (40% vs. healthy subjects)	↑ GDH activity	↓ Uptake (60% vs. healthy subjects. Further 50% reduction after HNE exposure, rescued by gluthathione/n-acetyl-cysteine)
Parkinson's disease	Possibly ↑ levels	↓ Uptake (48% vs. healthy subjects; correlation with disease severity)	= Levels of glutamine synthetase	Not investigated
Amyotrophic lateral sclerosis	Conflicting results Possibly ↑ levels	↓ Uptake (43% vs. healthy subjects) = Activity of glutamate dehydrogenase, glutaminase	↓ Levels of metabotropic glutamate receptor 2 mRNA = Activity of glutamate dehydrogenase, glutaminase	= Uptake
LHON	Not investigated	Not investigated	Not investigated	In cybrids, ↓ uptake
MELAS	Not investigated	In cybrids, ↓ uptake	Not investigated	Not investigated

hypotheses on the pathogenesis of the disease; possibly, these markers may be suitable secondary endpoints for clinical studies. Although beyond the scopes of this review, recent molecular biological analyses clearly show a novel function for glutamate as an extracellular autocrine/paracrine signal mediator in different peripheral tissues including, bone, testis, pancreas, and the adrenal, pituitary and pineal glands (reviewed in ref.[28]).

Here we will describe more in detail some of the most accessible tissues where, in our experience, the study of peripheral biochemical markers of the glutamatergic system is extremely promising and feasible: plasma, platelets, leukocytes and fibroblasts.

Glutamate Plasma Levels: A Tricky Issue

Despite its apparent simplicity, the task of assessing glutamate levels by HPLC in biological fluids raises some important technical issues concerning the reliability and reproducibility of the assay. As a matter of fact, considerable (up to 100-fold) variability in CSF glutamate levels has been observed among different studies, and sample processing procedures might be responsible for these discrepancies. Because it is probable that glutamate levels in biological fluids do not depend exclusively on tissue release but also on ongoing enzymatic activity, it is possible that spurious modifications may occur in vitro. Several methods have been proposed to inactivate enzymatic activity in biological fluids, but some of these, such as sample acidification without neutralization, may actually be the source of further bias. Our group developed a method of sample processing for glutamate detection that has been shown to yield stable values.[69] Other inactivation techniques, including sample boiling or filtration, may also be used, but their reliability has not been extensively tested. In light of this evidence, we believe that great care with regard to methodological procedures should be taken in the critical evaluation of the literature on this topic.[70] However, despite this very relevant methodological issue, glutamate level analysis in CSF and plasma has certainly contributed to our understanding of different neurological disorders, such as stroke, AIDS-dementia complex, or ALS (see below).

The Glutamatergic System in Human Platelets

Human platelets have been repeatedly considered as very promising tools for modeling CNS biochemical dysfunctions in neurological patients.[71,72] As an example, in platelets obtained from Alzheimer's disease patients, several groups reported dysfunction of MAO-B activity,[73] increased phenosulfotransferase activity,[74] decreased serotonin uptake,[75] increased membrane fluidity,[76] abnormal pattern of the ratio of the APP isoforms,[77,78] decreased benzodiazepine binding,[79] among numerous others parameters mirroring dysfunctions operative into the CNS of these patients. Moreover, platelets also store, release, and uptake several neurotransmitters, among which glutamate,[80] at the same time expressing fully functional cognate receptors.[81–86] Notably, our and other laboratory data indicate that platelets might represent an extremely suitable model to study the glutamatergic system dysfunction in neuropsychiatric disorders.[82,87–89] In fact, platelets display all the components of a functional glutamatergic system, although its role is not fully understood. For example, NMDA and AMPA receptors have been repeatedly shown both in platelets and megakaryocytes,[11,86,90,91] and a role for glutamate in platelet activation and/ or aggregation has already been proposed.[85,86,92,93] Moreover, we demonstrated in human platelets[88] the presence of the three major glutamate transporters, EAAT1, EAAT2, and EAAT3, while a functional high-affinity sodium-dependent glutamate reuptake system, similar to that expressed in synaptosomes, has been successfully described more than 20 years ago.[82] Apparently this reuptake system represents a major source of the amino acid removal from blood, and also displays very similar regulatory mechanisms and toxin susceptibility with respect to the CNS homologue.[94,95] Besides, we showed that human platelets express the two major vesicular glutamate transporters (VGLUT-1, VGLUT-2) as well, packing glutamate (possibly in dense bodies) and releasing it following aggregation.[96] Although the possible significance of these findings still need to be definitively addressed regarding to platelet and haemostasis physiology, it is conceivable that the mechanisms involved in glutamate homeostasis in platelets may represent suitable windows for studying analogous CNS alterations.

Leukocytes and Glutamate

White blood cells, with respect to platelets, offer the clear advantage of a nuclear structure. Moreover, leukocytes may, especially under conditions of inflammation, pass through the blood-brain barrier, possibly playing a role at least in some neurological diseases (such as multiple sclerosis; see, as an example regarding the glutamatergic system, ref.[97]). Also leukocytes, and especially lymphocytes, express both ionotropic[98,99] and metabotropic glutamate receptors,[100] for which several evidences show that they may be playing a role in the regulation of the immune function of these cells,[100–102] with possible important implication in the regulation of the neuroimmune homeostasis. Moreover, we previously showed that lympho-monocytes express the three major transporters as well, namely EAAT1, EAAT2, and EAAT3,[88] although the functional significance of this data still needs to be addressed.

Glutamate in Ex Vivo Fibroblasts

Another peripheral model that the scientific panorama is currently investigating is represented by fibroblasts, obtained through a very small skin biopsy with almost no nuisance for the patients and propagated by simple cell-culture procedures.[103] Fibroblasts, with respect to platelets, offer the advantage of surviving in culture and of displaying a functioning nucleus, allowing a better approximation of what is the complex neuronal situation. Hence, collectively, the available data strongly support a possible use of fibroblasts as useful model for exploring the role of excitotoxicity and other potential pathogenic mechanisms in neurological diseases, allowing compounds, or combination therapies to be tested. As a matter of fact, fibroblasts possess a sodium-dependent high affinity uptake of L-glutamate similar to that observed in brain tissue,[104,105] and express at least moderate levels of the three major glutamate transporters, EAAT1, EAAT2, and EAAT3.[105]

In the following sections we will briefly review the studies on the glutamatergic system in these peripheral markers, focusing on stroke, migraine and the three major neurodegenerative diseases (Alzheimer's disease, amyotrophic lateral sclerosis and Parkinson's disease), concluding with few data about the recently shown dysfunction of glutamate uptake in cybrids (derived from fibroblasts) of mitochondrial DNA disorders (LHON and MELAS).

Stroke and Migraine

There is convincing evidence that ischemia leads to an increase in the extracellular concentrations of excitatory amino acids, among which especially glutamate. The accumulation of this neurotransmitter, which can reach up to about 100 times normal at the ischemic core, is believed to be an important factor for the premature death of neurons that would otherwise survive the ischemic conditions and recover when flow is restored. Increased CSF and plasma glutamate levels were reported in ischemic stroke patients 24 h following symptom onset, and they were higher in patients with large cortical lesions or with a more severe clinical involvement.[106,107] Furthermore, a difference between patients with stable and progressive stroke was evident.[108] Also in animal models plasma concentrations of glutamate begin to rise 4–6 h following the occlusion of the middle cerebral artery, reaching a peak at about 8–24 h.[109] These data, besides supporting the key role of excitotoxicity in stroke, have brought to suggest that the detection of glutamate in CSF and/or plasma might be a useful prognostic tool in predicting the clinical outcome of patients. In fact, it is conceivable that having the possibility of dosing such putative biological marker may be of great practical value in selecting those patients who are more likely to benefit from neuroprotective therapies. For example Mallolas and colleagues[110] observed an increased prevalence of a polymorphism in the promoter of the glutamate transporter EAAT2 gene that abolishes a putative regulatory site for AP-2 creating a new consensus binding site for the repressor transcription factor GCF2. The authors found that the mutant genotype is associated with increased concentrations of glutamate in plasma and with a higher frequency of early neurological worsening in stroke patients.[110]

In any case, the issue of the origin of glutamate in both CSF and plasma still remains to be addressed. Raised glutamate in the CSF is most likely consequent to cellular necrosis in the lesion core and to energy failure, transport reversal, and amino acid release at the level of the surrounding ischemic penumbra. More arduous is deciding which sources are responsible for the increased glutamate levels in plasma. Obviously addressing the issue of the significance of glutamate in plasma is extremely important when considering that obtaining plasma samples raises much fewer concerns about patient safety and compliance with respect to CSF ones. Moreover, plasma and CSF glutamate levels have been shown to be correlated, raising the possibility that peripheral glutamate might derive, at least in part, from the brain lesion, leaking through a damaged blood–brain barrier. However, recent experimental evidence obtained by our group on ischemic stroke patients shows that glutamate levels in plasma may remain significantly elevated up to 2 weeks following symptom onset, when the blood–brain barrier function is likely restored.[111] Hence, it is probable that factors other than simple diffusion from the CNS are involved in the reported sustained elevation of glutamate plasma levels. Platelets might actually be one of the alternative sources of glutamate release.[88,112] As a matter of fact, they have been shown to be contributing to the clearance of the amino acid from the circulation, and, since we showed a reduced platelet glutamate uptake in stroke patients, it is possible that plasma glutamate might be increased as a result of a putative reduced clearance from the bloodstream.[111] Moreover, our results show that platelets express vesicular glutamate transporters and secrete glutamate upon stimulation. Hence, considering this data and the key role played by platelets in the homeostasis of the "neurovascular compartment", it is not surprising that also other authors have previously suggested that platelets might contribute to the pathogenesis of stroke by releasing soluble factors (among which we may list glutamate), which may both alter the permeability of the blood/brain barrier and possibly induce a direct damage to neurons.[113–115]

Not only glutamate is known to determine neuronal damage following cerebral ischaemia[116] and also to play a major role in the pathophysiology of epilepsy[117]: considerable evidence indicates that this neurotransmitter may play a role in the pathogenesis of migraine as well. In fact, glutamate release and NMDA receptor activation may be involved in the initiation, propagation and duration of the spreading depression, a phenomenon shown to be implicated in the pathophysiology of migraine attacks[118]; to further sustain this hypothesis it has been shown that the cerebral concentration of

magnesium decreases during a migraine attack,[119] causing enhancement of the sensitivity of NMDA receptors. Interestingly, excessive oral intake of glutamate may induce migraine-like episodes in susceptible individuals,[120] allowing to hypothesize a possible contribution of "peripheral glutamate" to this phenomenon. Indeed, increased concentrations of glutamate have been reported in plasma in migraine patients with respect to controls,[121,122] and in migraine patients during attacks if compared with the resting state[122]; platelet levels of glutamate have been shown to be raised in patients with migraine with aura.[123] Higher concentrations of glutamate have been found also in the CSF of migraine patients with respect to healthy controls.[124] Platelet activation and release of granule constituents, such as serotonin, have been already postulated to play an important role in the pathophysiology of migraine.[125] Moreover, while glutamate concentrations have been consistently reported as increased in platelets obtained from patients affected by migraine with aura, platelets obtained from patients affected by migraine without aura displayed a glutamate platelet content similar to that of controls.[112,123] Recently we reported an increase of plasma glutamate in subjects affected by migraine with aura, during the resting state, with respect to healthy controls and subjects affected by migraine without aura (which displayed higher values with respect to controls), and a similar increase of glutamate release from platelets.[126] Platelet glutamate uptake was increased in migraine with aura patients, while reduced in migraine without aura patients, with respect to healthy matched controls. Our data suggest that an up-regulation of the glutamatergic metabolism might be operative in migraine with aura, possibly due to a putative genetic trait. Another possibility is that the increased platelet glutamate re-uptake may be a compensatory mechanism linked to the increased plasma concentration of this amino acid; consistently with this assumption, we reported that glutamate can stimulate its own transport in human platelets.[94] Similarly, Rainesalo and colleagues[127] reported that platelet glutamate uptake in patients affected by temporal lobe epilepsy and hippocampal sclerosis is increased, suggesting this up-regulation might be a compensatory phenomenon for the increased glutamate release caused by seizure activity, in order to maintain the homeostasis. Moreover, the increased concentrations of glutamate in plasma in migraine with aura patients may derive, at least in part, from a peripheral source, such as platelets (considering the reported increase of glutamate release) and, interestingly, lamotrigine, which is known to inhibit glutamate release, also reduces the frequency of migraine with aura attacks.[128] Hence, globally our data further support the idea that migraine with aura and migraine without aura have different pathogenic mechanisms and that glutamate stored and released from platelets might play a role in the genesis of the symptoms, at least in the case of migraine with aura.

Alzheimer's Disease

Considering the progressive aging of the population, dementias, and in particular the most frequent among them, Alzheimer's disease (AD), are going to increasingly represent a real cost for both society and families, affecting thousands of patients. As a result, it is not surprising that concepts such as early diagnosis and possibility of clinical-instrumental monitoring of patients represent extremely important goals toward which researchers are currently oriented. Evidence that brain changes may occur more than four decades before clinical onset has been reported in AD patients,[129,130] promoting the idea that suitable biomarkers might be able to help in detecting AD when the patients present the first clinical symptoms, identifying from the general population subjects at high risk of developing AD, monitoring drug treatment and disease progression. Such tools would have a major impact on AD control at different levels, possibly resulting in important cost savings.

In particular, glutamatergic parameters have been assessed in both blood cells and fibroblasts obtained from AD patients, reflecting the central role of this route in the pathology of AD. For example, we reported in platelets obtained from AD patients[87] a decrease in the maximal velocity of the high-affinity sodium-dependent glutamate uptake (without changes in affinity), mirroring the decreased glutamate uptake correlated to neurodegeneration that has been previously shown in AD brains.[131] Interestingly, the dysfunction of platelet glutamate uptake was not present in vascular dementia patients, suggesting the hypothesis that excitotoxicity plays an important and specific role in the neurodegenerative process operative in AD dementia, conceivably being a candidate target for therapeutics (as in the case of Memantine). Moreover, glutamate uptake was cor-

related neither with the duration of the disease nor the severity (both MMSE and GDS), indicating that the impairment of glutamate uptake and the resulting excitotoxicity may be considered as early events in AD.[87] Different mechanisms may contribute to this apparent systemic alteration of glutamate uptake in AD, being this homeostatic process strictly dependent on the bioenergetic and redox status. Indeed, specific transporter alterations linked to abnormal APP expression have been described in AD brains,[132] although other data seems to suggest that the soluble form of APP displays a neuroprotective effect, facilitating glutamate uptake.[133] Besides, a role for inflammatory mechanisms may also be proposed to explain the observed AD-associated impairment of platelet glutamate uptake. For example, several cytokines have been reported to affect glutamate uptake in neuronal and peripheral cell models,[41] and the plasma levels of some of these inflammatory mediators have been repeatedly found increased in AD patients.[134–136] Consistently with the reduction of the uptake, EAAT1 expression (both mRNA and protein) is reduced in platelets from sporadic AD patients when compared to age-matched controls,[88] while EAAT2 and EAAT3 are unaffected, suggesting that the disease might specifically target the former transporter, theoretically endowed of a primary role in the maintenance of the correct amount of extracellular glutamate, as previously suggested for the CNS of AD patients.[137] In the same work, EAAT1 expression was also reduced in older healthy controls when compared to young healthy controls, allowing to hypothesize that EAAT1 downregulation is a phenomenon normally occurring with aging, possibly decreasing the "glutamatergic" threshold for developing AD.[88]

Few data are available on a putative dysregulation of the glutamatergic system in peripheral blood leukocytes obtained from AD patients. Iwatsuji and colleagues[138] reported an increased activity of total and heat-stable glutamate dehydrogenase (GDH) in lymphocytes from control subjects with aging, while total GDH activity was lower in patients with AD (and other neurodegenerative diseases) with respect to age-matched controls. Furthermore, Peeters and colleagues[139] described a dysregulation of the glutamine/glutamate pathway in AD and Down syndrome lymphocytes, possibly mirroring a CNS process operative in the pathogenesis of these amyloid disorders.[140] Finally, considering the growing evidence on the link between cerebrovascular disease and AD, it is not surprising that some authors studied the interactions of homocysteine with glutamate receptors in lymphocytes obtained from AD patients,[141] discussing the route through which homocysteine and homocysteic acid determine increased calcium and free radicals intracellular levels.

In fibroblasts from AD patients, similarly to platelets, EAAT1 expression is reduced when compared to age-matched control fibroblasts, which display lower levels of this protein with respect to young healthy controls,[105] suggesting again that EAAT1 downregulation might be an early process in aging, possibly increasing the vulnerability to develop the disease. Moreover, EAAT1 downregulation appears to be correlated to the disease severity (MMSE), indicating that this biomarker might be a state marker, and not merely a trait one. In order to further investigate the putative mechanisms involved in this reduction of glutamate uptake (~50%), we exposed cultured fibroblasts from AD patients and age-matched healthy controls to different concentrations of 4-hydroxy-2-nonenal (HNE),[142] one of the products of Abeta-induced lipid peroxidation, which in particular has been found increased in brain,[143] and cerebrospinal fluid of AD patients.[144] On the other hand, there is evidence that HNE itself can trigger lipid peroxidation and increase the levels of free radical production in synaptosomes,[42] and in primary astrocytic cultures.[145] HNE can selectively impair the function of different proteins, covalently binding to their cysteine, lysine, and histidine residues. Interestingly, it has been reported that HNE is specifically conjugated to the glutamate transporter subtype EAAT2 to a greater extent in AD brains and that Abeta can enhance HNE binding.[146] HNE concentrations up to 50 μM for 2 h were not able to induce changes in glutamate uptake in fibroblasts obtained from healthy age-matched controls; on the other hand, AD fibroblasts were more prone to the decrease of the uptake and to produce oxygen free radicals following HNE exposure.[142] Pre-incubation with glutathione-ethylester (GEE) was able, not only to prevent HNE noxious effects on glutamate uptake, but also to upregulate baseline uptake levels in AD fibroblasts bringing them close to the control ones, suggesting that increased oxidative stress and impairment of antioxidant defenses might play an important role in the downregulation of glutamate uptake, at least in this cell model.[142] In order to further investigate the possible links between increased Abeta production and glutamatergic dysfunction, we performed preliminary experiments indicating that exposure to Abeta, probably directly interacting with the transporters, significantly impairs glutamate uptake (Zoia and colleagues, submitted work).

Jimenez-Jimenez and colleagues[147] measured glutamate concentrations in plasma samples deriving from 37 AD patients without finding significant differences, although glutamate was increased in the CSF. On the contrary, Miulli and colleagues[148] found a significant difference assessing the same parameter in AD plasma samples with respect to controls. Basun and colleagues[149] reported a decrease of plasma glutamate in AD patients when compared to controls. This divergence of results might be possibly explained by both technical issues (such as correct inactivation of the sample, see ref.[69]), and by the relatively small number of subjects recruited. A pilot Russian study also reported the finding of autoantibody against glutamate in AD patients, which level increased with the severity of the dementia.[150]

Besides, a suggested model of "excitotoxic dementia" is possibly represented by HIV-associated dementia (HAD), although the clinical picture is clearly different from AD. In HIV patients, a sixfold increase of plasma glutamate was observed regardless of the presence of cognitive dysfunction. On the other hand, glutamate levels in the CSF were increased fivefold only in HAD patients with respect to non-demented HIV patients.[151] CSF glutamate levels were also related to the degree of dementia and brain atrophy, suggesting a role for glutamate intrathecal production in HAD, although no value for the follow-up can be clearly assigned to the measurement of glutamate in plasma for these patients.

Parkinson's Disease

The cause of idiopathic Parkinson's disease (PD) remains unknown, and the mechanisms responsible for nigral dopaminergic cell loss are still obscure. Many studies have demonstrated the capacity of oxidative stress and oxidizing toxins to induce nigral cell degeneration, supporting the occurrence of oxidative stress in PD pathogenesis (for a review see ref.[152]). Increased levels of basal oxidative stress were found in the substantia nigra pars compacta (SNc) from PD patients. Furthermore, symptomatic treatment with L-DOPA may add to the oxidative load and play a role in disease progression.[153,154] However, it is well known that other factors –including excitotoxic mechanisms, mitochondrial dysfunction, alterations in protein ubiquitination and degradation, inflammation – are involved in the cell death that occurs in PD. In particular, the involvement of the excitatory amino acid glutamate in the pathogenesis of PD is supported by the fact that SN cells receive a rich glutamatergic innervation from the subtalamic nucleus.[155] In PD, nigrostriatal dopamine depletion leads to an overactivity of glutamatergic projection neurons, which further contributes to excitotoxic nigral cell death.[156] Furthermore, the beneficial effect of anti-glutamatergic substances in animal experiments suggests that excess supply of glutamate might contribute to PD pathogenesis.

Interestingly, several lines of evidence suggest that cellular damages typical of PD are not limited to the brain but occur also in extracerebral tissue of these patients. A significant reduction of glutamate uptake, correlated with the disease severity, was observed in platelets from idiopathic PD patients, respect to controls and secondary parkinsonian syndromes. Furthermore, an increase of glutamate levels was found in platelets of PD patients, without correlation with the uptake decrease.[157] The assessment of possible polymorphisms of alpha-synuclein and ApoE genes in this population did not show neither differences between PD and controls, nor correlations with platelet glutamate uptake.[158] Since a reduced activity of the glutamate metabolizing enzyme glutamine synthetase leads to decreased uptake of glutamate and thus abundant glutamate, the activity of this enzyme was assessed in PBMCs from PD patients. Comparable activity of glutamine synthetase was demonstrated in PBMCs from PD patients and age-matched controls, supporting the absence of a systemic dysregulation of this enzyme in PD patients.[159] Moreover, the mitochondrial function of skin fibroblast cultures from PD patients was evaluated and a deficiency of both complex I and IV was observed, supporting the presence of a generalized mitochondrial defect in PD patients. The treatment of fibroblasts with the potent antioxidant CoQ(10) restored the activity of impaired respiratory chain complexes in the fibroblast cultures of 9 out of 18 PD patients.[160] We currently plan to investigate the possible interplay existing between mitochondrial and glutamate uptake dysfunction in PD fibroblasts, as we already did in some mtDNA diseases, such as LHON, where a defect of complex I is present, and MELAS, where both complex I and IV results to be compromised (see below).

Although with the opportune methodological criticism, glutamate plasma levels have been reported to be elevated in 20 PD patients with respect to an equal number of controls.[161] On the other hand, no difference regarding the CSF/plasma ratio was found in 21 patients affected by dementia with Lewy body with respect to 26 matched controls.[162] Finally Hartai and colleagues[163] reported evidence in plasma and blood cells obtained from PD patients for a dysfunctional metabolism of kynurenine, which derivative, kynurenic acid, is an excitatory amino acid repeatedly claimed to be playing an important role in different neurodegenerative disorders.[164]

Amyotrophic Lateral Sclerosis

The above cited issues about the use of biological parameters reflecting glutamatergic dysfunction would be especially relevant in ALS, where the neurodegenerative process is extremely rapid and conceivably does not allow enough space for an effective therapeutic intervention. As a matter of fact, ALS is a devastating neurological syndrome, due to upper and lower motor neuron loss, which appears both in a sporadic and a familial form, the last one in about 10% of patients. Patients with the sporadic form are clinically undistinguishable from patients with the familial one. After more than 100 years from the first clinical description, the pathogenesis of the disease is still unknown and no effective therapies exist, apart from a symptomatic and supportive treatment. Anyway, at the beginning of the 1990s it has been demonstrated that about 20% of familial ALS patients shows point mutations in the gene encoding for antioxidant enzyme superoxide Cu, Zn superoxide dismutase (SOD1).[165] This discovery sustained the hypothesis that mechanisms of oxidative stress are crucial in motor neuron degeneration, acting both in familial and in sporadic forms. SOD1 has a typical antioxidant action, since it is assigned to remove superoxide anion generated during normal cell metabolism. However, after mutation, this enzyme acquires pro-oxidant properties, such as the ability to use his normal reaction product (hydrogen peroxide) to generate hydroxyl radical,[166,167] or the ability to catalyze the formation of peroxynitrite from superoxide anion and nitric oxide.[168] After these findings, mutant SOD1 animal and cellular models have been widely used to investigate the pathogenic mechanisms underlying not only familial, but also the sporadic forms of ALS.

Besides oxidative stress, glutamate-mediated excitotoxicity is among the more likely causes of motor neuron selective degeneration observed in ALS. Experimental evidences of glutamate transport and metabolism alterations in ALS patients and in animal models of this disease have been strengthened up in the last years.[169,170] 60 to 70% of sporadic ALS patients show loss of the principal transporter of this amino acid (EAAT2) probably due to an abnormal splicing of his mRNA.[171] Furthermore, it must be underlined that motor neurons seem to be highly sensitive to glutamate-induced toxicity;[172–174] finally, some anti-glutamate agents (riluzole and gabapentin) are the only drugs with a certain success, although very limited, in the treatment of the disease.[175] However, several issues should be addressed before confirming the role of glutamate homeostasis dysfunction in ALS, and the mechanisms of selective motor neuronal damage must be explained to be plausible. Several characteristics specific of motor neurons have been described that may account for this selectivity.[176] Motor neurons are big cells, with high oxidative demands in order to sustain their metabolism, with consequent possible increased production of free radicals, able to impair glutamate uptake and to modulate glutamate receptor activity. Moreover, the relative lack of GluR2 AMPA subunits has been reported in motor neurons.[177] As previously described, the latter phenomenon determines increased calcium permeability through AMPA receptors, leading to disruption of calcium intracellular homeostasis, with the already described important functional consequences for the neurons. Additionally, motor neurons relatively lack of calcium-binding proteins, such as parvalbumin and calbindin D28k,[172] determining a lower capacity of buffering calcium increases in the cytoplasm of these neurons. Lots of evidences show that excitotoxic mechanisms and oxidative stress could converge and concur to generate motor neuron injury: some glutamate transporter subtypes are inactivated by oxidative stress products such as peroxynitrite[178] and hydroxynonenal.[179] It has also been demonstrated that SOD1 mutant forms, typical of familial ALS, are able to oxidatively inactivate EAAT2 when expressed in Xenopus oocytes.[180] Moreover, glutamate uptake was significantly decreased in human neuroblastoma SH-SY5Y cells expressing mutant G93A SOD1, possibly due to an oxidative inactivation of glutamate transporters

induced by the G93A mutation.[181] In the same study it was shown that mutant cells displayed a marked sensitivity to oxidants, resulting in a more pronounced reduction of glutamate uptake not reversed by a short-term antioxidant treatment.[181]

In line with these assumptions, in platelets obtained from ALS patients we reported[182] a decrease in the maximal velocity of high-affinity sodium-dependent glutamate uptake (without changes in affinity), of the same order of magnitude previously observed into the CNS by other groups,[169] with respect to both healthy controls, and subjects affected by other neurological, non-neurodegenerative disorders. This data apparently indicate a more widespread pathology in ALS patients than previously believed. As a matter of fact, oxidative stress, endogenously generated or derived from xenobiotic exposure, may directly impair energy-dependent glutamate uptake and also can lead to the generation of lipid peroxidation products and free radicals, which in turn may damage glutamate transporters. A role for inflammatory mechanisms may also be proposed to explain the observed ALS-associated impairment of glutamate uptake, both in periphery, and CNS. For example, several cytokines have been reported to affect glutamate uptake[41] in neuronal and peripheral cell models, and plasma levels of some of this inflammatory mediators have been found increased in ALS patients.[183] An alternative hypothesis may be indicated in the existence of a genetic trait predisposing to the development of motor neuron disease, possibly reducing the neuronal threshold to ROS- or other agents-mediated injury. In line with the latter hypothesis, unpublished data from our group indicate that platelet uptake values are stable over time in individual ALS patients, thus suggesting that decreased transport is a disease trait rather than disease state marker. Although our data indicate a systemic involvement in ALS, other groups failed to report in platelets[184] and leukocytes[185] obtained from ALS patients a reduction of the activity of the enzymes cytochrome oxidase, glutamate dehydrogenase, glutaminase, and PKC histone H1 phosphotransferase, as described in motor neuron disease affected CNS areas. It is reasonable to conclude that only some ALS-associated markers may be usefully studied in platelets, possibly due to the implicit phenotypic differences among different cell types.

Also lymphocytes have been proposed as model for studying glutamatergic activation, oxidative metabolism and calcium homeostasis. While no difference was found in lymphocytes from ALS patients in basal oxygen consumption rate (QO_2), cytochrome c oxidase activity, catalase activity, and lactate production, a decrease in QO_2, induced by an uncoupler of oxidative phosphorylation, FCCP, was observed.[186] Moreover, the basal (resting) level of free cytosolic calcium was higher in lymphocytes from sporadic ALS patients. Further increase in free intracellular calcium challenged by a K^+ channel blocker or by FCCP was similar in ALS and control lymphocytes.[186] The levels of metabotropic glutamate receptor 2 mRNA were found to be markedly reduced in T lymphocytes of ALS patients, whereas the expression of other subtypes (1b, 3, 8) was similar to control levels, providing a reliable peripheral marker of the glutamatergic dysfunction that characterizes ALS.[187]

An early study indicated that fibroblasts from 5 patients with non-SOD1 familial ALS did not display signs of abnormal oxidative stress, based on the assay of ROS metabolism and of thiobarbituric-acid-reactive species (TBARS), a marker of lipid peroxidation.[188] On the other hand, other authors demonstrated an increased sensitivity to oxidative stress of fibroblasts from patients with sporadic and SOD1-associated familial ALS following treatment with free radical-generating agents.[189] Because fibroblasts possess a sodium-dependent high affinity uptake of L-glutamate mediated by a transport system similar to that observed in brain tissue[104] and express at least moderate levels of glutamate transporters,[190] we are currently using fibroblasts from sporadic ALS patients in order to verify whether this peripheral cells can be used to model the alterations of the glutamate uptake system already documented in the CNS. Kinetic analyses of sodium-dependent glutamate uptake, determined as [³H]-glutamate influx, showed no difference in fibroblasts from ALS patients with respect to controls (Gessica Sala et al. unpublished observations). Furthermore, no significant difference was observed in ROS production, protein carbonylation and nitration, susceptibility to hydrogen peroxide, p38 mitogen-activated protein kinase activation, and ATP intracellular content between patient and control cells, and no correlation with the disease severity was found.[191] Collectively, our data show no major alterations of the oxidative and bioenergetic status in cultured fibroblasts from sporadic ALS patients, suggesting that these cells do not represent a useful model to study the glutamatergic and oxidative dysfunctions associated to ALS.

Early reports in ALS showed increased glutamate levels in the CSF and/or in plasma of affected patients.[192–195] However, further studies by different groups failed to confirm this result,[196–198] possibly as a consequence of both, the already described methodological issues, and of the small numbers of recruited patients. Interestingly, a significant increase in CSF glutamate levels has been more recently reported on a larger number of ALS patients, being more pronounced in patients affected by the spinal form of the disease.[199] Although a possible role in the pathophysiology for glutamate into the CSF might be quite easily postulated, the interpretation of the finding of increased glutamate plasma levels, as reported in ALS patients by some groups, is more arduous, since extensive muscle atrophy along the course of the disease may be responsible for a putative elevation of serum amino acid content.[195] Very recently, Andreadou and colleagues reported that at baseline, increased plasma glutamate levels correlated with spinal onset and male gender in ALS patients, whereas glycine levels did not differ with respect to healthy controls.[200] No significant change was observed for both amino acids following riluzole treatment for 6 months. The same group reported in 65 untreated ALS patients increased plasma glutamate levels correlating with the duration of the disease and the spinal subtype, reinforcing the suggestion of a possible role for glutamate in the pathophysiology of ALS, and further suggesting that its "plasma window" might be more reliable than previously thought.[201]

Mitochondrial Diseases: Glutamate Uptake in LHON and MELAS Cybrids

Leber hereditary optic neuropathy (LHON) is a maternally inherited form of retinal ganglion cell (RGC) degeneration leading to optic atrophy which is caused by point mutations in the mitochondrial genome (mtDNA). Three pathogenic mutations (positions 11778/ND4, 3460/ND1 and 14484/ND6) account for the majority of LHON cases and they affect genes that encode for different subunits of mitochondrial complex I. Despite LHON-associated alterations of complex I have been extensively investigated,[202] it remains unknown why RGCs appear to specifically degenerate. Excitotoxic injury to RGCs and the optic nerve has been previously hypothesized, especially given the high susceptibility of this neural cell type to glutamate toxicity compared to other retinal cell types.[203] Furthermore, the pharmacological inhibition of glutamate transport causes selective degeneration of RGCs.[204] It is important to remember that EAAT1 is the major retinal glutamate transporter, expressed by glial Muller cells, and is known to be inhibited by ROS.[205] Increased levels of extracellular glutamate, coupled to complex I dysfunction, might promote selective injury to RGCs. A reduced activity of the excitatory amino acid transporter-1 (EAAT1), correlated with mitochondrial ROS levels, was demonstrated in osteosarcoma-derived cytoplasmic hybrids (cybrids) constructed using enucleated fibroblasts from LHON patients with respect to cybrids obtained from healthy controls.[206] These findings have been recently confirmed in primary rat retinal cultures treated with different concentrations of rotenone to induce gradual complex I inhibition,[207] suggesting that complex I-derived free radicals and disruption of glutamate transport might represent key elements for explaining the selective RGC death in LHON. Moreover, in a recent paper, glutamate transport impairment occurring in LHON cybrids has been demonstrated to be partially restored by exposure to the antioxidants Trolox and decylubiquinone. In the same study it has been reported that rotenone, a classic complex I inhibitor, does not worsen glutamate uptake defect present in LHON cybrids under basal conditions, while significantly reduces glutamate transport in control cybrids. An increased protein carbonylation under basal conditions, not further affected by rotenone, and partially counteracted by antioxidants, has been also observed in LHON cybrids, strengthening the hypothesis that the complex I defect associated with LHON causes free radical overproduction responsible for glutamate transport inhibition.[208] Furthermore, in the same experimental model it has been found that all three different LHON mutations impair ATP synthesis and the respiratory control ratio driven by complex I substrates. In contrast, succinate-driven ATP synthesis, respiration rates, and respiratory control ratios were not affected. However, the defective ATP synthesis with complex I substrates did not result in reduced ATP cellular content, indicating a compensatory mechanism.[209]

Although LHON is characterized by quite a selective expression of the pathology, there are several other mitochondrial diseases characterized by dramatic clinical phenotypes, involving different levels of the nervous system. These are the mitochondrial encephalomyopathies, which are a diverse range of disorders caused by a number of different mutations of either the mitochondrial or nuclear genomes.[210,211] One of the most important, the MELAS phenotype may be caused by several different mitochondrial DNA (mtDNA) mutations of which the A3243G is the most common.[212] The clinical effects are diverse and encompass not only the full MELAS phenotype, but also monosymptomatic features such as deafness or diabetes mellitus. Moreover, the stroke-like episodes represent one of the most important clinical features of MELAS. Their pathogenesis is likely to be multifactorial and it is still debated: the ischemic vascular hypothesis suggests they are caused by "mitochondrial angiopathy," and the generalized cytopathic theory proposes that neuronal hyperexcitability may initiate, then maintain, and develop the cascade of stroke-like events caused by "mitochondrial cytopathy." Once neuronal hyperexcitability is developed in a localized brain region, this could depolarize the adjacent neurons, spreading at the surrounding cortex, in agreement with the non vascular distribution of the stroke-like events.[213] The molecular consequences of the A3243G mutation are not completely understood, but may include effects on both transcription and translation of mtDNA. In fact, this mutation has been linked to a marked decrease in both the rates of synthesis and the steady-state levels of the mitochondrial translation products; moreover a small but consistent increase in the levels of an unprocessed RNA containing the tRNALeu(UUR) sequence (RNA 19) has been reported and it could contribute to the observed inhibition of mitochondrial protein synthesis.[214] Other authors have shown a diminished steady-state level of the tRNALeu(UUR) in mutant cellular models. This has been linked to posttranscriptional modifications such as diminution of methylation, which could be responsible for a slower processing rate of the precursor transcript and may accelerate the rate of degradation of the tRNA, lowering its steady-state level.[215] Other results suggest that the primary biochemical defect in cells with high levels of A3243G mutated mtDNA is the inability to translate the UUR leucine codons.[216] Nevertheless, all these different mechanisms seem to have in common a decreased ability to translate the tRNA of leucine. While the exact mechanism is still debated, the A3243G mutation results in dysfunction of the respiratory chain, which can lead to cellular energy impairment and cell death by failure of different key homeostatic processes and activation of the apoptotic cascade.[217] Notably, glutamate uptake is a cellular process strictly dependent upon energy supply and a mitochondrial respiratory chain defect may induce a reduction of glutamate transport,[29] leading to excitotoxic damage. Moreover, the potential contribution of abnormal glutamate handling in the pathogenesis of mtDNA diseases has recently been highlighted in LHON, as mentioned above.[206,208] Interestingly, LHON-MELAS overlap syndromes have been described in association with mtDNA mutations G13513A[218] and G3376A.[219] Glutamate uptake was found to be significantly reduced in osteosarcoma-derived cybrids (same ρ^0 as for the LHON cybrids described in ref.[206]) expressing high levels of the A3243G mutation; moreover, an inverse relationship between A3243G mutation load and mitochondrial ATP synthesis, without any evidence of increased cellular or mitochondrial free radical production, was also observed.[57] Collectively, our results strength the hypothesis of the possible role played by glutamate excitotoxicity in the pathological conditions characterized by a mitochondrial respiratory chain deficiency. However, we previously demonstrated that the reduction of glutamate uptake in LHON cybrids might relate to an excess of mitochondria-derived free radicals, from the complex I defect in this disorder.[206,220] Even if MELAS and LHON syndromes appear to have pathogenic aspects in common,[218,219] the biochemical consequences of the different mtDNA mutations diverge in some respects. The various mitochondrial mutations linked to LHON (11778/ ND4, 3460/ND1 and 14484/ND6) all selectively target complex I with a consequent loss of function in addition to an increased production of free radicals which may be considered as a toxic gain of function.[206,207,220] However, while the A3243G MELAS mutation predominantly affects complex I activity, in contrast to the LHON mutations, there is also an impairment of complex III and IV activities at high mutant loads[221] with no evidence of increased free radical generation. In fact, even in different cybrid lines harboring high mutant loads of the A3243G mtDNA mutation causing decreased mitochondrial oxidative phosphorylation, no evidence of increased free radical production or oxidative damage was found. Our results suggest that cybrid models carrying the A3243G mutation do not acquire a toxic gain of function resulting in an enhancement of free radicals production. This is in general agreement with previous studies showing no direct evidence of increased free radical generation in glioblastoma cybrids carrying the A3243G mutation,[222] although these clones were more susceptible to pro-oxidant stress stimuli.[223] However, in contrast to these data, a reduced GSH/GSSG ratio[224] and increased activity of antioxidant enzymes[225] have been shown in cellular models harboring the A3243G mutation, although a direct demonstration of higher ROS production is still missing. Our results allow suggesting a putative role for excitotoxicity in MELAS. We might in fact hypothesize that in conditions of elevated energetic demand, a reduced ATP synthesis might result in a dysfunction of those homeostatic systems strictly dependent on the energetic supply for they correct functioning, such as glutamate uptake. This is very interesting when thinking that, to date, no drug intervention has been demonstrated to be conclusively neuroprotective in mitochondrial encephalomyopathies. However, the high frequency of stroke-like episodes in MELAS suggests that the use of an effective anti-glutamatergic therapy may be worth evaluating, at least in those patients exhibiting the more severe phenotypes. It may be that such therapy could be more effective than in ischemic stroke where there is no underlying bioenergetic defect influencing glutamate uptake. The observations that mitochondrial respiratory chain dysfunction leads to decreased glutamate uptake has a relevance to the pathogenesis of neurodegenerative disorders associated with mitochondrial dysfunction, including both diseases with primary mtDNA mutations and those where the defect may be secondary. However, more studies are certainly required to understand the mechanisms involved and their relevance to the pathogenesis, in particular the role of ATP synthesis, free radical generation and glutamate transporter regulation, especially focusing on the reasons of the described difference of the oxidative status between LHON and MELAS.

Conclusion and Future Directions

Up to now, a huge number of putative peripheral biochemical markers of neurological dysfunction have been proposed, but among the dozens of them, only few display a level of sensitivity and specificity good enough to be useful in the clinical setting. Moreover, it is conceivable that when facing the need of choosing a suitable peripheral marker, only those related to the pathogenesis of the disease might be more closely reflecting the processes operative directly into the CNS, possibly not merely being unspecific secondary effects. However, nowadays only few of these biomarkers display both, the good reliability, and the good enough accessibility necessary to start thinking to a large-scale clinical use. It is mandatory, hence, organizing future experiments further investigating the utility of these biomarkers, opening the way to a more rational use of the available therapeutic resources.

As we have been discussing, peripheral markers of excitotoxicity offers (see the criteria for a diagnostic biomarker in Chapter 1) the clear advantage of reflecting a "core" process involved in the pathology of various neurological diseases, and not just their symptoms.[3] Although this data argues for excitotoxicity possibly as being a downstream pathologic mechanism common to different disorders, at same time stresses its importance, clearly implying the possibility of successfully interfering with a disease in a significant way by targeting excitotoxicity with specific therapies. However, although peripheral biomarkers of glutamatergic dysfunction do not yet offer a good enough specificity,[2] they are reproducible,[4] also in different laboratories[7] (with the discussed exception of plasma glutamate when the opportune sample inactivation is not performed), and most important, quite inexpensive[8] and very accessible[5] (for most of them by a simple blood withdraw[6]), especially when compared to more "central" procedures, such as, for example, neuroimaging or a lumbar puncture.

Looking back at the growing body of evidence implicating excitotoxicity in neurology during these past years, it is conceivable that together with the development of novel research tools, future lines of research will focus on identifying more specific biomarkers of glutamatergic homeostasis dysfunction, possibly endowed of a higher level of reliability, that is certainly needed when treating with patients. For this reason, considering all the reported evidence, we would like to conclude this review by stating that peripheral markers of glutamatergic dysfunction may become in the near future reliable tools used by the clinicians in the setting of different neurological disorders as a consistent help for determining diagnosis, prognosis, and therapeutic choices, significantly increasing the quality of patient care.

References

1. Curtis DR, Phillis JW, Watkins JC. Chemical excitation of spinal neurones. Nature. 1959;183:611–612
2. Curtis DR, Watkins JC. The excitation and depression of spinal neurones by structurally related amino acids. J Neurochem. 1960;6:117–141
3. Lucas DR, Newhouse JP. The toxic effect of sodium L-glutamate on the inner layers of the retina. AMA Arch Ophthalmol. 1957;58:193–201
4. Mody I, MacDonald JF. NMDA receptor-dependent excitotoxicity: the role of intracellular Ca2+ release. Trends Pharmacol Sci. 1995;16:356–359
5. Beal MF. Does impairment of energy metabolism result in excitotoxic neuronal death in neurodegenerative illnesses? Ann Neurol. 1992;31:119–130
6. Beal MF. Mechanisms of excitotoxicity in neurologic diseases. FASEB J. 1992;6:3338–3344
7. Ferrarese C, Beal MF, eds. Excitotoxicity in neurological diseases: new therapeutic challenge, Kluwer, Boston, 2003
8. Olney JW. Brain lesions, obesity and other disturbances in mice treated with monosodium glutamate. Science. 1969;164:719–721
9. Olney JW. Excitatory amino acids and neuropsychiatric disorders. Biol Psychiatry. 1989;26:505–525
10. Cotman CW, Monaghan DT, Ganong AH. Excitatory amino acid neurotransmission: NMDA receptors and Hebb-type synaptic plasticity. Annu Rev Neurosci. 1988;11:61–80
11. Skerry TM, Genever PG. Glutamate signalling in non-neuronal tissues. Trends Pharmacol Sci. 2001;22:174–181
12. Hisano S. Vesicular glutamate transporters in the brain. Anat Sci Int. 2003;78:191–204
13. Torgner I, Kvamme E. Synthesis of transmitter glutamate and the glial-neuron interrelationship. Mol Chem Neuropathol. 1990;12:11–17
14. Broer S, Brookes N. Transfer of glutamine between astrocytes and neurons. J Neurochem. 2001;77:705–719
15. Heath PR, Shaw PJ. Update on the glutamatergic neurotransmitter system and the role of excitotoxicity in amyotrophic lateral sclerosis. Muscle Nerve. 2002;26:438–458
16. Daw NW, Stein PS, Fox K. The role of NMDA receptors in information processing. Annu Rev Neurosci. 1993;16:207–222
17. Massieu L, Garcia O. The role of excitotoxicity and metabolic failure in the pathogenesis of neurological disorders. Neurobiology. 1998;6:99–108
18. Williams K. Interactions of polyamines with ion channels. Biochem J. 1997;325:289–297
19. Li YV, Hough CJ, Sarvey JM. Do we need zinc to think? Sci STKE. 2003;182:19
20. Lipton SA, Singel DJ, Stamler JS. Nitric oxide in the central nervous system. Prog Brain Res. 1994;103:359–364
21. Mattson MP. Excitotoxic and excitoprotective mechanisms: abundant targets for the prevention and treatment of neurodegenerative disorders. Neuromolecular Med. 2003;3:65–94
22. Lazarewicz JW, Wroblewski JT, Palmer ME, et al. Activation of N-methyl-D-aspartate-sensitive glutamate receptors stimulates arachidonic acid release in primary cultures of cerebellar granule cells. Neuropharmacology. 1988;27:765–769

23. Garthwaite J, Garthwaite G, Palmer RM, et al. NMDA receptor activation induces nitric oxide synthesis from arginine in rat brain slices. Eur J Pharmacol. 1989;172:413–416

24. Melcher T, Geiger JR, Jonas P, et al. Analysis of molecular determinants in native AMPA receptors. Neurochem Int. 1996;28:141–144

25. Bredt DS, Nicoll RA. AMPA receptor trafficking at excitatory synapses. Neuron. 2003;40:361–379

26. Huettner JE. Kainate receptors and synaptic transmission. Prog Neurobiol. 2003;70:387–407

27. Mori H, Mishina M. Roles of diverse glutamate receptors in brain functions elucidated by subunit-specific and region-specific gene targeting. Life Sci. 2003;74:329–336

28. Hinoi E, Takarada T, Ueshima T, et al. Glutamate signaling in peripheral tissues. Eur J Biochem. 2004;271:1–13

29. Danbolt NC. Glutamate uptake. Prog Neurobiol. 2001;65: 1–105

30. Beretta S, Begni B, Ferrarese C. Pharmacological manipulation of glutamate transport. Drug News Perspect. 2003; 16:435–445

31. O'Shea RD. Roles and regulation of glutamate transporters in the central nervous system. Clin Exp Pharmacol Physiol. 2002;29:1018–1023

32. Amara SG, Fontana AC. Excitatory amino acid transporters: keeping up with glutamate. Neurochem Int. 2002;41:313–318

33. Sato H, Tamba M, Ishii T, et al. Cloning and expression of a plasma membrane cystine/glutamate exchange transporter composed of two distinct proteins. J Biol Chem. 1999;274:11455–11458

34. Flynn J, McBean GJ. Kinetic and pharmacological analysis of L-[35 S] cystine transport into rat brain synaptosomes. Neurochem Int. 2000;36:513–521

35. Baker DA, Shen H, Kalivas PW. Cystine/glutamate exchange serves as the source for extracellular glutamate: modifications by repeated cocaine administration. Amino Acids. 2002;23:161–162

36. Bannai S. Exchange of cystine and glutamate across plasma membrane of human fibroblasts. J Biol Chem. 1986;261: 2256–2263

37. Danbolt NC, Holmseth S, Skar A, et al. Glutamate uptake transporters. In: Ferrarese C, Beal MF, eds. Excitotoxicity in neurological diseases: new therapeutic challenge, Kluwer, Boston 2003

38. Fiermonte G, Palmieri L, Todisco S, et al. Identification of the mitochondrial glutamate transporter. Bacterial expression, reconstitution, functional characterization, and tissue distribution of two human isoforms. J Biol Chem. 2002;277: 19289–19294

39. Munir M, Correale DM, Robinson MB. Substrate-induced up-regulation of Na(+)-dependent glutamate transport activity. Neurochem Int. 2000;37:147–162

40. Duan S, Anderson CM, Stein BA, et al. Glutamate induces rapid upregulation of astrocyte glutamate transport and cell-surface expression of GLAST. J Neurosci. 1999; 19:10193–10200

41. Fine SM, Angel RA, Perry SW, et al. Tumor necrosis factor alpha inhibits glutamate uptake by primary human astrocytes. Implications for pathogenesis of HIV-1 dementia. J Biol Chem. 1996;271:15303–15306

42. Keller JN, Mark RJ, Bruce AJ, et al. 4-hydroxynonenal, an aldehydic product of membrane lipid peroxidation, impairs

43. Springer JE, Azbill RD, Mark RJ, et al. 4-hydroxynonenal, a lipid peroxidation product, rapidly accumulates following traumatic spinal cord injury and inhibits glutamate uptake. J Neurochem. 1997;68:2469–2476

44. Chen W, Aoki C, Mahadomrongkul V, et al. Expression of a variant form of the glutamate transporter GLT1 in neuronal cultures and in neurons and astrocytes in the rat brain. J Neurosci. 2002;22:2142–2152

45. Schmitt A, Asan E, Lesch KP, et al. A splice variant of glutamate transporter GLT1/EAAT2 expressed in neurons: cloning and localization in rat nervous system. Neuroscience. 2002;109:45–61

46. Northington FJ, Traystman RJ, Koehler RC, et al. GLT1, glial glutamate transporter, is transiently expressed in neurons and develops astrocyte specificity only after midgestation in the ovine fetal brain. J Neurobiol. 1999; 39:515–526

47. Rauen T, Fischer F, Wiessner M. Glia-neuron interaction by high-affinity glutamate transporters in neurotransmission. Adv Exp Med Biol. 1999;468:81–95

48. Lehre KP, Danbolt NC. The number of glutamate transporter subtype molecules at glutamatergic synapses: chemical and stereological quantification in young adult rat brain. J Neurosci. 1998;18:8751–8757

49. Conti F, DeBiasi S, Minelli A, et al. EAAC1, a high-affinity glutamate transporter, is localized to astrocytes and gabaergic neurons besides pyramidal cells in the rat cerebral cortex. Cereb Cortex. 1998;8:108–116

50. Peghini P, Janzen J, Stoffel W. Glutamate transporter EAAC-1-deficient mice develop dicarboxylic aminoaciduria and behavioral abnormalities but no neurodegeneration. EMBO J. 1997;16:3822–3832

51. Tanaka K, Watase K, Manabe T, et al. Epilepsy and exacerbation of brain injury in mice lacking the glutamate transporter GLT-1. Science. 1997;276:1699–1702

52. Dehnes Y, Chaudhry FA, Ullensvang K, et al. The glutamate transporter EAAT4 in rat cerebellar Purkinje cells: a glutamate-gated chloride channel concentrated near the synapse in parts of the dendritic membrane facing astroglia. J Neurosci. 1998;18:3606–3619

53. Marie H, Billups D, Bedford FK, et al. The amino terminus of the glial glutamate transporter GLT-1 interacts with the LIM protein Ajuba. Mol Cell Neurosci. 2002;19:152–164

54. Lin CI, Orlov I, Ruggiero AM, et al. Modulation of the neuronal glutamate transporter EAAC1 by the interacting protein GTRAP3–18. Nature. 2001;410:84–88

55. Jackson M, Song W, Liu MY, et al. Modulation of the neuronal glutamate transporter EAAT4 by two interacting proteins. Nature. 2001;410:89–93

56. Massieu L, Morales-Villagran A, Tapia R. Accumulation of extracellular glutamate by inhibition of its uptake is not sufficient for inducing neuronal damage: an in vivo microdialysis study. J. Neurochem. 1995;64:2262–2272

57. DiFrancesco JC, Cooper JM, Lam A, et al. MELAS mitochondrial DNA mutation A3243G reduces glutamate transport in cybrids cell lines. Exp Neurol. 2008;212:152–156

58. Tremolizzo L, Beretta S, Ferrarese C. Peripheral markers of glutamatergic dysfunction in neurological diseases: focus on ex vivo tools. Crit Rev Neurobiol. 2004;16:141–146

glutamate transport and mitochondrial function in synaptosomes. Neuroscience. 1997;80:685–696

59. Shulman RG. Functional imaging studies: linking mind and basic neuroscience. Am J Psychiatry. 2001;158:11–20

60. Petroff OA, Errante LD, Rothman DL, et al. Glutamate-glutamine cycling in the epileptic human hippocampus. Epilepsia. 2002;43:703–710

61. Pioro EP, Majors AW, Mitsumoto H, et al. 1H-MRS evidence of neurodegeneration and excess glutamate + glutamine in ALS medulla. Neurology. 1999;53:71–79

62. Taylor-Robinson SD, Turjanski N, Bhattacharya S, et al. A proton magnetic resonance spectroscopy study of the striatum and cerebral cortex in Parkinson's disease. Metab Brain Dis. 1999;14:45–55

63. Berger C, Annecke A, Aschoff A, et al. Neurochemical monitoring of fatal middle cerebral artery infarction. Stroke. 1999;30:460–463

64. Berger C, Schabitz WR, Georgiadis D, et al. Effects of hypothermia on excitatory amino acids and metabolism in stroke patients: a microdialysis study. Stroke. 2002;33:519–524

65. Schulz MK, Wang LP, Tange M, et al. Cerebral microdialysis monitoring: determination of normal and ischemic cerebral metabolisms in patients with aneurysmal subarachnoid hemorrhage. J Neurosurg. 2000;93:808–814

66. Hutchinson PJ, O'Connell MT, Al-Rawi PG, et al. Increases in GABA concentrations during cerebral ischaemia: a microdialysis study of extracellular amino acids. J Neurol Neurosurg Psychiatry. 2002;72:99–105

67. Vespa P, Prins M, Ronne-Engstrom E, et al. Increase in extracellular glutamate caused by reduced cerebral perfusion pressure and seizures after human traumatic brain injury: a microdialysis study. J Neurosurg. 1998;89:971–982

68. Stahl N, Mellergard P, Hallstrom A, et al. Intracerebral microdialysis and bedside biochemical analysis in patients with fatal traumatic brain lesions. Acta Anaesthesiol Scand. 2001;45:977–985

69. Ferrarese C, Pecora N, Frigo M, et al. Assessment of reliability and biological significance of glutamate levels in cerebrospinal fluid. Ann Neurol. 1993;33:316–319

70. Espey MG, Basile AS, Heaton RK, et al. Increased glutamate in CSF and plasma of patients with HIV dementia. Neurology. 2002;58:1439

71. Da Prada M, Cesura AM, Launay JM, et al. Platelets as a model for neurones? Experientia. 1988;44:115–126

72. Di Luca M, Colciaghi F, Pastorino L, et al. Platelets as a peripheral district where to study pathogenetic mechanism of Alzheimer disease: the case of amyloid precursor protein. Eur J Pharmacol. 2000;405:277–283

73. Parnetti L, Reboldi GP, Santucci C, et al. Platelet MAO-B activity as a marker of behavioural characteristics in dementia disorders. Aging. 1994;6:201–207

74. Bongioanni P, Donato M, Castagna M, et al. Platelet phenol-sulphotransferase activity, monoamine oxidase activity and peripheral-type benzodiazepine binding in demented patients. J Neural Transm. 1996;103:491–501

75. Inestrosa NC, Alarcón R, Arriagada J, et al. Platelet of Alzheimer patients: increased counts and subnormal uptake and accumulation of [14C]5-hydroxytryptamine. Neurosci Lett. 1993;163:8–10

76. Zubenko GS, Teply I, Winwood E, et al. Prospective study of increased platelet membrane fluidity as a risk factor for Alzheimer's disease: results at 5 years. Am J Psychiatry. 1996;153:420–423

77. Rosenberg RN, Baskin F, Fosmire JA, et al. Altered amyloid protein processing in platelets of patients with Alzheimer disease. Arch Neurol. 1997;54:139–144

78. Di Luca M, Pastorino L, Bianchetti A, et al. Differential level of platelet amyloid beta precursor protein isoforms: an early marker for Alzheimer disease. Arch Neurol. 1998;55:1195–1200

79. Bongioanni P, Castagna M, Mondino C, et al. Platelet and lymphocyte benzodiazepine binding in patients with Alzheimer's disease. Exp Neurol. 1997;146:560–566

80. D'Andrea G, Canazi AR, Ferro-Milone F, et al. Platelet levels of glutamate and aspartate in normal subjects. Stroke. 1989;20:299–300

81. Yang M, Srikiatkhachorn A, Anthony M, et al. Serotonin stimulates megakaryocytopoiesis via the 5-HT2 receptor. Blood Coagul Fibrinolysis. 1996;7:127–133

82. Mangano RM, Schwarcz R. The human platelet as a model for the glutamatergic neuron: platelet uptake of L-Glutamate. J Neurochem. 1981;36:1067–1076

83. Sherif FM. GABA-transaminase in brain and blood platelets: basic and clinical aspects. Prog Neuropsycopharmacol Biol Psychiatry. 1994;18:1219–1233

84. Altamura CA, Mauri MC, Ferrara A, et al. Plasma and platelet excitatory amino acids in psychiatric disorders. Am J Psychiatry. 1993;150:1731–1733

85. Franconi F, Miceli M, De Montis MG, et al. NMDA receptors play an antiaggregating role in human platelets. Thromb Haemost. 2000;76:84–87

86. Morrell CN, Sun H, Ikeda M, et al. Glutamate mediates platelet activation through the AMPA receptor. J Exp Med. 2008;205:575–584

87. Ferrarese C, Begni B, Canevari C, et al. Glutamate uptake is decreased in platelets from Alzheimer's disease patients. Ann Neurol. 2000;47:641–643

88. Zoia C, Cogliati T, Tagliabue E, et al. Glutamate transporters in platelets: EAAT1 decrease in aging and in Alzheimer's disease. Neurobiol Aging. 2004;25:149–157

89. do Nascimento CA, Nogueira CW, Borges VC, et al. Changes in [(3)H]-glutamate uptake into platelets from patients with bipolar I disorder. Psychiatry Res. 2006;141:343–347

90. Genever PG, Wilkinson DJP, Patton AJ, et al. Expression of a functional N-methyl-D-aspartate-type glutamate receptor by bone marrow megakaryocytes. Blood. 1999;23:2876–2883

91. Reilmann R, Rolf LH, Lange HW. Huntington's disease: N-methyl-D-aspartate receptor coagonist glycine is increased in platelets. Exp Neurol. 1997;144:416–419

92. Franconi F, Micele M, De Montis MG, et al. NMDA receptors play an anti-aggregating role in human platelets. Thromb Haemost. 1996;76:84–87

93. Franconi F, Miceli M, Alberti L, et al. Further insights into the anti-aggregating activity of NMDA in human platelets. Br J Pharmacol. 1998;124:35–40

94. Begni B, Tremolizzo L, D'Orlando C, et al. Substrate-induced modulation of glutamate uptake in human platelets. Br J Pharmacol. 2005;145:792–799

95. Borges VC, Santos FW, Rocha JB, et al. Heavy metals modulate glutamatergic system in human platelets. Neurochem Res. 2007;32:953–958

96. Tremolizzo L, DiFrancesco JC, Rodriguez-Menendez V, et al. Human platelets express the synaptic markers VGLUT1 and 2 and release glutamate following aggregation. Neurosci Lett. 2006;404:262–265

97. Sarchielli P, Di Filippo M, Candeliere A, et al. Expression of ionotropic glutamate receptor GLUR3 and effects of glutamate on MBP- and MOG-specific lymphocyte activation and chemotactic migration in multiple sclerosis patients. J Neuroimmunol. 2007;188:146–158

98. Lombardi G, Dianzani C, Miglio G, et al. Characterization of ionotropic glutamate receptors in human lymphocytes. Br J Pharmacol. 2001;133:936–944

99. Ganor Y, Besser M, Ben-Zakay N, et al. Human T cells express a functional ionotropic glutamate receptor GluR3, and glutamate by itself triggers integrin-mediated adhesion to laminin and fibronectin and chemotactic migration. J Immunol. 2003;170:4362–4372

100. Chiocchetti A, Miglio G, Mesturini R, et al. Group I mGlu receptor stimulation inhibits activation-induced cell death of human T lymphocytes. Br J Pharmacol. 2006;148:760–768

101. Miglio G, Varsaldi F, Dianzani C, et al. Stimulation of group I metabotropic glutamate receptors evokes calcium signals and c-jun and c-fos gene expression in human T cells. Biochem Pharmacol. 2005;70:189–199

102. Miglio G, Varsaldi F, Lombardi G. Human T lymphocytes express N-methyl-D-aspartate receptors functionally active in controlling T cell activation. Biochem Biophys Res Commun. 2005;338:1875–1883

103. Connolly GP. Fibroblast models of neurological disorders: fluorescence measurement studies. Trends Pharmacol Sci. 1998;19:171–177

104. Balcar VJ, Shen J, Bao S, et al. Na(+)-dependent high affinity uptake of L-glutamate in primary cultures of human fibroblasts isolated from three different types of tissue. FEBS Lett. 1994;339:50–54

105. Zoia CP, Tagliabue E, Isella V, et al. Fibroblast glutamate transport in aging and in AD: correlations with disease severity. Neurobiol Aging. 2005;26:825–832

106. Castillo J, Davalos A, Naveiro J, et al. Neuroexcitatory amino acids and their relation to infarct size and neurological deficit in ischemic stroke. Stroke. 1996;27:1060–1065

107. Davalos A, Castillo J, Serena J, et al. Duration of glutamate release after acute ischemic stroke. Stroke. 1997;28:708–710

108. Serena J, Leira R, Castillo J, et al. Neurological deterioration in acute lacunar infarctions: the role of excitatory and inhibitory neurotransmitters. Stroke. 2001;32:1154–1161

109. Davalos A, Shuaib A, Wahlgren NG. Neurotransmitters and pathophysiology of stroke: evidence for the release of glutamate and other transmitters/mediators in animals and humans. J Stroke Cerebrovasc Dis. 2000;9:2–8

110. Mallolas J, Hurtado O, Castellanos M, et al. A polymorphism in the EAAT2 promoter is associated with higher glutamate concentrations and higher frequency of progressing stroke. J Exp Med. 2006;203:711–717

111. Aliprandi A, Longoni M, Stanzani L, et al. Increased plasma glutamate in stroke patients might be linked to altered platelet release and uptake. J Cereb Blood Flow Metab. 2005;25:513–519

112. Cananzi AR, D'Andrea G, Perini F, et al. Platelet and plasma levels of glutamate and glutamine in migraine with and without aura. Cephalalgia. 1995;15:132–135

113. Sharp CD, Hines I, Houghton J, et al. Glutamate causes a loss in human cerebral endothelial barrier integrity through activation of NMDA receptor. Am J Physiol Heart Circ Physiol. 2003;285:2592–2598

114. Fujimoto T, Suzuki H, Tanque K, et al. Cerebrovascular injuries induced by activation of platelets in vivo. Stroke. 1985;16:245–250

115. Ballabh P, Braun A, Nedergaard M. The blood–brain barrier: an overview: structure, regulation, and clinical implications. Neurobiol Dis. 2004;16:1–13

116. Dirnagl U, Iadecola C, Moskowitz MA. Pathobiology of ischaemic stroke: an integrated view. Trends Neurosci. 1999;22:391–397

117. Sherwin A, Robitaille Y, Quesney F, et al. Excitatory amino acids are elevated in human epileptic cerebral cortex. Neurology. 1988;38:920–923

118. Marrannes R, Willems R, De Prins E, et al. Evidence for a role of the N-methyl-D-aspartate (NMDA) receptor in cortical spreading depression in the rat. Brain Res. 1998;457:226–240

119. Ramadan NM, Halvorson H, Vande-Linde A, et al. Low brain magnesium in migraine. Headache. 1989;29:416–419

120. Schaumburg HH, Byck R, Gerstl R, et al. Monosodium 1-glutamate: its pharmacology and role in the Chinese restaurant syndrome. Science. 1969;163:826–828

121. Alam Z, Coombes N, Waring RH, et al. Plasma levels of neuroexcitatory amino acids in patients with migraine or tension headache. J Neurol Sci. 1998;156:102–106

122. Ferrari MD, Odink K, Bos KD, et al. Neuroexcitatory plasma amino acids are elevated in migraine. Neurology. 1990;40:1582–1586

123. D'Andrea G, Cananzi AR, Joseph R, et al. Platelet glycine, glutamate and aspartate in primary headache. Cephalalgia. 1991;11:197–200

124. Martìnez F, Castillo J, Rodrìguez JR, et al. Neuroexcitatory amino acid levels in plasma and cerebrospinal fluid during migraine attacks. Cephalalgia. 1993;13:89–93

125. Hanington E. The platelet and migraine. Headache. 1986;26:411–415

126. Vaccaro M, Riva C, Tremolizzo L, et al. Platelet glutamate uptake and release in migraine with and without aura. Cephalalgia. 2007;27:35–40

127. Rainesalo S, Keränen T, Peltola J, et al. Glutamate uptake in blood platelets from epileptic patients. Neurochem Int. 2003;43:389–392

128. D'Andrea G, Nordera GP, Allais G. Treatment of aura: solving the puzzle. Neurol Sci. 2006;27(2, Suppl):96–99

129. Morris JC. Clinical dementia rating: a reliable and valid diagnostic and staging measure for dementia of the Alzheimer type. Int Psychogeriatr. 1997;9(1, Suppl): 173–176

130. Small GW, Komo S, La Rue A, et al. Early detection of Alzheimer's disease by combining apolipoprotein E and neuroimaging. Ann N Y Acad Sci. 1996;802:70–78

131. Masliah E, Alford M, DeTeresa R, et al. Deficient glutamate transport is associated with neurodegeneration in Alzheimer's disease. Ann Neurol. 1996;40:759–766

132. Li S, Mallory M, Alford M, et al. Glutamate transporter alterations in Alzheimer disease are possibly associated with abnormal APP expression. J Neuropathol Exp Neurol. 1997;56:901–991

133. Masliah E, Raber J, Alford M, et al. Amyloid protein precursor stimulates excitatory amino acid transport. Implications for roles in neuroprotection and pathogenesis. J Biol Chem. 1998;273:12548–12554

134. Galimberti D, Fenoglio C, Lovati C, et al. Serum MCP-1 levels are increased in mild cognitive impairment and mild Alzheimer's disease. Neurobiol Aging. 2006;27:1763–1768
135. Licastro F, Grimaldi LM, Bonafè M, et al. Interleukin-6 gene alleles affect the risk of Alzheimer's disease and levels of the cytokine in blood and brain. Neurobiol Aging. 2003;24:921–926
136. Sala G, Galimberti G, Canevari C, et al. Peripheral cytokine release in Alzheimer patients: correlation with disease severity. Neurobiol Aging. 2003;24:909–914
137. Scott HL, Pow DV, Tannenberg AE, et al. Aberrant expression of the glutamate transporter excitatory amino acid transporter 1 (EAAT1) in Alzheimer's disease. J Neurosci. 2002;22:RC206
138. Iwatsuji K, Nakamura S, Kameyama M. Lymphocyte glutamate dehydrogenase activity in normal aging and neurological diseases. Gerontology. 1989;35:218–224
139. Peeters MA, Salabelle A, Attal N, et al. Excessive glutamine sensitivity in Alzheimer's disease and Down syndrome lymphocytes. J Neurol Sci. 1995;133:31–41
140. Walton HS, Dodd PR. Glutamate-glutamine cycling in Alzheimer's disease. Neurochem Int. 2007;50:1052–1066
141. Boldyrev AA, Johnson P. Homocysteine and its derivatives as possible modulators of neuronal and non-neuronal cell glutamate receptors in Alzheimer's disease. J Alzheimers Dis. 2007;11:219–228
142. Begni B, Brighina L, Sirtori E, et al. Oxidative stress impairs glutamate uptake in fibroblasts from patients with Alzheimer's disease. Free Radic Biol Med. 2004;37:892–901
143. Markesbery WR, Lovell MA. Four-hydroxynonenal, a product of lipid peroxidation, is increased in the brain in Alzheimer's disease. Neurobiol Aging. 1998;19:33–36
144. Lovell MA, Ehmann WD, Mattson MP, et al. Elevated 4-hydroxynonenal in ventricular fluid in Alzheimer's disease. Neurobiol Aging. 1997;18:457–461
145. Blanc EM, Keller JN, Fernandez S, et al. 4-Hydroxynonenal, a lipid peroxidation product, impairs glutamate transport in cortical astrocytes. Glia. 1998;22:149–160
146. Lauderback CM, Hackett JM, Huang FF, et al. The glial glutamate transporter, GLT-1, is oxidatively modified by 4-hydroxy-2-nonenal in the Alzheimer's disease brain: the role of Abeta1–42. J. Neurochem. 2001;78:413–416
147. Jiménez-Jiménez FJ, Molina JA, Gómez P, et al. Neurotransmitter amino acids in cerebrospinal fluid of patients with Alzheimer's disease. J Neural Transm. 1998;105:269–277
148. Miulli DE, Norwell DY, Schwartz FN. Plasma concentrations of glutamate and its metabolites in patients with Alzheimer's disease. J Am Osteopath Assoc. 1993;93:670–676
149. Basun H, Forssell LG, Almkvist O, et al. Amino acid concentrations in cerebrospinal fluid and plasma in Alzheimer's disease and healthy control subjects. J Neural Transm Park Dis Dement Sect. 1990;2:295–304
150. Davydova TV, Voskresenskaya NI, Fomina VG, et al. Induction of autoantibodies to glutamate in patients with Alzheimer's disease. Bull Exp Biol Med. 2007;143:182–183
151. Ferrarese C, Aliprandi A, Tremolizzo L, et al. Increased glutamate in CSF and plasma of patients with HIV dementia. Neurology. 2001;57:671–675
152. Jenner P. Oxidative stress in Parkinson's disease. Ann Neurol. 2003;53(3, Suppl):S26–S38

153. Shulman LM. Levodopa toxicity in Parkinson disease: reality or myth? Reality – practice patterns should change. Arch Neurol. 2000;57:406–407
154. Weiner WJ. Is levodopa toxic? Arch Neurol. 2000;57: 408–410
155. Greenamyre JT. Glutamatergic influences on the basal ganglia. Clin Neuropharmacol. 2001;24:65–70
156. Rodriguez MC, Obeso JA, Olanow CW. Subthalamic nucleus-mediated excitotoxicity in Parkinson's disease: a target for neuroprotection. Ann Neurol. 1998;44(1, Suppl):S175–S188
157. Ferrarese C, Zoia C, Pecora N, et al. Reduced platelet glutamate uptake in Parkinson's disease. J Neural Transm. 1999;106:685–692
158. Ferrarese C, Tremolizzo L, Rigoldi M, et al. Decreased platelet glutamate uptake and genetic risk factors in patients with Parkinson's disease. Neurol Sci. 2001;22:65–66
159. Zipp F, Demisch L, Derouiche A, et al. Glutamine synthetase activity in patients with Parkinson's disease. Acta Neurol Scand. 1998;97:300–302
160. Winkler-Stuck K, Wiedemann FR, Wallesch CW, et al. Effect of coenzyme Q10 on the mitochondrial function of skin fibroblasts from Parkinson patients. J Neurol Sci. 2004;220:41–48
161. Iwasaki Y, Ikeda K, Shiojima T, et al. Increased plasma concentrations of aspartate, glutamate and glycine in Parkinson's disease. Neurosci Lett. 1992;145:175–177
162. Molina JA, Gómez P, Vargas C, et al. Neurotransmitter amino acid in cerebrospinal fluid of patients with dementia with Lewy bodies. J Neural Transm. 2005;112:557–563
163. Hartai Z, Klivenyi P, Janaky T, et al. Kynurenine metabolism in plasma and in red blood cells in Parkinson's disease. J Neurol Sci. 2005;239:31–35
164. Klivényi P, Toldi J, Vécsei L. Kynurenines in neurodegenerative disorders: therapeutic consideration. Adv Exp Med Biol. 2004;541:169–183
165. Deng HX, Hentati A, Tainer JA, et al. Amyotrophic lateral sclerosis and structural defects in Cu, Zn superoxide dismutase. Science. 1993;261:1047–1051
166. Wiedau-Pazos M, Goto JJ, Rabizadeh S, et al. Altered reactivity of superoxide dismutase in familial amyotrophic lateral sclerosis. Science. 1996;271:515–518
167. Yim HS, Kang JH, Chock PB, et al. A familial amyotrophic lateral sclerosis-associated A4 V Cu, Zn-superoxide dismutase mutant has a lower Km for hydrogen peroxide. Correlation between clinical severity and the Km value. J Biol Chem. 1997;272:8861–8863
168. Estevez AG, Crow JP, Sampson JB, et al. Induction of nitric oxide-dependent apoptosis in motor neurons by zinc-deficient superoxide dismutase. Science. 1999;286:2498–2500
169. Rothstein JD, Martin LJ, Kuncl RW. Decreased glutamate transport by the brain and spinal cord in amyotrophic lateral sclerosis. N Engl J Med. 1992;326:1464–1468
170. Shaw PJ, Chinnery RM, Ince PG. [3H]D-aspartate binding sites in the normal human spinal cord and changes in motor neuron disease: a quantitative autoradiographic study. Brain Res. 1994;655:195–201
171. Lin CL, Bristol LA, Jin L, et al. Aberrant RNA processing in a neurodegenerative disease: the cause for absent EAAT2, a glutamate transporter, in amyotrophic lateral sclerosis. Neuron. 1998;20:589–602

172. Ince P, Stout N, Shaw P, et al. Parvalbumin and calbindin D-28k in the human motor system and in motor neuron disease. Neuropathol Appl Neurobiol. 1993;19:291–299

173. Urushitani M, Shimohama S, Kihara T, et al. Mechanism of selective motor neuronal death after exposure of spinal cord to glutamate: involvement of glutamate-induced nitric oxide in motor neuron toxicity and nonmotor neuron protection. Ann Neurol. 1998;44:796–807

174. Takuma H, Kwak S, Yoshizawa T, et al. Reduction of GluR2 RNA editing, a molecular change that increases calcium influx through AMPA receptors, selective in the spinal ventral gray of patients with amyotrophic lateral sclerosis. Ann Neurol. 1999;46:806–815

175. Louvel E, Hugon J, Doble A. Therapeutic advances in amyotrophic lateral sclerosis. Trends Pharmacol Sci. 1997;18:196–203

176. Shaw PJ, Kuncl RW. Current concepts in the pathogenesis of ALS. In: Kuncl RW (ed.) Motor Neuron Disease. W.B. Saunders, London, 2002

177. Williams TL, Day NC, Ince PG, et al. Calcium-permeable alpha-amino-3-hydroxy-5-methyl-4-isoxazole propionic acid receptors: a molecular determinant of selective vulnerability in amyotrophic lateral sclerosis. Ann Neurol. 1997;42:200–207

178. Trotti D, Rossi D, Gjesdal O, et al. Peroxynitrite inhibits glutamate transporter subtypes. J Biol Chem. 1996;271:5976–5979

179. Pedersen WA, Fu W, Keller JN, et al. Protein modification by the lipid peroxidation product 4-hydroxynonenal in the spinal cords of amyotrophic lateral sclerosis patients. Ann Neurol. 1998;44:819–824

180. Trotti D, Rolfs A, Danbolt NC, et al. SOD1 mutants linked to amyotrophic lateral sclerosis selectively inactivate a glial glutamate transporter. Nat Neurosci. 1999;2:427–433

181. Sala G, Beretta S, Ceresa C, et al. Impairment of glutamate transport and increased vulnerability to oxidative stress in neuroblastoma SH-SY5Y cells expressing a Cu, Zn superoxide dismutase typical of familial amyotrophic lateral sclerosis. Neurochem Int. 2005;46:227–234

182. Ferrarese C, Sala G, Riva R, et al. Decreased platelet glutamate uptake in patients with amyotrophic lateral sclerosis. Neurology. 2001;56:270–272

183. Houi K, Kobayashi T, Kato S, et al. Increased plasma TGF-beta1 in patients with amyotrophic lateral sclerosis. Acta Neurol Scand. 2002;106:299–301

184. Gluck MR, Thomas RG, Sivak MA. Unaltered cytochrome oxidase, glutamate dehydrogenase and glutaminase activities in platelets from patients with sporadic amyotrophic lateral sclerosis – a study of potential pathogenetic mechanisms in neurodegenerative diseases. J Neural Transm. 2000;107:1437–1447

185. Lanius RA, Paddon HB, Mezei M, et al. A role for amplified protein kinase C activity in the pathogenesis of amyotrophic lateral sclerosis. J Neurochem. 1995;65:927–930

186. Curti D, Malaspina A, Facchetti G, et al. Amyotrophic lateral sclerosis: oxidative energy metabolism and calcium homeostasis in peripheral blood lymphocytes. Neurology. 1996;47:1060–1064

187. Poulopoulou C, Davaki P, Koliaraki V, et al. Reduced expression of metabotropic glutamate receptor 2mRNA in T cells of ALS patients. Ann Neurol. 2005;58:946–949

188. Jansen GA, Wanders RJ, Jobsis GJ, et al. Evidence against increased oxidative stress in fibroblasts from patients with non-superoxide-dismutase-1 mutant familial amyotrophic lateral sclerosis. J Neurol Sci. 1996;139(Suppl):91–94

189. Aguirre T, Van Den Bosch L, Goetschalckx K, et al. Increased sensitivity of fibroblasts from amyotrophic lateral sclerosis patients to oxidative stress. Ann Neurol. 1998;43:452–457

190. Cooper B, Chebib M, Shen J, et al. Structural selectivity and molecular nature of L-glutamate transport in cultured human fibroblasts. Arch Biochem Biophys. 1998;353:356–364.

191. Sala G, Trombin F, Mattavelli L, et al. Lack of Evidence for Oxidative Stress in Sporadic Amyotrophic Lateral Sclerosis Fibroblasts. Neurodegener Dis. 2008 Mar 18 [Epub ahead of print]

192. Rothstein JD, Tsai G, Kuncl RW, et al. Abnormal excitatory amino acid metabolism in amyotrophic lateral sclerosis. Ann Neurol. 1990;28:18–25

193. Iwasaki Y, Ikeda K, Kinoshita M. Plasma amino acid levels in patients with amyotrophic lateral sclerosis. J Neurol Sci. 1992;107:219–222

194. Gredal O, Møller SE. Effect of branched-chain amino acids on glutamate metabolism in amyotrophic lateral sclerosis. J Neurol Sci. 1995;129:40–43

195. Plaitakis A, Caroscio JT. Abnormal glutamate metabolism in amyotrophic lateral sclerosis. Ann Neurol. 1987;22: 575–579

196. Perry TL, Krieger C, Hansen S, et al. Amyotrophic lateral sclerosis: amino acid levels in plasma and cerebrospinal fluid. Ann Neurol. 1990;28:12–17

197. Camu W, Billiard M, Baldy-Moulinier M. Fasting plasma and CSF amino acid levels in amyotrophic lateral sclerosis: a subtype analysis. Acta Neurol Scand. 1993;88:51–55

198. Iłzecka J, Stelmasiak Z, Solski J, et al. Plasma amino acids concentration in amyotrophic lateral sclerosis patients. Amino Acids. 2003;25:69–73

199. Spreux-Varoquaux O, Bensimon G, Lacomblez L, et al. Glutamate levels in cerebrospinal fluid in amyotrophic lateral sclerosis: a reappraisal using a new HPLC method with coulometric detection in a large cohort of patients. J Neurol Sci. 2002;193:73–78

200. Andreadou E, Kapaki E, Kokotis P, et al. Plasma glutamate and glycine levels in patients with amyotrophic lateral sclerosis. In Vivo. 2008;22:137–141

201. Andreadou E, Kapaki E, Kokotis P, et al. Plasma glutamate and glycine levels in patients with amyotrophic lateral sclerosis: the effect of riluzole treatment. Clin Neurol Neurosurg. 2008;110:222–226

202. Brown MD. The enigmatic relationship between mitochondrial dysfunction and Leber's hereditary optic neuropathy. J Neurol Sci. 1999;165:1–5

203. Luo X, Heidinger V, Picaud S, et al. Selective excitotoxic degeneration of adult pig retinal ganglion cells in vitro. Invest Ophthalmol Vis Sci. 2001;42:1096–1106

204. Vorwerk CK, Naskar R, Schuettauf F, et al. Depression of retinal glutamate transporter function leads to elevated intravitreal glutamate levels and ganglion cell death. Invest Ophthalmol Vis Sci. 2000;41:3615–3621

205. Muller A, Maurin L, Bonne C. Free radicals and glutamate uptake in the retina. Gen Pharmacol. 1998;30:315–318

206. Beretta S, Mattavelli L, Sala G, et al. Leber hereditary optic neuropathy mtDNA mutations disrupt glutamate transport in cybrid cell lines. Brain. 2004;127:2183–2192

207. Beretta S, Wood JP, Derham B, et al. Partial mitochondrial complex I inhibition induces oxidative damage and perturbs glutamate transport in primary retinal cultures. Relevance to Leber Hereditary Optic Neuropathy (LHON). Neurobiol Dis. 2006;24:308–317

208. Sala G, Trombin F, Beretta S, et al. Antioxidants partially restore glutamate transport defect in leber hereditary optic neuropathy cybrids. J Neurosci Res. 2008 July 9 [Epub ahead of print]

209. Baracca A, Solaini G, Sgarbi G, et al. Severe impairment of complex I-driven adenosine triphosphate synthesis in leber hereditary optic neuropathy cybrids. Arch Neurol. 2005;62:730–736

210. DiMauro S, Schon EA. Mitochondrial respiratory-chain diseases. N Engl J Med. 2003;348:2656–2668

211. Schapira AH. Mitochondrial disease. Lancet. 2006;368: 70–82

212. Goto Y, Nonaka I, Horai S. A mutation in the tRNA (Leu)(UUR) gene associated with the MELAS subgroup of mitochondrial encephalomyopathies. Nature. 1990;348: 651–653

213. Iizuka T, Sakai F. Pathogenesis of stroke-like episodes in MELAS: analysis of neurovascular cellular mechanisms. Curr Neurovasc Res. 2005;2:29–45

214. King MP, Koga Y, Davidson M, et al. Defects in mitochondrial protein synthesis and respiratory chain activity segregate with the tRNA(Leu(UUR)) mutation associated with mitochondrial myopathy, encephalopathy, lactic acidosis, and strokelike episodes. Mol Cell Biol. 1992; 12:480–490

215. Helm M, Florentz C, Chomyn A, et al. Search for differences in post-transcriptional modification patterns of mitochondrial DNA-encoded wild-type and mutant human tRNALys and tRNALeu(UUR). Nucleic Acids Res. 1999;27:756–763

216. El Meziane A, Lehtinen SK, Hance N, et al. A tRNA suppressor mutation in human mitochondria. Nat Genet. 1998;18:350–353

217. Liu CY, Lee CF, Hong CH, et al. Mitochondrial DNA mutation and depletion increase the susceptibility of human cells to apoptosis. Ann N Y Acad Sci. 2004;1011:133–145

218. Pulkes T, Eunson L, Patterson V, et al. The mitochondrial DNA G13513A transition in ND5 is associated with a LHON/MELAS overlap syndrome and may be a frequent cause of MELAS. Ann Neurol. 1999;46:916–919

219. Blakely EL, de Silva R, King A, et al. LHON/MELAS overlap syndrome associated with a mitochondrial MTND1 gene mutation. Eur J Hum Genet. 2005;13:623–627

220. Wong A, Cavelier L, Collins-Schramm HE, et al. Differentiation-specific effects of LHON mutations introduced into neuronal NT2 cells. Hum Mol Genet. 2002;11:431–438

221. Yoneda M, Tanaka M, Nishikimi M, et al. Pleiotropic molecular defects in energy-transducing complexes in mitochondrial encephalomyopathy (MELAS). J Neurol Sci. 1989;92:143–158

222. Sandhu JK, Sodja C, McRae K, et al. Effects of nitric oxide donors on cybrids harbouring the mitochondrial myopathy, encephalopathy, lactic acidosis and stroke-like episodes (MELAS) A3243G mitochondrial DNA mutation. Biochem J. 2005;391:191–202

223. Wong A, Cortopassi G. mtDNA mutations confer cellular sensitivity to oxidant stress that is partially rescued by calcium depletion and cyclosporin A. Biochem Biophys Res Commun. 1997;239:139–145

224. Pang CY, Lee HC, Wei YH. Enhanced oxidative damage in human cells harboring A3243G mutation of mitochondrial DNA: implication of oxidative stress in the pathogenesis of mitochondrial diabetes. Diabetes Res Clin Pract. 2001;54(2, Suppl):S45–S56

225. Rusanen H, Majamaa K, Hassinen IE. Increased activities of antioxidant enzymes and decreased ATP concentration in cultured myoblasts with the 3243A–>G mutation in mitochondrial DNA. Biochim Biophys Acta. 2000;1500:10–16

Chapter 35
Melatonin as a Biological Marker in Schizophrenia

Armando L. Morera, Pedro Abreu-Gonzalez, and Manuel Henry

Abstract It is widely accepted that schizophrenia is not a single disease but different subgroups of biological and clinically heterogeneous entities. There are no routine laboratory tests that could help clinicians in its diagnosis as in other diseases. The association between pineal gland and mental functions stems from Descartes. Melatonin is the main hormone of the pineal gland. The use of melatonin as a possible marker in schizophrenia has gone parallel with the development of laboratory techniques. It is not until the late seventies of the past century that such techniques were generalised. A common drawback of researching on this topic consists of studying melatonin levels in schizophrenic patients without having a clear hypothesis that would have had linked their results with the pathophysiology of schizophrenia. In this chapter the biosynthesis, receptor subtypes and functions of melatonin are highlighted and especially the role of melatonin as a possible biological marker in schizophrenia is reviewed. General research in this topic lacks of homogeneity concerning researching methodology. Several recommendations are made when researching in this area. Clear conclusions are drawn regarding the results from studies relating melatonin to schizophrenia. With the recent advent of new dregs targeting melatonin receptors we will assist to an explosion affecting basic as well as clinical research, rendering plausible important applications in the near future.

Keywords Schizophrenia • melatonin • pineal gland • biological markers

Abbreviations BPRS Brief psychiatric rating scale; Camp Cyclic adenosine monophosphate CNS Central nervous system; CRSD Circadian rhythms sleep disorder; CSF Cerebrospinal fluid; DMPEA Dimethoxyphenylethylamine; DOP Dopamine; HIOMT Hydroxyindole-*O*-methyltransferase; MAO Monoamine oxidase; MESOR Midline estimating statistic of rhythm; MLT Melatonin; NAS *N*-acetylserotonin; NADMPEA *N*-acetyl-3,4-dimethoxyphenethylamine; NE Norepinephrine; PANSS Positive and negative syndrome scale; PCK Protein kinase C; PVN Paraventricular nucleus; SAM *S*-adenosyl methionine; SANS Scale for the assessment of negative symptoms; SAPS Scale for the assessment of positive symptoms; SCG Superior cervical ganglia; SCN Suprachiasmatic nucleus; SER Serotonin; SNAT Serotonin *N*-acetyltranferase; 6SMT 6-sulfatoxymelatonin; 6HMT 6-Hydroxymelatonin; TRP Tryptophan; TRPH Tryptophan hydroxylase; 5OHTRPD 5-Hydroxytryptophan decarboxylase;

Introduction

Schizophrenia is a chronic mental illness with still unknown aetiology. It is widely accepted that schizophrenia is not a single disease but different subgroups of biologically and clinically heterogeneous entities.[1] There are no routine laboratory tests as in other diseases which could help clinicians in its diagnosis. The association between pineal gland and psychiatry stems from Descartes when he placed within the pineal gland

A. L. Morera and M. Henry
Department of Internal Medicine, Dermatology and Psychiatry

P. Abreu-Gonzalez
Department of Physiology. La Laguna University. Santa Cruz de Tenerife. Canary Islands. Spain.

M.S. Ritsner (ed.), *The Handbook of Neuropsychiatric Biomarkers, Endophenotypes and Genes*,
© Springer Science + Business Media B.V. 2009

the seat of rational though and the link between body and soul. During the first 50 years of the twentieth century, there was a renewed interest in the therapeutic efficacy of pineal extracts,[2] which were administered as a treatment for psychotic states.[3]

The main hormonal product of the pineal gland is MLT. In as much as 80% of pineal secretion is in the form of MLT, though several other products such as 5-methoxytryptophol and *O*-acetyl-5-methoxytryptophenol are also secreted by the pineal gland.[4]

The use of MLT as a possible marker in schizophrenia has gone parallel with the development of laboratory techniques that allow researchers to quantify MLT in a valid and reliable fashion. It is not until the late seventies of the past century that such techniques were generalised.[5–7] Research on MLT levels in schizophrenia has reported controversial results. It has been found that schizophrenic patients have a decreased,[8–19] unaffected[18–22] or increased[8,9] MLT level.

A common drawback of researching on this topic consists of studying MLT levels in schizophrenic patients without having a clear hypothesis which would have linked their results with the pathophysiology of schizophrenia. Most hypotheses have tried to link MLT with the pathophysiology of schizophrenia positive symptoms. Studies on the other two main symptoms subgroups, negative and cognitive/disorganised, are very scanty or inexistent.[15,16,19,22]

Clinical applications of MLT in psychiatry have been reviewed elsewhere.[23] In this chapter the role of MLT as a possible biological marker in schizophrenia is reviewed.

Melatonin Biosynthesis

The steps of MLT synthesis in the pineal gland are well known. MLT and other potential pineal hormones are products of TRP metabolism.[24] In humans MLT synthesis also depends upon TRP availability and is reduced by acute TRP depletion.[25] TRP is actively transported into the pinealocyte and the enzyme TRPH catalyzes the conversion of TRP to 5OHTRP. TRPH activity is inhibited by *p*-chlorophenylalanine which reduces SER levels in the pineal gland.[26] The 5OHTRPD removes the terminal carboxyl group from 5-OHTRP leading to the formation of SER. The activity of 5OHTRPD in the pineal gland is higher than in

most tissues and its activity remains unchanged throughout the light-dark cycle.[27] The SNAT enzyme catalyzes the transfer of an acetyl group from acetyl-CoA to SER, resulting in the formation of NAS. This is the rate-limiting step in the MLT biosynthesis. SNAT activity in the pineal gland exhibits a 24h rhythm with a nocturnal peak 20–100 times greater than daytime levels; this results in an increased pineal MLT content during the night.[28] The HIOMT enzyme catalyzes the *O*-methylation of NAS by SAM to form MLT. In mammals, pineal HIOMT activity is high and appears to exhibit no diurnal variation.[29] MLT synthesis steps are summarized in Fig. 35.1.

Since there is no evidence of MLT storage in the pineal, the hormone is thought to be released directly into the bloodstream and/or the CSF, where up to 70% is bound to albumin.[30] MLT half life in blood after intravenous infusion is about 30 min[31]; but a biphasic elimination pattern with half-life of about 3 and 45 min has also been observed following oral administration.[31] Inactivation of MLT occurs in the liver, where it is converted to 6HMT by the P-450-dependent microsomal mixed-function oxidase enzyme system. Most of 6HMT is excreted in the urine and faeces as sulphate conjugate (6SMT), and a much smaller amount as a glucuronide.[32] Liver and kidney pathologies alter clearance rate.[33] In humans, the main metabolite is 6SMT. Its urinary concentration accounts for up to 90% of administrated MLT.[32] Apart from blood, saliva and urine, MLT has also been detected in CSF of mammalians and in the anterior chamber of the eye.[34] MLT is also found in many fluids related to reproduction.[35] In the brain, at least in animal models, MLT has been reported to be concentrated in several regions of the cortex, cerebellum, thalamus and the paraventricular nuclei of the hypothalamus.[36]

Regulation of Melatonin Biosynthesis

The pineal gland is a photoneuroendocrine organ, converting external luminous stimuli into a hormone secretion, being responsible for the synchronization between internal homeostasis and the environment.[37] The circadian activity of the SCN is synchronised to the light/dark cycle mainly by light perceived by the retina. The signal generated in the retina is transmitted to the SCN through the retino-hypothamic tract that

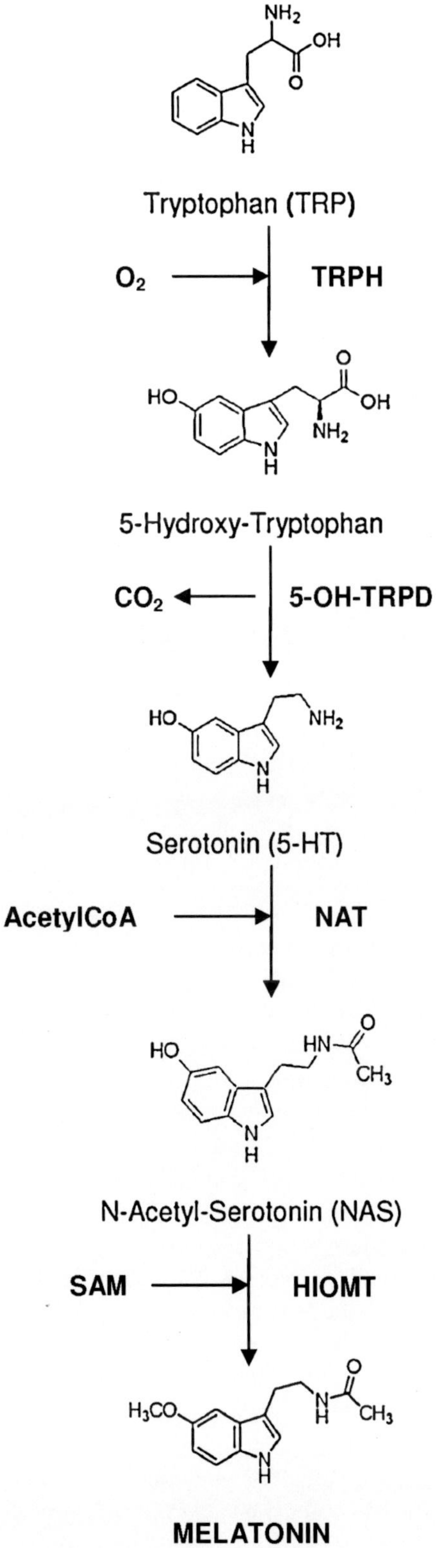

Tryptophan (TRP)

$O_2 \longrightarrow$ **TRPH**

5-Hydroxy-Tryptophan

$CO_2 \longleftarrow$ **5-OH-TRPD**

Serotonin (5-HT)

AcetylCoA $\longrightarrow$ **NAT**

N-Acetyl-Serotonin (NAS)

SAM $\longrightarrow$ **HIOMT**

MELATONIN

Fig. 35.1 Steps of melatonin biosynthesis

has its origin in the ganglion cell layer of the retina.[38] In the absence of light during the night, an increase in MLT biosynthesis in the pineal gland is stimulated by electrical signals originating from neurons in the SCN.[39] These neurons receive inputs from the retina and send inputs via PVN and then make synaptic connections with the intermediolateral column of the cervical spinal cord to the SCG. The postganglionic sympathetic fibres coming from SCG terminate adjacent to the pinealocytes.[40] The neurotransmitter at the postganglionic sympathetic nerve terminal is NE, which comes from catecholamine synthesis. Activation of the β-adrenoceptors present on the pinealocytes results in increases in cAMP concentration that promotes the biosynthesis of MLT.[41] Alpha-1-adrenoceptors potentiate β-adrenergic activity that increases intracellular Ca^{++} and the activation of the PKC.[42] During the day, NE release from the sympathetic fibres is suppressed by an increase in electrical activity in the SCN. At night, when SCN activity is inhibited, the release of NE is enhanced.[43]

Natural or artificial light, by influencing SCN output, suppresses MLT secretion in a dose dependent fashion.[44] Minima suppressive effects are observed with full spectrum light intensities of 200–300 lux.[45] Complete MLT suppression is obtained with light intensities above 2,000–2,500 lux.[44] The response to light is rapid and only 15 min of bright light exposure (1,500 lux) is sufficient to shunt down MLT production.[46] Experimental evidence indicates that, beside light, weak electromagnetic fields[47] and temperature may influence the endogenous production of MLT.[48] Figure 35.2 shows schematically the regulation of MLT biosynthesis.

Melatonin Receptors

At present, three mammalian MLT receptors have been cloned: MT_1,[49] MT_2[50] and an affinity purified MT_3.[51] Only two of them, MT_1 and MT_2, have been identified in humans, based on their binding affinity (picomolar or nanomolar) and chromosomal localization (chromosome 4q35 or 11p21–22).[50]

The MT_1 receptor has been detected in the SCN of hypothalamus, which controls the rhythmic production of MLT in the pineal gland.[52] Both MT_1 and

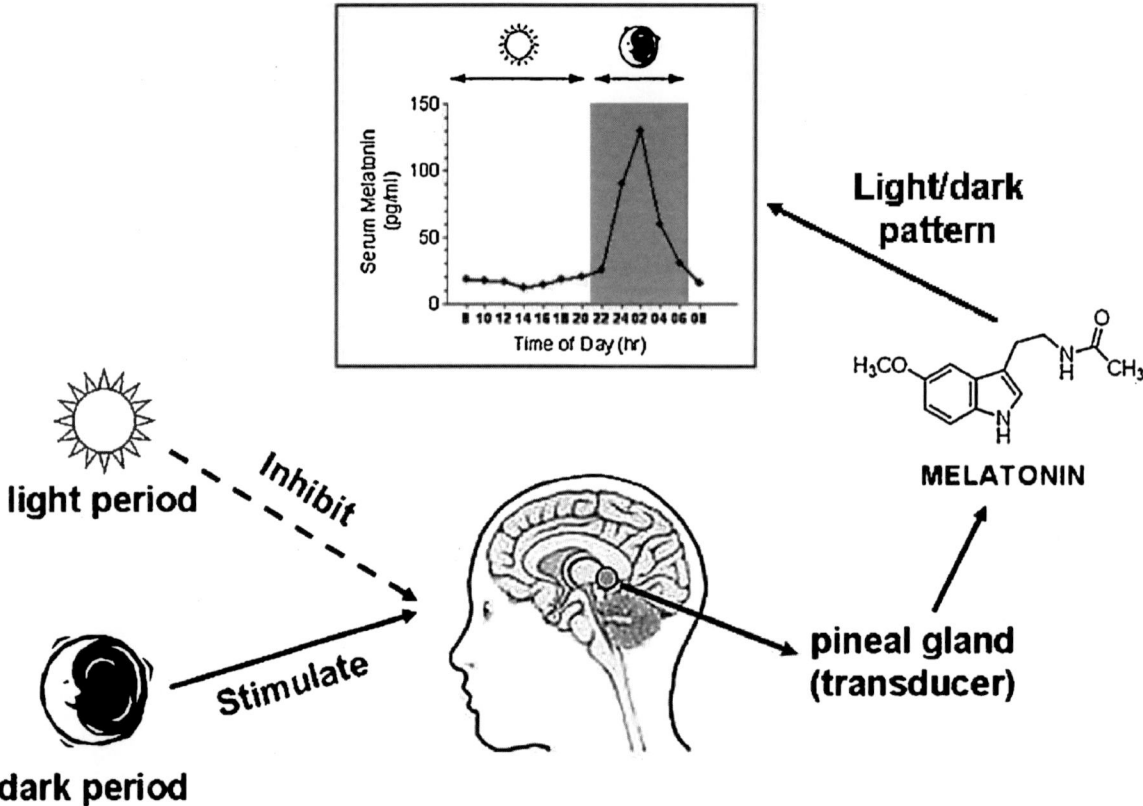

Fig. 35.2 Regulation of melatonin biosynthesis

MT_2 receptors have been found in the cerebellum[53] and retina.[54] Outside of the CNS, human MLT receptors have been localized in lymphocytes,[55] granulosa cells,[56] spermatozoa,[57] colon,[58] platelets[59] and cardiac vessels.[60]

The MT_1 receptor can couple to a variety of G proteins which may explain its diversity of response within the body.[61] When the MLT is acting through MT_1 receptor, it has been shown to produce inhibitory response on the cAMP signal transduction cascade.[62] This is a more generalized signalling mechanism proposed for MT_1 receptor. Besides, MT_1 receptors are responsible for the sleep promoting and circadian effects of MLT (amplitude of circadian rhythmicity) and the suppression of firing in the neurons of the CNS.[63] The expression of MT_1 in the human CNS decrease with advancing age as well as in the late stage of Alzheimer's disease. Sleep disruptions, nightly restlessness and circadian rhythm disturbances seen in the elderly and in patients with Alzheimer's disease may be due to an alteration of MT_1 receptor expression found in the CNS.[64]

The role of MT_2 receptor in mammalian physiology as well as its signalling properties is now becoming clearer with the recent development of MT_2-selective ligands.[65] It is known that MT_2 receptors are involved in retina physiology,[66] in modulating circadian rhythm,[67] in dilating cardiac vessels[60] and are involved in the inflammatory response of the microcirculation.[68] The MT_2 receptor is coupled to an inhibition of cAMP and coupled to the stimulation of phosphoinositide hydrolysis.[69]

Recently, a protein that displays a binding profile similar to the MT_2 receptor, now denoted as MT_3, was affinity-purified from Syrian hamster kidney. It was shown a 95% homology to the human quinine reductase 2, an enzyme involved in detoxification process.[51] This protein is widely distributed in hamster tissues, mouse, dog and monkey.[70] In RMPI cells (Syrian hamster melanoma), activation of MT_3 sites increases phosphatidylinositol turnover.[71]

Even in the absence of receptors, the highly diffusible MLT molecule exerts systemic effects at the most basic cellular levels, by modulating cytoskeletal and mitotic functions through binding with calmodulin.[72] Through binding to cytosolic calmodulin, MLT may directly affect calcium signalling by interacting with target enzymes, such as adenylate cyclase and phosphodiesterase, as well as with structural proteins.[73] MLT has recently been identified as a ligand for two orphan receptors, α and β, in the family of nuclear retinoid Z receptors.[74] The binding was in the low nanomolar range, suggesting that these receptors may be involved in nuclear signalling by the hormone.

MLT analogues that act as agonists or antagonists, with increased selectivity for MT_1 and MT_2 receptors, have been examined as pharmacological agents for the treatment of insomnia and CRSD. Among the most used in psychiatry are:

- Ramelteon. In vitro studies showed that the selectivity of ramelteon for MT_1/MT_2 binding sites is 1,000-fold greater that MLT. Ramelteon has been used in the treatment of insomnia characterised by difficulty with sleep onset in patients with DSM-IV primary chronic insomnia.[75]
- Agomelatine is another selective MLT agonist which show a high affinity for MT_1 and MT_2 receptor in the same level as MLT.[76] Additionally, it has antagonistic affinity for $5-HT_{2c}$ receptor with antidepressant capacity in humans.[77]
- VEC-162 is a selective MT_1 and MT_2 receptor agonist that is currently being studied for its efficacy in sleep disorders and depression. VEC-162 improves sleep efficiency, reduced sleep latency and attenuates REM reduction.[78]
- LY156735 is a MLT agonist under development for CSRD[78] treatment. LY156735 has also shown a reduction in subjective sleep latency compared to placebo.

Functions of Melatonin

MLT posses several biological properties. Many of these properties have been shown in animals and most of them have also been observed in humans. In the next paragraph we will shortly review the main biological actions of MLT.

Immunomodulatory effects: Using various pharmacological interventions, Maestroni et al.[79] provided the first evidence of a possible involvement of MLT on humoral and T-cell immune reactions as well as on spleen and cellularity in mice. Chronic MLT administration increases T-helper cell activity and IL-2 production and suppress 5-lypoxigenase gene expression in human lymphocytes.[80,81] Endogenous MLT regulates production of IL-1 in blood mononuclear cells.[82] The mechanisms of MLT action on the immune system have been confirmed in a variety of immune parameters.[83]

Antioxidant properties: The antioxidant potential of MLT was described the first time by Ianas et al.[84] The antioxidant effect of MLT has been confirmed in numerous subsequent in vitro investigations.[85,86] Its action has been documented on: hydroxyl radicals/hydrogen peroxide,[87] peroxyl radicals,[88] singlet oxygen,[89] nitric oxide/peroxinitrite anion,[90] and inhibition of nitric oxide synthase,[91] stimulation of antioxidant enzymes[92,93] and lipid peroxidation.[94] The mechanism of the antioxidant action of MLT involves an electron donation[95] or the reaction of singlet oxygen radical with MLT to produce N1-acetyl-N2-formyl-5-methoxykynurenaine,[96] a derivative MLT compound.

Cellular protector: MLT has revealed to have a protective effect to physical agents such as UV-light. MLT has been shown to protect the skin and other cells from oxidative damage by UV radiation.[97] In vitro studies using MLT as a protector agent for ionizing radiation exposure were conduced by Vijayalaxmi et al.[98] Those authors proposed that MLT may influence the stimulation/activation of enzymes involved in the repair of lesions in cellular DNA.[99]

Antitumoral actions: An important property of MLT is its influence on cell division. MLT exhibits oncostatic activity by inhibiting the mitosis, with a partial delay occurring at the metaphase stage.[100] Several studies have been published regarding the inhibitory action of MLT on the growth of steroid hormone-dependent cells as well as on the development of experimental malignant tumours, both, in vitro and in vivo conditions.[101,102] MLT as a therapeutic agent, promotes the stabilization of the disease and influences objective tumour regression in synergy with a chemotherapeutic agent.[103]

Thermoregulation effects: The administration of pharmacological doses of MLT tends to lower core body temperature and increases distal skin temperature, indicating increased heat loss. Similar changes in core and skin temperature occur at sleep onset, which suggest that the effect of MLT on sleep may be mediated by thermoregulatory mechanisms.[104,105] The relationship between MLT, core body temperature and polysomnographic sleep parameters was explored in a study by van den Heuvel et al.,[106] in which β-adrenergic transmission blocked by atenolol produced a blunting of the night-time decrease in core body temperature and the suppression of nocturnal MLT levels. These effects were reversed by the administration of 5 mg of MLT.

Reproductive functions: The demonstration of MLT receptors in reproductive organs[107] and the localization of sex hormone receptors in the pineal gland,[108] support the idea that MLT plays an important role in the reproductive functions. Besides, several studies have concluded that neurons found in the pre-optic area and/or the mediobasal hypothalamus and pituitary are the main sites through which MLT exerts its reproductive actions.[109,110] High MLT levels in amenorrhea women have been reported. This fact would support a causal relationship between high MLT concentration and hypothalamic-pituitary-gonadal hypofunction.[111]

Resetting/entraining actions: Non 24h sleep/wake disorders is commonly encountered in totally blind individuals since their sleep/wake cycles are not synchronised to the 24h light/dark cycle. The circadian rhythm of sleepiness is shifted out of phase with the desired time for sleeping. The most frequently reported symptoms in these patients are recurrent insomnia and daytime sleepiness.[112] In several blind individuals with no conscious light perception, administration of MLT at dosage of 3–5 mg/day was found effective not only in normalising the sleep/wake cycle but also in reducing the variability in the time on night sleep onset and increasing sleep duration.[113] Disturbed night-time sleep with impaired daytime alertness and performance are symptoms of jet lag frequently seen in intercontinental travellers (jet lag syndrome).[114] MLT taken during the evening at the local time of the new time zone (after transmeridian flight) has been shown to alleviate the symptoms of jet lag.[115] A comparative study to determine the optimal MLT dosage form for the alleviation of jet lag syndrome has been performed by Suhner et al.[116] The lower physiological dose of 0.5 mg was almost as effective as the pharmacological dose of 5 mg. Shift work is associated with a number of health problems. Insomnia or excessive sleepiness are common complications among shift workers. Endogenous factors such as MLT levels and increased circulating cortisol levels are factors that contribute to the frag-

mented/reduction sleep in night shift workers.[117] Some studies indicate that both MLT production and sleep patterns are altered in shift workers.[118] It was found that shift workers who took MLT, ranged 0.5–3 mg, adapted to a shift sleep schedule more quickly than those that took placebo.[119] It was found that the ability to phase-shift endogenous MLT rhythm is associated with improve shift work tolerance.[120] MLT has also been proposed as a treatment for the advanced and delayed sleep phase syndrome,[121,122] as well as a time resetter in pinealectomised patients.[123]

Cardiovascular system: There is a well-know diurnal variation in human cardiovascular functions, with lower blood pressure, heart rate and cardiac output and higher peripheral vascular resistance at night, when the MLT is high, relative to the day.[124] Increased risk of myocardial infarction coincides with the fall in nocturnal MLT levels.[125] The suprachiasmatic nucleus of hypothalamus and, possibly, the melatonergic system modulate the cardiovascular rhythm. It has been observed that patients with coronary heart disease have a low MLT production rate, which correlates with the stage of the disease, e.g., greater reductions are observed in patients with a higher risk of cardiac infarction and/or sudden death.[126] Acute coronary syndrome is an inflammatory disorder that results from the interaction between various components of the immune system and inflammatory proteins.[127] Previous studies from Dominguez-Rodriguez et al. have reported a relationship between MLT and light/dark variations in the production of inflammatory systemic markers, such as interleukin-6,[128] C-reactive protein,[129] matrix metalloproteinase-9[130] and soluble vascular cell adhesion molecule-1.[131] In the same way, Dominguez-Rodriguez et al.[132] demonstrated the prognostic role of MLT in a prospective cohort of patients who were admitted into the coronary care unit within 6h of symptom onset in patients with ST-segment elevation myocardial infarction. These findings confirm that patients who had developed adverse events (heart failure or cardiac death), during follow-up (6 months) have significantly lower nocturnal MLT levels than patients who do not exhibit these events.

Hypnotic actions: MLT administration in healthy subjects have shown to increase total sleep time, sleep efficiency index, stage 2 of sleep, and reduced wakefulness.[133] Due to its hypnotic activity MLT has been used as a treatment for primary insomnia and secondary insomnia.[134–136]

Melatonin as a Biological Marker in Schizophrenia

The first hypothesis that linked MLT with schizophrenia was posed by McIsaac[137] in 1961. This investigator based his hypothesis on the structural similarity between MLT and the hallucinogenic harmala alkaloids harmine and harmaline. Previous researches studying the effect of parental administration of beef pineal gland extracts in schizophrenic patients produced controversial results.[138,139] McIsaac[137] proposed that the formation of harmala alkaloids from SER metabolism could cause hallucinations. The formation of such alkaloid, 10-methoxyharmalan, could be produced by the MLT removal of one molecule of water by cyclodehydration. Harmala alkaloids are potent MAO inhibitors and could prevent the normal breakdown of SER, which would cause SER metabolism to shunt down the pathways which would lead to the production of more 5-methoxytriptamine, MLT and 10-methoxyharmalan. By doing this, once 10-methoxyharmalan is formed, it tends to perpetuate (positive feedback) its own formation. As far as we know, there is no research in humans that has confirmed this hypothesis. Besides, the fact that most researches have reported low serum MLT levels in schizophrenic patients points clearly against this hypothesis. The only two papers that have reported high blood levels of MLT in schizophrenic patients[8,9] were carried out in small samples, five patients, that were taking chlorpromazine; as it has been reported that chlorpromazine increases MLT blood levels[140] the results of these two papers are therefore seemingly expected.

Kunz et al.[141] found that the size of uncalcified pineal was positively associated with the total amount of 24 h 6SMT. Sandyk and Kay[142] studied the relationship between degree of pineal calcification and psychopathology in a sample of antipsychotic treated schizophrenic patients. Patients with no pineal calcification had more hallucinations, measured by means of the BPRS, than patients with minimal or prominent pineal calcification. Assuming that patients with no pineal calcification secrete more MLT, it could be concluded that an increased MLT level is related with hallucinations. This would indicate that in a step of the MLT metabolism, this MLT would be transformed into a compound with hallucinogenic characteristics. Sandyk and Kay[142] did not measure MLT in any

biological fluid, therefore, our assumption is merely speculative. In a sample of nine schizophrenic patients Monteleone et al.[16] studied the relation between MLT levels and positive and negative psychopathology. Positive symptoms were measured by means of the SAPS while negative symptoms were measured by means of the SANS. They did not find correlations between MLT and SAPS/SANS total scores. Bersani et al.[19] also studied the relation between MLT secretion and positive and negative symptoms, measured by means of the SAPS, SANS and PANSS. Again there was no correlation between MLT and psychopathology. In our opinion the negative results may be due to the fact that the scales that psychiatrists use in researching are more focused on the study of the schizophrenic subsyndromes than on the study of specific symptoms.

Hartley and Smith,[143] based on the in vitro formation of NADMPEA in bovine pineal gland cultures, proposed that if the rhythm of HIOMT would comes out of phase with its normal substrate (NAS), it would be free to act on abnormal substrates such as the isomeric methyl ethers of DOP to produce hallucinogenic methylated compounds (NADMPEA), as a consequence MLT production would be reduced. Though the final reduction of MLT would be explained by means of this hypothesis, NADMPEA has not been isolated from schizophrenic biological tissues. A similar related hallucinogenic compound, the DMPA, has also been related to schizophrenia, but experimental results have not consistently found such component in the urine of healthy subjects, schizophrenic patients or non-schizophrenic psychiatric patients.[144]

Another line of research has been centred on the study of the enzymes involved in MLT production from SER, SNAT and HIOMT. Smith et al.[145] studied the pineal enzyme activity of SNAT and HIOMT from the brain autopsy of 11 schizophrenic patients and 67 non schizophrenic subjects. They found that the schizophrenic group had an elevated HIOMT activity by about 25% when compared to controls. They suggest that a lack of substrate or an abnormally low activity of an enzyme prior to HIOMT in the biosynthesis of MLT could explain the observed low serum MLT concentrations. Rao et al.[14] studied the blood levels of TRP, SER and MLT in a sample of 33 drug-naïve patients, 35 drug-free patients, 34 drug-treated (antipsychotics) patients and 25 controls. They found that drug-naïve and drug-treated patients had significantly lower levels of TRP than control subjects. Besides, SER levels were not significantly different among the four groups. The only bias that we find in this research is that patients were not a "pure" sample of schizophrenics; they were diagnosed of schizophrenia or schizoaffective psychosis. In a later study of the same group,[21] they reported no significant differences in the MESOR of TRP and SER blood levels among healthy subjects, drug-free schizophrenics and antipsychotic treated schizophrenics. Several studies[146,147] have found that schizophrenic patients had a mean blood SER concentration, significantly higher than that of controls. In spite of the differences among those researches, they indicate that SER blood levels (one of the substrates of MLT) are not decreased in schizophrenia, so the final product of SER metabolism should not be decreased.

Crow[148] proposed that schizophrenia type II was characterised by negative symptoms, cerebral atrophy and poor response to antipsychotics. Sandyk and Kay[149] suggested that a subgroup of schizophrenics was characterised by presenting diminished MLT secretion, cerebral atrophy, ventricular enlargement, negative symptoms, and impaired cognitive and psychosexual development. From this study, we could predict that schizophrenic patients with negative symptoms should have low MLT levels. Mann et al.[22] related the MLT secretion profile with negative psychopathology, measured by the PANSS, in a sample of schizophrenic patients with predominantly negative symptoms before and after the administration of olanzapine. Schizophrenic patients had normal circadian rhythms of MLT secretion on both conditions and there was no relation between negative PANSS subscores and MLT. Again, our prediction has not been confirmed by the scanty research on this topic.

MLT has direct and indirect antioxidant properties.[150,151] Some schizophrenic patients have altered their oxidant-antioxidant equilibrium,[152–154] so MLT might be directly involved in the pathophysiology of schizophrenia through its antioxidant properties. Because schizophrenic patients are biochemically oxidised,[155,156] the low MLT levels might be the result of the body reaction trying to compensate this hyperoxidative status. Earlier works of Altschule et al.,[157,158] reported that schizophrenic patients had low levels of glutathione, and that the injection of pineal extracts corrected these deficits. Based on the results of in vitro studies, in which MLT reduced by about 83% the enzymatic oxidation of DOP and by about 35.7% the autoxidation of DOP, Hartley and Smith[159] proposed that

MLT may act as a free radical quenching agent, thus slowing down the autoxidation rate.

MLT and DOP have an inverse relation. The "low MLT syndrome" in schizophrenia might be biochemically related to elevated DOP[14] since low MLT disinhibits hypothalamic DOP release[160] and thus might contribute to the hyperdopaminergia in schizophrenia. As the hyperdomaminergic hypothesis was one of the first biochemical hypothesis formulated in schizophrenia,[161] it seem expected that drug-naïve and drug-free patients had high levels of DOP causing an inhibition of MLT secretion. Research results on drug-naïve and drug-free patients seem to support this hypothesis.[11,12,15]

Diagnostic biomarkers need to meet certain criteria to be considered as useful. Validity, reliability, sensitivity and specificity are paramount as in any other medical biomarker. Though MLT was discovered 50 years ago,[162] its study in medicine is not generalised, and it is still mainly focused on the realm of investigation. Hence, it is too premature to confirm or deny that MLT complies with the general requirements that are to be taken when considered as a diagnostic biomarker. Further information about additional criteria concerning diagnostic biomarkers can be seen in chapter 1. If conditions would be met, MLT then, as a valuable biomarker, would render an important tool related to schizophrenia as a specific disease.

At present MLT, can be measured with easy and inexpensive biochemical techniques. As far as it is known no harmful effects have been reported in humans when MLT has been administered in clinical use. Probably when its use would be commonly generalised in clinical practice techniques would become a routine test and costs would be even cheaper.

Conclusions and Future Directions

In general, the results from studies relating MLT to schizophrenia appear to be rather inconclusive. From our point of view this may be due to the great variety of methodologies applied in the different researches. When researching on this topic we would like to make the following recommendations and statements:

1. Small samples are common in this area of research, and it would be advisable, when possible, to recruit bigger samples.

2. "Purity" of clinical diagnosis is a must in this type of research. It is difficult to draw valid conclusions about the relation between MLT and schizophrenia when samples have a "mixture" of psychiatric diagnosis.

3. MLT in healthy subjects has a circadian rhythm, with lower serum levels during the day and higher levels at night. Therefore, studies that sample subjects at one point during the day do not reflect the physiological rhythm of MLT secretion. Because sampling at several times of the day is difficult, we may propose that at least two samples should be carried out, at night, when MLT levels are high and during the day, when MLT levels are low. In order to preserve as much as possible the sleeping hours, sampling times at midday (12:00) and midnight (24:00) are proposed.

4. The effect of antipsychotics on MLT production/secretion is not clear. Studies with drug-naïve and drug-free patients and comparisons among different antipsychotics should be promoted.

5. Biasing effects should be taken into account: body position, wearing sunglasses, transoceanic flights, caffeine containing beverages, smoking status and menstrual cycle phase among others.

6. Schizophrenia is a complex medical condition. Though important progress about its diagnosis has been achieved, we believe that future investigations should focus not only on the main psychopathology syndromes but also on individual symptoms, e.g. hallucinations, delusions, blunt affect, etc.

With the recent advent of new drugs targeting MLT receptors (ramelteon, agomelatine, VEC-162, etc.) we will assist in the coming years to an explosion affecting this area of research, in basic research as well as in clinical trials and their future applications.

References

1. Sadock BJ, Sadock VA. Schizophrenia. In: Sadock BJ, Sadock VA (eds) Kaplan & Sadock's Synopsys of Psychiatry, 10th ed. Lippincott Williams & Wilkins, Baltimore, MD; 2007;467–497
2. Mullen PE, Silman RE. The pineal and psychiatry: a review. Psychol Med 1977;7:407–417
3. Becker WH. Epiglandol bei dementia praecox. Therapeust Halbmonast 1920;34:667–668
4. Hofman MA, Skene DJ, Swaab DF. Effect of photoperiod on the diurnal melatonin and 5-methoxytryptophol rhythms in the human pineal gland. Brain Res 1995;671:254–260
5. Lynch HJ, Ozaki Y, Wurtman RJ. The measurement of melatonin in mammalian tissues and body fluids. J Neural Transm 1978;Suppl 13:251–264
6. Lewy AJ, Markey SP. Analysis of melatonin in human plasma by gas chromatography negative chemical ionization mass spectrometry. Science 1978;201:741–743
7. Thoresen TS. Radioimmmunoassay for melatonin in human serum. Scand J Clin Lab Invest 1978;38:687–692
8. Smith JA, Mee TJX, Barnes JLC. Elevated melatonin serum concentrations in psychiatric patients treated with chlorpromazine. J Pharm Pharmacol 1977;Suppl. 29:30P
9. Smith JA, Mee TJ, Barnes JD. Increased serum melatonin levels in chlorpromazine-treated psychiatric patients. J. Neural Transm 1978;Suppl 13:397
10. Smith JA, Barnes JL, Mee TJ. The effect of neuroleptic drugs on serum and cerebrospinal fluid melatonin concentrations in psychiatric subjects. J Pharm Pharmacol 1979;31:246–248
11. Ferrier IN, Johnstone EC, Crow TJ, et al. Melatonin/cortisol ratio in psychiatric illness. Lancet 1982;8:1070
12. Ferrier IN, Arendt J, Johnstone EC, et al. Reduced nocturnal melatonin secretion in chronic schizophrenia: relationship to body weight. Clin Endocrinol (Oxf) 1982;17:181–187
13. Fanget F, Claustrat B, Dalery J, et al. Nocturnal plasma melatonin levels in schizophrenic patients. Biol Psychiatry 1989;25:499–501
14. Rao ML, Gross G, Strebel B, et al. Serum amino acids, central monoamines, and hormones in drug-naive, drug-free, and neuroleptic-treated schizophrenic patients and healthy subjects. Psychiatry Res 1990;34:243–257
15. Monteleone P, Maj M, Fusco M, et al. Depressed nocturnal plasma melatonin levels in drug-free paranoid schizophrenics. Schizophr Res 1992;7:77–84
16. Monteleone P, Natale M, La Rocca A, et al. Decreased nocturnal secretion of melatonin in drug-free schizophrenics: no change after subchronic treatment with antipsychotics. Neuropsychobiology 1997;36:159–163
17. Jiang HK, Wang JY. Diurnal melatonin and cortisol secretion profiles in medicated schizophrenic patients. J Formos Med Assoc 1998;97:830–837
18. Vigano D, Lissoni P, Rovelli F, et al. A study of light/dark rhythm of melatonin in relation to cortisol and prolactin secretion in schizophrenia. Neuro Endocrinol Lett 2001;22:137–141
19. Bersani G, Mameli M, Garavini A, et al. Reduction of night/day difference in melatonin blood levels as a possible disease-related index in schizophrenia. Neuro Endocrinol Lett 2003;24:181–184
20. Beckmann H, Wetterberg L, Gattaz WF. Melatonin immunoreactivity in cerebrospinal fluid of schizophrenic patients and healthy controls. Psychiatry Res 1984;11:107–110
21. Rao ML, Gross G, Strebel B, et al. Circadian rhythm of tryptophan, serotonin, melatonin, and pituitary hormones in schizophrenia. Biol Psychiatry 1994;35:151–163
22. Mann K, Rossbach W, Muller MJ, et al. Nocturnal hormone profiles in patients with schizophrenia treated with olanzapine. Psychoneuroendocrinology 2006;31:256–264
23. Morera A, Henry M, Abreu P, et al. Melatonin therapeutic use in psychiatry: a 39 year bibliographic study. Actas Esp Psiquiatr 2006;34:344–351

24. Ebadi M. Regulation of the synthesis of MLT and its significance to neuroendocrinology. In: Reiter RJ (ed.) The Pineal Gland. Raven Press, New York, 1984;1–38

25. Zimmermann RC, McDougle CJ, Schumacher M, et al. Effects of acute tryptophan depletion on nocturnal MLT secretion in humans. J Clin Endocrinol Metab 1993;76: 1160–1164

26. Underwood H. The pineal and melatonin: regulators of circadian function in lower vertebrates. Experientia 1990;46: 120–128

27. Hall JC. Cycling transcript and the circadian clock. Curr Biol 1991;1:89–90

28. Reiter RJ. Pineal melatonin: cell biology of its synthesis and of its physiological interactions. Endocrine Rev 1991;12:151–180

29. Sugden D. Melatonin biosynthesis in the mammalian pineal gland. Experientia 1989;45:922–932

30. Cardinali DP, Lynch HJ, Wurtman RJ. Binding of melatonin to human and rat plasma protein. Endocrinology 1972;91:1213–1218

31. Mallo C, Zaidan R, Galy G, et al. Pharmacokinetics of MLT in man after intravenous infusion and bolus injection. Eur J Clin Pharmacol 1990;38:297–301

32. Arendt J. Melatonin and the Mammalian Pineal Gland. Chapman & Hall, London, 1995

33. Viljoen M, Steyrn ME, van Rensburg BW, et al. Melatonin in chronic renal failure. Nephron 1992;60:138–142

34. Martin XD, Malina HZ, Brennan MC, et al. The ciliary body the third organ found to synthesize indoleamines in humans. Eur J Ophthalmol 1992;2:67–72

35. Cagnacci A. Melatonin in relation to physiology in adult humans. J Pineal Res 1996;21:200–213

36. Menendez-Pelaez A, Reiter RJ. Distribution of melatonin in mammalian tissues: the relative importance of nuclear versus cytosolic localization. J Pineal Res 1993;15:59–69

37. Turek FW, Dugovic C, Zee PC. Current understanding of the circadian clock and the clinical implications for neurological disorders. Arch Neurol 2001;58:1781–1787

38. Morin LP, Allen CN. The circadian visual system. Brain Res Rev 2006;51:1–60

39. Saper CB, Lu J, Chou TC, et al. The hypothalamic integrator for circadian rhythms. Trends Neurosci 2005;28:152–157

40. Moore RY. Neural control of the pineal gland. Behav Brain Res 1996;73:125–130

41. Klein DC, Weller JL, Moore RY. Melatonin metabolism: neural regulation of pineal serotonin: acetyl coenzyme A N-acetyl-transferase activity. Proc Natl Acad Sci USA 1971;68:3107–3110

42. Ho AK, Klein DC. Activation of alfa1-adrenoceptors, protein kinase C, or treatment with intracellular free Ca^{++} elevating agent's increases pineal phospholipase A_2 activity: evidence that protein kinase C may participate in Ca^{++}-dependent alpha 1-adrenergic stimulation of pineal phospholipase A_2 activity. J Biol Chem 1987;262: 11764–11770

43. Klein DC, Schaad NL, Namboordiri MA, et al. Regulation of pineal serotonin N-acetyltransferase activity. Biochem Soc Trans 1992;20:299–304

44. Cagnacci A, Soldani R, Yen SSC. The effect of light on core body temperature is mediated by MLT in women. J Clin Endocrinol Metab 1993;76:1036–1038

45. Dollins AB, Lynch HJ, Wurtman RJ, et al. Effects of illumination on human nocturnal serum MLT levels and performance. Physiol Behav 1993;53:153–160

46. Petterborg LJ, Kjelamn BF, Thalen BE, et al. Effects of 15 minute light pulse on nocturnal serum MLT levels in human volunteers. J Pineal Res 1991;10:9–13

47. Reiter RJ. Static and extremely low frequency electromagnetic field exposure: reported on the circadian production of melatonin. J Cell Biochem 1993;51: 394–403

48. Shanahan TL, Kronauer RE, Duffy JF, et al. Melatonin rhythm observed throughout a three-cycle bright-light stimulus designed to reset the human circadian pacemaker. J Biol Rhythms 1999;14:237–253

49. Reppert SM, Weaver DR, Ebisawa T. Cloning and characterization of a mammalian melatonin receptor that mediates reproductive and circadian response. Neuron 1994;13: 1177–1185

50. Reppert SM, Godson C, Mahle CD, et al. Molecular characterization of a second melatonin receptor expressed in human retina and brain: the Mel 1b melatonin receptor. Proc Natl Acad Sci USA 1995;92:8734–8738

51. Nosjean O, Ferro M, Coge F, et al. Identification of the melatonin-binding site MT3 as the quinone reductase 2. J Biol Chem 2000;275:31311–31317

52. Reppert SM, Weaver DR, Rivkees SA, et al. Putative melatonin receptors in a human biological clock. Science 1988;242:78–81

53. Al-Ghoul WM, Herman MD, Dubocovich, ML. Melatonin receptor subtype expression in human cerebellum. Neuroreport 1988;9:4063–4068

54. Scher J, Wankiewicz E, Brown GM, et al. Melatonin receptor in the human retina: expression and localization. Invest Ophthalmol Vis Sci 2002;43:889–897

55. Lopez-Gonzalez MA, Calvo JR, Ossuna C, et al. Interaction of melatonin with human lymphocytes: evidence for binding sites coupled to potentiation of cyclic AMP stimulated by vasoactive intestinal peptide and activation of cyclic GMP. J Pineal Res 1992;12:97–104

56. Yie SM, Niles LP, Youglai EV. Melatonin receptors on human granulose cell membranes. J Clin Endocrinol Metab 1995;80:1747–1749

57. van Vuuren RJ, Pitout MJ, van Aswegen, et al. Putative melatonin receptor in human spermatozoa. Clin Biochem 1992;25:125–127

58. Poon AMS, Mak ASY, Luk HT. Melatonin and 2[125] iodomelatonin binding sites in the human colon. Endocrinol Res 1996;22:77–94

59. Vacas MI, Del Zar MM, Martinuzzo DP, et al. Binding sites for [3H]-melatonin in human platelets. J Pineal Res 1992;13:60–65

60. Doolen S, Krause DN, Dubocovich ML, et al. Melatonin mediates two distinct responses in vascular smooth muscle. Eur J Pharmacol 1998;345:67–69

61. Brydon L, Roka F, Petit L, et al. Dual signalling of human Mel1a a melatonin receptors via G(i2), G(i3), and G(q/11) proteins. Mol Endocrinol 1999;13:2025–2038

62. Brydon L, Petit L, de Coppet P, et al. Polymorphism and signalling of melatonin receptors. Reprod Nutr Dev 1999;39:315–324

63. Liu C, Weaver DR, Jin X, et al. Molecular dissection of two distinct actions of melatonin on the suprachiasmatic circadian clock. Neuron 1997;19:91–102

64. Wu YH, Zhou JN, van Heerikhuize J, et al. Decrease MT1 melatonin receptor expression in the suprachiasmatic nucleus in aging and Alzheimer's disease. Neurobiol Aging 2007;28:1239–1247

65. Sugden D, Yeh LK, The MT. Design of subtype selective melatonin receptor agonist. Reprod Nutr Dev 1999;39:335–344

66. Duvocovich ML, Masana MI, Jacob S, et al. Melatonin receptor antagonists that differentiate between the human Mel1a and Mel1b recombinant subtypes are used to asses the pharmacological profile of the rabbit retina ML1 presynaptic heteroreceptor. Naunyn-Schmiedebergs Arch Pharmacol 1997;355:365–375

67. Duvocovich ML, Yun K, Al-Ghoul WM, et al. Selective MT2 melatonin receptor antagonist bock melatonin-mediated phase advances in circadian rhythms. FASEB J 1998;12:1211–1220

68. Lotufo CM, Lopes C, Duvocovich ML, et al. Melatonin and N-acetylserotonin inhibits leucocytes rolling and adhesion to rat microcirculation. Eur J Pharmacol 2001;430:351–457

69. McKenzy RS, Melan MA, Passey DK, et al. Dual coupling of MT(1) and MT(2) melatonin receptors to cyclic AMP and phosphoinositide signal transduction cascade their regulation following melatonin exposure. Biochem Pharmacol 2002;63:587–595

70. Nosjean O, Nocolas JP, Klupsch F, et al. Comparative pharmacological studies of melatonin receptors: MT1, MT2 and MT3/QR2. Biochem Pharmacol 2001;61:1369–1379

71. Eison AS, Mullins UL. Melatonin binding sites are functionally coupled to phosphoinositide hydrolysis in Syrian hamsters RPMI 1846 melanoma cells. Life Sci 1993;53:PL393–PL398

72. Fjaerly O, Lund T, Osterud B. The effect of melatonin on cellular activation processes in human blood. J Pineal Res 1999;26:50–55

73. Benitez-King G, Anton-Tay F. Calmodulin mediates melatonin cytoskeletal effects. Experientia 1993;49:635–641

74. Becker-Andre M, Wiesenberg I, Schaeren-Wiemers N, et al. Pineal gland hormone melatonin binds and activates an orphan of the nuclear receptor superfamily. J Biol Chem 1994;269:28531–2834

75. Erman M, Seiden D, Zammit G, et al. An efficacy, safety, and dose-response study of Ramelteon in patients with chronic primary insomnia. Sleep Med 2006;7:17–24

76. Yous S, Andrieux J, Howell HE, et al. Novel naphthalenic ligands with high affinity for the melatonin receptor. J Med Chem 1992;35:1484–1486

77. Loo H, Hale A, D'Haenen H. Determination of the dose of agomelatine, a melatoninergic agonist and selective 5-HT(2C) antagonist, in the treatment of major depressive disorder: a placebo-controlled dose range study. Int Clin Psychopharmacol 2002;17:239–247

78. Doghramji K. Melatonin and its receptors: a new class of sleep-promoting agents. J Clin Sleep Med 2007;3(5 Suppl):S17–S23

79. Maestroni GJM, Conti A, Pierpaoli W. Role of the pineal gland in immunity: circadian synthesis and release of melatonin modulates the antibody response and antagonize the immunosuppressive effect of corticosterone. J Neuroimmunol 1986;13:19–30

80. Caroleo MC, Frasce D, Nistico G, et al. Melatonin as immunomodulator in immunodeficient mice. Immunopharmacology 1992;2:81–89

81. Steinhilber D, Brungs M, Werz O, et al. The nuclear receptor for melatonin represses 5-lypoxigenase gene expression in human B lymphocytes. J Biol Chem 1995;270:7037–7040

82. Garcia-Mauriño S, Gonzalez-Haba MG, Calvo JR, et al. Involvement of nuclear binding sites for melatonin in the regulation of IL-2 and IL-6 production by human blood mononuclear cells. J Neuroimmunol 1998;92:76–84

83. Giordano M, Palermo MS. Melatonin-induced enhancement of antibody dependent cellular cytotoxicity. J Pineal Res 1991;10:117–121

84. Guerrero JM, Reiter RJ. Melatonin-immune system relationship. Curr Top Med Chem 2002;2:167–179

85. Ianas O, Olnescu R, Badescu I. Melatonin involvement in oxidative stress. Rom J Endocrinol 1991;29:147–153

86. Marshall KA, Reiter RJ, Poeggeler B, et al. Evaluation of the antioxidant activity of melatonin in vitro. Free Radic Biol Med 1996;21:307–315

87. Chan TY, Tang PL. Characterization of the antioxidant effects of melatonin and related indoleamines in vitro. J Pineal Res 1996;20:187–191

88. Tan DX, Manchester LC, Reiter RJ, et al. A novel melatonin metabolite, cyclic 3-hydroxymelatonin: a biomarker of in vivo hydroxyl radical generation. Biochem Biophys Res Commun 1998;253:614–620

89. Pieri C, Marra M, Moroni F, et al. Melatonin: a peroxyl radical scavenger more effective than vitamin E. Life Sci 1994;55:PL271–PL276

90. Sewerynek E, Reiter RJ, Melchiorri D, et al. Oxidative damage in the liver induced by ischemia-reperfusion: protection by melatonin. Hepatogastroenterology 1996;43:898–905

91. Blanchard B, Pompon D, Ducrocq C. Nitrosation of melatonin by nitric oxide and peroxynitrite. J Pineal Res 2000;29:184–192

92. Pozo D, Reiter RJ, Calvo JR, et al. Inhibition of cerebellar nitric oxide synthase and cyclic GMP production by melatonin via complex formation with calmodulin. J Cell Biochem 1997;65:430–442

93. Kotler M, Rodríguez C, Sainz RM, et al. Melatonin increases gene expression for antioxidant enzymes in rat brain cortex. J Pineal Res 1998;24:83–89

94. Antolin I, Rodriguez C, Sainz RM, et al. Neurohormone melatonin prevents cell damage: effect on gene expression for antioxidant enzymes. FASEB J 1996;10:882–890

95. Melchiorri D, Reiter RJ, Sewerynek E, et al. Melatonin reduces kainate-induced lipid peroxidation in homogenates of different brain regions. FASEB J 1995;9:1205–1210

96. Poeggeler B, Saarela S, Reiter RJ, et al. Melatonin-a highly potent endogenous radical scavenger and electron donor: new aspects of the oxidation chemistry of this indole accessed in vitro. Ann N Y Acad Sci 1994;738:419–420

97. Hardeland R, Reiter RJ, Poeggeler B, et al. The significance of the metabolism of the neurohormone melatonin: antioxidative protection and formation of bioactive substances. Neurosci Biobehav Rev 1993;17:347–357

98. Dreher F, Gabard B, Schwindt DA, et al. Topical melatonin in combination with vitamins E and C protects skin from ultraviolet-induced erythema: a human study in vivo. Br J Dermatol 1998;139:332–339

99. Vijayalaxmi, Reiter RJ, Leal BZ, et al. Effect of melatonin on mitotic and proliferation indices, and sister chromatid exchange in human blood lymphocytes. Mutat Res 1996;351:187–192

100. Vijayalaxmi, Reiter RJ, Meltz ML, et al. Melatonin: possible mechanisms involved in its 'radioprotective' effect. Mutat Res 1998;404:187–189

101. Banerjee S, Margulis L. Mitotic arrest by melatonin. Exp Cell Res 1973;78:314–318

102. Hill SM, Spriggs LL, Simon MA, et al. The growth inhibitory action of melatonin on human breast cancer cells is linked to the estrogen response system. Cancer Lett 1992;64:249–256

103. Cos S, Fernández F, Sánchez-Barceló EJ. Melatonin inhibits DNA synthesis in MCF-7 human breast cancer cells in vitro. Life Sci 1996;58:2447–2453

104. Ravindra T, Lakshimi NK, Ahuja YR. Melatonin in the pathogenesis and therapy of cancer. Ind J Med Scim 2006;60:523–535

105. Dawson D, Encel N. Melatonin and sleep in humans. J Pineal Res 1993;15:1–12

106. Kräuchi K, Cajochen C, Wirz-Justice A. A relationship between heat loss and sleepiness: effects of postural change and melatonin administration. J Appl Physiol 1997;83: 134–139

107. Van Den Heuvel CJ, Reid KJ, Dawson D. Effect of atenolol on nocturnal sleep and temperature in young men: reversal by pharmacological doses of melatonin. Physiol Behav 1997;61:795–802

108. Soares JM Jr, Masana MI, Erşahin C, et al. Functional melatonin receptors in rat ovaries at various stages of the oestrous cycle. J Pharmacol Exp Ther 2003;306:694–702

109. Sanchez JJ, Abreu P, Gonzalez-Hernandez T, et al. Estrogens modulation of adrenoceptor responsiveness in the female rat pineal gland: differential expression of intracellular estrogen receptors. J Pineal Res 2004;37:26–35

110. Vanecek J, Klein DC. Melatonin inhibits gonadotropin-releasing hormone-induced elevation of intracellular Ca2 + in neonatal rat pituitary cells. Endocrinology 1992;130: 701–707

111. Zemková H, Vaněček J. Inhibitory effect of melatonin on gonadotropin-releasing hormone-induced Ca2 + oscillations in pituitary cells of newborn rats. Neuroendocrinology 1997;65:276–283

112. Berga SL, Mortola JF, Yen SS. Amplification of nocturnal melatonin secretion in women with functional hypothalamic amenorrhea. J Clin Endocrinol Metab 1988;66: 242–244

113. Lockley SW, Skene DJ, Tabandeh H, et al. Relationship between napping and melatonin in the blind. J Biol Rhythms 1997;12:16–25

114. Palm L, Blennow G, Wetterberg L. Long-term melatonin treatment in blind children and young adults with circadian sleep-wake disturbances. Dev Med Child Neurol 1997;39:319–325

115. Cardinali DP, Brusco LI, Lloret SP, et al. Melatonin in sleep disorders and jet-lag. Neuro Endocrinol Lett 2002;23(1 Suppl):9–13

116. Oxenkrug GF, Requintina PJ. Melatonin and jet lag syndrome: experimental model and clinical implications. CNS Spectr 2003;8:139–148

117. Suhner A, Schlagenhauf P, Johnson R, et al. Comparative study to determine the optimal melatonin dosage form for the alleviation of jet lag. Chronobiol Int 1998;15:655–666

118. Sack RL, Blood ML, Lewy AJ. Melatonin rhythms in night shift workers. Sleep 1992;15:434–441

119. Roden M, Koller M, Pirich K, et al. The circadian melatonin and cortisol secretion pattern in permanent night shift workers. Am J Physiol 1993;265(1 Pt 2):R261–R267

120. Quera-Salva MA, Guilleminault C, Claustrat B, et al. Rapid shift in peak melatonin secretion associated with improved performance in short shift work schedule. Sleep 1997;20:1145–1150

121. Cavallo A, Ris MD, Succop P, et al. Melatonin treatment of pediatric residents for adaptation to night shift work. Ambul Pediatr 2005;5:172–177

122. Kayumov L, Brown G, Jindal R, et al. A randomized, double-blind, placebo-controlled crossover study of the effect of exogenous melatonin on delayed sleep phase syndrome. Psychosom Med 2001;63:40–48

123. Morgenthaler TI, Lee-Chiong T, Alessi C, et al. Standards of Practice Committee of the American Academy of Sleep Medicine. Practice parameters for the clinical evaluation and treatment of circadian rhythm sleep disorders. An American Academy of Sleep Medicine report. Sleep 2007;30:1445–1459

124. Jan JE, Tai J, Hahn G, et al. Melatonin replacement therapy in a child with a pineal tumor. J Child Neurol 2001;16: 139–140

125. Veerman DP, Imholz BP, Wieling W, et al. Circadian profile of systemic hemodynamics. Hypertension 1995;26: 55–59

126. Dominguez-Rodriguez A, Abreu-Gonzalez P, Garcia M, et al. Decreased nocturnal melatonin levels during acute myocardial infarction. J Pineal Res 2002;33:248–252

127. Sewerynek E. Melatonin and the cardiovascular system. Neuroendocrinol Lett 2002;23:79–83

128. Armstrong EJ, Morrow DA, Sabatine MS. Inflammatory biomarkers in acute coronary syndromes: part II: acute-phase reactants and biomarkers of endothelial cell activation. Circulation 2006;113:152–155

129. Dominguez-Rodriguez A, Abreu-Gonzalez P, Garcia M, et al. Light/dark patterns of interleukin-6 in relation to the pineal hormone melatonin in patients with acute myocardial infarction. Cytokine 2004;26:89–93

130. Dominguez-Rodriguez A, Garcia-Gonzalez M, Abreu-Gonzalez P, et al. Relation of nocturnal melatonin levels to C-reactive protein concentration in patients with ST-segment elevation myocardial infarction. Am J Cardiol 2006;97:10–12

131. Dominguez-Rodriguez A, Abreu-Gonzalez P, Garcia-Gonzalez MJ, et al. Relation of nocturnal melatonin levels

to serum matrix metalloproteinase-9 concentrations in patients with myocardial infarction. Thromb Res 2007;120:361–366

132. Dominguez-Rodriguez A, Abreu-Gonzalez P, Garcia-Gonzalez MJ, et al. Light/dark patterns of soluble vascular cell adhesion molecule-1 in relation to melatonin in patients with ST-segment elevation myocardial infarction. J Pineal Res 2008;44:65–69

133. Dominguez-Rodriguez A, Abreu-Gonzalez P, Garcia-Gonzalez M, et al. Prognostic value of nocturnal melatonin levels as a novel marker in patients with ST-segment elevation myocardial infarction. Am J Cardiol 2006;97: 1162–1164

134. Stone BM, Turner C, Mills SL, et al. Hypnotic activity of melatonin. Sleep. 2000;23:663–669

135. Buscemi N, Vandermeer B, Hooton N, et al. The efficacy and safety of exogenous melatonin for primary sleep disorders. A meta-analysis. J Gen Intern Med 2005;20: 1151–1158

136. Suresh Kumar PN, Andrade C, Bhakta SG, et al. Melatonin in schizophrenic outpatients with insomnia: a double-blind, placebo-controlled study. J Clin Psychiatry 2007;68: 237–241

137. Dolberg OT, Hirschmann S, Grunhaus L. Melatonin for the treatment of sleep disturbances in major depressive disorder. Am J Psychiatry. 1998;155:1119–1121

138. McIsaac WM. A biochemical concept of mental disease. Postgrad Med 1961;30:111–118

139. Altschule MD. Some effects of aqueous extracts of acetone-dried beef-pineal substance in chronic schizophrenia. N Engl J Med 1957;257:919–922

140. Eldred SH, Bell NW, Sherman LJ. A pilot study comparing the effects of pineal extract and a placebo in patients with chronic schizophrenia. N Engl J Med 1960;263: 1330–1335

141. Wurtman RJ, Axelrod J. Effect of chlorpromazine and other drugs on the disposition of circulating melatonin. Nature 1966;212:312

142. Kunz D, Schmitz S, Mahlberg R, et al. A new concept for melatonin deficit: on pineal calcification and melatonin excretion. Neuropsychopharmacology 1999;21:765–772

143. Sandyk R, Kay SR. Pineal calcification and its relationship to hallucinations in schizophrenia. Int J Neurosci 1992;64: 217–219

144. Hartley R, Smith JA. Formation in vitro of N-acetyl-3,4-dimethoxyphenethylamine by pineal hydroxyl-indole-O-methyl transferase. Biochem Pharmacol 1973;22: 2425–2428

145. Hempel K, Ullrich H, Philippu G. Quantitative investigation on the urinary excretion and metabolism of 3,4-dimethoxyphenylethylamine in schizophrenics and normal individuals. Biol Psychiatry 1982;17:49–59

146. Smith JA, Mee TJ, Padwick DJ, et al. Human post-mortem pineal enzyme activity. Clin Endocrinol (Oxf) 1981;14: 75–81

147. Garelis E, Gillin JC, Wyatt RJ, et al. Elevated blood serotonin concentrations in unmediated chronic schizophrenic patients: a preliminary study. Am J Psychiatry 1975;132: 184–186

148. DeLisi LE, Neckers LM, Weinberger DR, et al. Increased whole blood serotonin concentrations in chronic schizophrenic patients. Arch Gen Psychiatry 1981;38:647–650

149. Crow TJ. Molecular pathology of schizophrenia: more than one disease process? Br Med J 1980;280:66–68

150. Sandyk R, Kay SR. Pineal melatonin in schizophrenia: a review and hypothesis. Schizophr Bull 1990;16: 653–662

151. Tan DX, Manchester LC, Reiter RJ, et al. Cyclic 3-hydroxymelatonin: a melatonin metabolite generated as a result of hydroxyl radical scavenging. Biol Signals Recept 1999;8:70–74

152. Antolin I, Rodriguez C, Sainz RM, et al. Neurohormone melatonin prevents cell damage: effect on gene expression for antioxidant enzymes. FASEB J 1996;10:882–890

153. Gama CS, Salvador M, Andreazza AC, et al. Elevated serum superoxide dismutase and thiobarbituric acid reactive substances in schizophrenia: a study of patients treated with haloperidol or clozapine. Prog Neuropsychopharmacol Biol Psychiatry 2006;30:512–515

154. Zhang XY, Tan YL, Cao LY, et al. Antioxidant enzymes and lipid peroxidation in different forms of schizophrenia treated with typical and atypical antipsychotics. Schizophr Res 2006;81:291–300

155. Morera AL, Henry M, García-Hernández A, et al. Acute phase proteins as biological markers of negative psychopathology in paranoid schizophrenia. Actas Esp Psiquiatr 2007;35:249–252

156. Dietrich-Muszalska A, Olas B, Rabe-Jablonska J. Oxidative stress in blood platelets from schizophrenic patients. Platelets 2005;16:386–391

157. Yao JK, Leonard S, Reddy R. Altered glutathione redox state in schizophrenia. Dis Markers 2006;22:83–93

158. Altschule MD, Siegel EP, Goncz RM, et al. Effect of pineal extracts on blood glutathione level in psychotic patients. AMA Arch Neurol Psychiatry 1954;71:615–618

159. Alstchule MD, Siegel EP, Henneman DH. Blood glutathione level in mental disease before and after treatment. AMA Arch Neurol Psychiatry 1952;67:64–68

160. Hartley R, Smith JA. Inhibition of catecholamine oxidation by indoles. Biochem Pharmacol 1972;21:2007–2012

161. Zisapel N. Melatonin-dopamine interactions: from basic neurochemistry to a clinical setting. Cell Mol Neurobiol 2001;21:605–616

162. Laruelle M, Abi-Dargham A, Gil R, et al. Increased dopamine transmission in schizophrenia: relationship to illness phases. Biol Psychiatry 1999;46:56–72

163. Lerner AB, Case JD, Takahashi Y, et al. Isolation of melatonin, the pineal gland factor that lightens melanocytes. J Am Chem Soc 1958;80:2587

Chapter 36
Peripheral Biological Markers for Mood Disorders

Ghanshyam N. Pandey and Yogesh Dwivedi

Abstract The biogenic amine hypothesis of affective disorders is based primarily on the observation that antidepressant drugs may produce their therapeutic effects by interacting with the noradrenergic (NA) or the serotonergic (5HT) systems. For example, several antidepressants cause inhibition of the reuptake of norepinephrine (NE) or serotonin (5HT), thus increasing their levels in the synaptic cleft. Antidepressants could also decrease the metabolism of NE or 5HT by inhibiting enzymes, such as monoamine oxidase (MAO). However, direct evidence in support of the MAO hypothesis has been lacking. Initial studies to validate this hypothesis focused primarily on the determination of amines and their metabolites in the CSF, urine, and/or plasma. Further studies revealed that only are these abnormalities of biogenic amine levels and of their metabolites but that affective disorders may be associated with abnormalities of several receptors and the receptor-linked signaling systems, such as abnormalities of α- and β-adrenergic receptors and abnormalities of signaling systems, namely, the phosphoinositide (PI) and the adenylyl cyclase (AC) signaling systems. Some more recent studies indicate abnormalities of several transcription factors, such as cyclic-AMP response element binding (CREB) and of their target genes, such as brain-derived neurotrophin factor (BDNF), in affective disorders. These studies are primarily based on either postmortem brain tissue obtained from deceased subjects or on the use of peripheral tissues, such as platelets or white cells obtained from patients. Because of the inaccessibility of the living human brain, these peripheral sources have been used both for studying the pathophysiology of mood disorders (MD) and for their possible use as diagnostic and prognostic markers. Although the use of peripheral tissues, such as platelets and lymphocytes, as possible central markers seems questionable, several lines of evidence support their relationship to the central nervous system (CNS).

In this chapter, we describe the studies of several neurotransmitter receptors, namely, the α_1- and α_2-adrenergic receptors in platelets and the β_2-adrenergic receptors in neutrophils, as well as the $5HT_{2A}$ receptors in platelets of depressed and bipolar patients. We also describe the studies of the receptor-linked signaling system, namely, the phosphoinositide (PI) and the adenylyl cyclase (AC) signaling systems, in platelets and/or lymphocytes of these patients. In addition, we have also reviewed the studies of various components of these receptor-linked signaling pathways, namely, studies of protein kinase C (PKC) in depression and bipolar illness, studies of phospholipase C (PLC) and protein kinase A (PKA) studies of some of the G proteins in platelets and/or lymphocytes. Finally, we have discussed the studies of CREB and other transcription factors, as well as the role of some of their target genes, such as BDNF, which have been shown to play an important role in the pathophysiology of MD. We have also reviewed the studies on the role of cytokines in MD. In this review we also discuss the similarities or dissimilarities in the findings between peripheral tissues and postmortem brains and the usefulness of these receptors and signaling molecules as possible biological markers for the identification of patients with MD and also their possible role as prognostic markers. Finally, uses of some of these markers in identifying suicidal patients have also been discussed.

Keywords Adenylyl cyclase (AC) • α-adrenergic receptors • β-adrenergic receptors • BDNF • bipolar illness • CREB • CSF • cytokines • depression • G proteins • 5HIAA • $5HT_{2A}$ receptors • 5HT

G.N. Pandey and Y. Dwivedi
Department of Psychiatry, University of Illinois at Chicago, USA

M.S. Ritsner (ed.), *The Handbook of Neuropsychiatric Biomarkers, Endophenotypes and Genes,*
© Springer Science+Business Media B.V. 2009

transporter • MARCKS • MHPG • monoamines • mood disorders • norepinephrine • phosphoinositide (PI) • phospholipase c (PLC) • protein kinase A (PKA) • protein kinase C (PKC) • serotonin (5HT) • signal transduction • suicide

Abbreviations AC (signaling system): Adenylate cyclase; ACTH: **Adrenocorticotropic hormone;** AGP: α_1-Acid glycoprotein; BDNF: Brain-derived neurotrophic factor; BPRS: Brief Psychiatric Rating Scale; cAMP, cyclic AMP: Cyclic adenosine monophosphate; CREB: cAMP response element-binding; CNS: Central nervous system; CPR: C-reactive protein; CSF: Cerebrospinal fluid; CYP: Cyanopindolol; DAG: Ciacylglycerol; DHE: ^{3}H-dihydroergocryptine; DHA: ^{3}H-dihydroalprenolol; EP: Epinephrine; G_i: Inhibitory guanine nucleotide binding protein; GH: Growth hormone; G_s: Stimulatory guanine nucleotide binding protein; GTP or G Proteins: Guanosine 5-triphosphate binding proteins; HDRS: Hamilton Depression Rating Scale; HPA axis: Hypothalamic–pituitary–adrenal axis; 5HIAA: 5-Hydroxyindoleacetic acid; 5HT: 5-Hydroxytryptamine; $5HT_{2A}$: 5-Hydroxytryptamine receptor 2A; 5HTT: 5-Hydroxytryptamine transporter; IFN: Interferons; IL-1-rA: IL-1 receptor antagonist; IP_3: Inositol 1,4,5 triphosphate; ISOP: Isoproterenol; kD: Kilo Dalton; MAO: Monoamine oxidase; MARCKS: Myristoylated alanine-rich protein kinase C substrate; MHPG: 3-Methoxy-4-hydroxyphenylglycol; MCPP: Metachlorophenyl piperzine; MD: Mood disorders; MDD: Major depressive disorders; NA: Noradrenergic; NE: Norepinephrine; NGF: Nerve growth factor; NT: Neurotrophin; PAC: ^{3}H-para-amino-clonidine; PBP: Pediatric bipolar patients; PFC: Prefrontal cortex; PGE1: Prostaglandin El; PI: Phosphoinositide; PIP_2: Phosphatidylinositide-4,5 bisphosphate; PKA: Protein kinase A; PKC: Protein kinase C; PLC: Phospholipase C; RDC: Research diagnostic criteria; TfR: Transferrin receptors; TNF: Tumor necrosis factor

Introduction

Mood disorders (MD), which include major depressive disorders (MDD) and bipolar illness, are a major public health concern. It is estimated that about 17% of the population suffers from major depression at some point in their life.[1,2] Of these, about 4% of men and 8% of women have a lifetime risk of developing a severe form of depression. Also, about 1–2% of the population is at risk of developing a bipolar disorder. A mood disorders is a major risk factor for suicidal behavior and suicide, which in itself is a large public health problem. About 30,000 people die of suicide each year within the United States alone.[3–5] In teenagers, it is the second or third leading cause of death, after auto accidents and homicide.[6] Early identification of MD, as well as of suicidal behavior, is therefore very crucial, not only in the treatment of these disorders, but also in the prevention of suicide. Clinical, behavioral, and psychological risk factors are important in identifying these patients, but many times they may provide a false/positive or a false/negative diagnosis. A combination of biological factors with psychosocial factors might more accurately predict the diagnosis of MD and suicidal behavior. MDD, in particular, and MD, in general, are not homogeneous illnesses. There are many different types of MDD, based on several behavioral criteria, and nonresponse to a particular antidepressant medication may be related to the heterogeneity of these disorders. Biological factors may thus help in identifying the more homogenous forms and subgroups of these disorders, and there have been numerous attempts to determine potential biological markers, some of which seem promising.

Initial biological theories postulated the involvement of neurotransmitters norepinephrine (NE)[7,8] and serotonin (5-hydroxytryptamine [5HT])[9–11] in MD. The initial theories of MD have been followed by more specific biological theories of depression, such as the neurotrophin hypothesis of depression[12] or the signal transduction abnormality hypothesis of depression[13–20] The initial theories of MD were primarily based either on animal studies or on the mode of action of antidepressant drugs. For example, it was observed that treatment with reserpine, a drug that depletes catecholamines, produced symptoms similar to depression.[7,21,22] Effective antidepressant drugs, such as tricyclics and monoamine oxidase (MAO) inhibitors, presumably produce their therapeutic effects by either inhibiting the reuptake of biogenic amines, such as 5HT or NE, or by inhibiting their metabolism. Attempts have therefore been made to validate the initial biological theories of MD in humans, because if we know the specific biological abnormalities associated with these disorders, it may be possible to identify patients with these disorders at an early stage using biochemical or biological methods. However, because of the inaccessibility of the human brain, initial attempts to validate these hypotheses in humans were primarily focused on studies in peripheral tissues, such as cerebrospinal fluid (CSF), plasma, blood cells, or urine, and also on studies of neurotransmitters, such as 5HT, NE, and their metabolites (for review, see Maes and Meltzer,[23] Pandey and Dwivedi,[24,25] Kopin,[26] Schatzberg et al.[27]). More recent studies have therefore focused on the functional abnormalities of these neurotransmitters and biogenetic amines. 5HT and NE, for example, produce their biological or physiological effects by interacting with their own specific receptors. The interactions of these neurotransmitters with their receptors initiate a series of events known as a signaling cascade, or intracellular signaling. In general, these signal transduction pathways include the following components: receptors, guanine nucleotide binding protein (G protein), effectors, and second messengers. The second messengers trigger the signaling cascade further downstream and cause a cellular response. This downstream intracellular signaling coordinates the behavior of individual cells and results in various physiological responses. Later studies, therefore, focused on the responsiveness of these receptor-agonist interactions and on the components of these signaling pathways in peripheral cells or in postmortem brain tissues obtained from patients with MD. The downstream pathways include activation of phosphorylating enzymes, such as protein kinase C (PKC), of cyclic adenosine monophosphate response element binding protein (CREB), and of the transcription factors that finally cause the transcription of some target genes.

In this chapter we will focus on studies of the serotonergic and the adrenergic systems, primarily in peripheral cells and in plasma or CSF in patients with MD with or without suicidal behavior. This will be followed by studies of the various receptor subtypes for 5HT and NE that may be present in these peripheral cells. Some of these studies examined the roles of phosphorylating enzymes, such as PKC and protein kinase A (PKA), as well as of some transcription factors, e.g., CREB. Target genes, such as neurotrophins and neuropeptide Y (NPY), will also be discussed.

The intent of this chapter is not to exhaustively review all the studies pertaining to these two monoamines (5HT and NE), their metabolites, their receptors, and the related signal transduction mechanisms in all peripheral tissues but to instead focus on those studies

in peripheral tissues that found peripheral markers that show promise as potential biological markers for either MD or suicidal behavior.

Serotonergic Markers

5HT and 5HIAA in Patients with Mood Disorders and in Suicide

Initial studies of the serotonergic system in MD and suicide focused primarily on measuring the levels of 5HT and its major metabolite, known as 5-hydroxyindoleacetic acid (5HIAA), in the platelets, plasma, or CSF of patients with MD with or without suicidal behavior.[28–37] In the periphery, 5HT is primarily stored in the platelets, and 5HT in the blood is transported by an active 5HT uptake mechanism in the platelets. Because the level of 5HT in the plasma is very low, the level of 5HT has been studied primarily either in platelets, or in whole blood, which is representative of the content in the platelets, and to some extent in the CSF. Many studies showed decreased levels of 5HT in the platelets of depressed patients[38–42]; however, these results were not always confirmed or replicated.[43] The relationship between 5HT or other measures of 5HT, such as serotonin transporter (5HTT) sites, and $5HT_{2A}$ receptors in the platelets has also been studied in depression.[44,45] However, the major focus of serotonin studies has been 5HIAA in the CSF of patients with MD and in suicide.[28–35]

Although some studies suggest a decrease in 5HIAA,[28,29,46] others have reported no changes in 5HIAA[23,47,48] in the CSF of depressed patients compared with normal control subjects.[47] An important observation made by Asberg et al.[46] was that depressed patients with suicidal behavior had a significantly lower level of 5HIAA in the CSF compared with depressed patients without suicidal behavior or control subjects. This observation implicated an abnormal serotonergic system, primarily related to suicidal behavior. Asberg and colleagues[29,46] also made an observation with regard to the predictive power of a lower level of 5HIAA in the CSF for suicide risk. They measured 5HIAA levels in the CSF of 119 subjects who attempted subjects and found that the seven patients who subsequently completed suicide had levels of CSF 5HIAA that were lower compared with the levels of the nonsuicidal depressed patients. This occurred within 1 year of testing. Lower levels of CSF 5HIAA have been found in those subjects who were involved in a violent or impulsive suicide attempt rather than passive or premeditated attempts. This implicates abnormal serotonergic mechanisms and lower CSF 5HIAA in suicide as well as in impulsive aggressive behavior. To summarize the studies of 5HT and 5HIAA in depressed and suicidal patients, it appears that although abnormalities of 5HT and 5HIAA may exist in depressed patients, a lower level of CSF 5HIAA may be of greater value in predicting suicidal behavior in patients with MD.

5HT$_{2A}$ Receptors in Mood Disorders and Suicide

As discussed earlier, initial studies of serotonin function in MD and suicide were primarily carried out by investigating 5HIAA in CSF obtained from depressed patients with and without suicidal behavior. Since a decrease in 5HIAA was found in the CSF of depressed and/or suicidal patients compared with control subjects, this implied that the function and the turnover of 5HT may be altered in suicidal subjects. The alterations in 5HT turnover could result not only in an altered level of 5HT in the synaptic cleft but also in changes in the postsynaptic receptors for 5HT. Therefore, it is not surprising that 5HT receptor subtypes have been studied in the platelets and the postmortem brain of depressed and suicidal subjects.[49,50] However, determining 5HT receptors is a complex process because several types of 5HT receptors have been identified.[51–54] Serotonin receptors have been classified into several subtypes, known as 5HT$_1$, 5HT$_2$, 5HT$_3$, 5HT$_4$, 5HT$_5$, 5HT$_6$, and 5HT$_7$. These receptors have been further subdivided: 5HT$_1$ into 5HT$_{1A}$ and 5HT$_{1B}$; 5HT$_2$ into 5HT$_{2A}$, 5HT$_{2B}$, and 5HT$_{2C}$ receptors (for review, see Humphrey et al.[51]). Of these receptors only the 5HT$_{2A}$ receptor subtype has been unequivocally shown to be present in peripheral blood cells (specifically the platelets) of human subjects,[55–59] and therefore it is not surprising that platelet 5HT$_{2A}$ receptors have been extensively studied in patients with MD and in suicide.

5HT$_{2A}$ receptors in platelets have generally been studied by radioligand binding techniques using various ligands, such as [125]I- or ^{3}H-LSD and ^{3}H-ketanserin. The presence of 5HT$_{2A}$ receptors in human platelets has been reported by several investigators.[50,60–62]

The pharmacological profile of the 5HT$_{2A}$ receptors in platelets appears to be very similar to that observed in the brain.[55–57] For example, Elliott and Kent[58] studied rabbit platelet and brain 5HT$_{2A}$ receptors and showed that agonist and antagonist displacement appeared to be very similar in both tissues. Similar observations were made by Ostrowitzki et al.[63], who observed that the binding characteristics and the expression of 5HT$_{2A}$ receptors in the platelets and the brain of pigs appear to be very similar. Additional investigators have studied 5HT$_{2A}$ receptors in the platelets of depressed or bipolar patients with and/or without suicidal behavior.[25,61,64,65] The studies of platelet 5HT$_{2A}$ receptors in depression have been reviewed by Mendelson et al.[66]

In our initial studies of 5HT$_{2A}$ receptors in platelets, we studied 23 depressed patients during a drug-free period and 20 normal control subjects using ^{125}I-LSD as the ligand.[61] We found that the ^{125}I-LSD binding B$_{max}$ was higher in depressed patients compared with normal control subjects. However, when we divided the depressed patients into those with suicidal behavior and those with no suicidal behavior, we found that the 5HT$_{2A}$ receptors were even more elevated in the depressed patients with suicidal behavior compared with normal control subjects and nonsuicidal depressed patients. To examine if the increase in 5HT$_{2A}$ receptors in platelets of depressed suicidal patients is independent of diagnosis, we studied 5HT$_{2A}$ receptors in patients with different diagnoses, for example, depression, bipolar depression, mania, schizoaffective disorder, or schizophrenia.[50] We observed that the mean B$_{max}$ of the total group of suicidal patients was significantly higher than of nonsuicidal patients or normal control subjects. We also found that B$_{max}$ was significantly higher in 23 suicidal depressed patients, 5 suicidal patients with bipolar illness, 5 suicidal patients with schizophrenia, and 9 suicidal patients with schizoaffective disorders, compared with normal control subjects. This study thus suggested a significant increase in platelet 5HT$_{2A}$ receptors in suicidal patients independent of diagnosis. To examine the predictive value of platelet 5HT$_{2A}$ receptors, we determined the sensitivity and the specificity of platelet 5HT$_{2A}$ receptors. Since the B$_{max}$ of 5HT$_{2A}$ receptors for 90% of the comparison subjects was less than 70 fmol/mg protein, this value was arbitrarily assigned as a cut-off point for the upper limit of the normal. Hence, the specificity of normal comparison subjects was 90% according to this cut-off point. This cut-off point has a sensitivity of 55% in predicting suicidal behavior (that is, it correctly identifies 55% of patients as having suicidal behavior) with a specificity of 76%, that is 76% of patients whose B_{max} was below 70 fmol/mg protein were correctly identified as not having suicidal behavior.

Platelet $5HT_{2A}$ receptors have been studied by several investigators. At least nine investigators[60–62,64,65,67–70] found an increase in platelet $5HT_{2A}$ receptors in depressed patients. However, there are several studies that did not find differences in the $5HT_{2A}$ receptors between depressed patients and normal control subjects.[50,71–75] A relationship between an increase in $5HT_{2A}$ receptors and suicide was observed by Bigeon et al.,[76] who found that $5HT_{2A}$ receptors were significantly increased in suicidal men. Others have generally found an increase in $5HT_{2A}$ receptors associated with depression.[73,74,77]

There have been several studies of $5HT_{2A}$ receptors in the postmortem brain of suicide victims, half of them indicating no change[78–85] and half of them indicating an increase in $5HT_{2A}$ receptors[49,86–91] very similar to that in the platelets. Most of these studies of $5HT_{2A}$ receptors were carried out using radioligands such as ^{125}I-LSD or ^{3}H-ketanserin none of which were specific in labeling the $5HT_{2A}$ receptors. More recently we investigated $5HT_{2A}$ receptors in the postmortem brains of teenage suicide victims. We measured the LSD binding, as well as the protein and the mRNA expression, of $5HT_{2A}$ receptors in the prefrontal cortex (PFC) and hippocampus and found an increase, not only in the binding, but also in the protein and the mRNA expression of $5HT_{2A}$ receptors in teenage suicide subjects compared with normal control subjects.

The increase in $5HT_{2A}$ receptors in depression and/or suicide has also been substantiated by functional studies of $5HT_{2A}$ receptors. $5HT_{2A}$ receptors are linked to the phosphoinositide (PI) system. Mikuni et al.[92] determined the 5HT-stimulated IP_3 formation in platelets of depressed patients and found that $5HT_{2A}$ receptor responsiveness was increased, which suggests an increase in $5HT_{2A}$ receptor number, as well as in $5HT_{2A}$ receptor function in depression.

In summary, it appears that there is a similarity in the binding, as well as in other pharmacological characteristics, between $5HT_{2A}$ receptors in the platelets and in the human postmortem brain. In general, there appears to be an increase in $5HT_{2A}$ receptors in the platelets of suicidal patients compared with normal control subjects; whereas there may or may not be an increase in $5HT_{2A}$ receptors in depressed patients without suicidal

behavior. Increased $5HT_{2A}$ receptors in bipolar patients have also been reported by our lab but appear to be related more to suicidal behavior.[50] The specificity and predictive value of the $5HT_{2A}$ receptor is high in predicting suicidal behavior and it may be a useful marker for suicide rather than depression.

Neuroendocrine Studies of Serotonin Function

Neuroendocrine studies, often called the window to the brain, provide another useful method for studying central serotonergic function using peripheral sources. The procedure involves the administration of serotonergic probes, such as 5HT precursors like 5-hydroxltriptophan, or agents that cause the release of 5HT, or antagonist, such as fenfluramine, or a 5HT agonist. The 5HT agonist/antagonist, such as metachlorophenyl piperzine (mCPP), buspirone, or ipsapirone, acts on and stimulates 5HT receptor subtypes. Certain hormones, such as prolactin, ACTH, or cortisol that are released as a result of 5HT acting on the serotonergic system can then be measured. This strategy has been used extensively to study 5HT function in depressed patients. Using 5HT precursor studies Meltzer et al.[93–96] measured the 5HT-induced cortisol levels of 40 patients with MD compared with control subjects and found that the cortisol response was significantly increased in patients who had made a suicide attempt or who had a history of suicidal behavior.

Another method used to study central serotonergic function has been the measurement of fenfluramine-induced prolactin or cortisol release in patients with depression. The administration of fenfluramine causes a release of 5HT from presynaptic neurons and the blockade of 5HT uptake into the presynaptic terminal. The released 5HT acts on the postsynaptic 5HT receptors and causes the release of anterior pituitary hormones, such as prolactin and ACTH. ACTH in turn causes the release of cortisol from the adrenals. Using this method of study several investigators have observed a decrease in prolactin release following the administration of DL-fenfluramine in depressed patients, although some studies have not observed such differences. Caccaro et al.[97] found that the prolactin response to DL-fenfluramine was significantly decreased in those depressed patients and personality disorder patients who had a history of suicide attempts.

Similar results were reported by Malone et al.,[98] who also observed decreased prolactin response to fenfluramine in patients with a history of suicide attempts. O'Keane and Dinan[99] studied the fenfluramine-induced prolactin response in depressed patients and control subjects and found reduced prolactin and cortisol response in patients with depression compared with control subjects. More recently Cleare et al.[100] found a decreased prolactin and cortisol response in patients with depression compared with normal control subjects. However, they also observed that patients with a history of suicide attempts had a lower cortisol response than patients without a history of suicide attempts or normal control subjects. Both mCPP and ipsaperone have also been studied in children with depression with mixed results.

5HT Transporter Studies in Mood Disorders

Serotonin produces its physiological effects by interacting with postsynaptic receptors, such as $5HT_{2A}$, as discussed earlier. This primarily occurs after the release of 5HT from the presynaptic neurons. The 5HT in the synaptic cleft needs to achieve a certain level before it can produce postsynaptic effects by interacting with the postsynaptic receptors. The 5HT in the synaptic cleft is primarily inactivated by two processes: by its degradation by the enzyme MAO and by reuptake in presynaptic neurons. The reuptake in presynaptic neurons is achieved by 5HT transporter sites. The mechanism by which 5HT is transported, along with sodium ions, and the cloning of transporters have been reviewed by Owens and Nemeroff.[45] The antidepressant imipramine, and to a greater extent, 5HT reuptake inhibitors, such as paroxetine or fluoxetine, appear to produce their therapeutic effects through the inhibition of 5HT uptake in presynaptic neurons, thus increasing the level of 5HT in the synaptic cleft. It has been suggested that the abnormalities in 5HT function in depressed patients may be related to abnormalities in 5HT uptake. Platelets are a very suitable model for studying 5HT uptake and 5HT transport, since the properties of 5HT uptake in the platelets and in the brain appear to be very similar, as is the case with $5HT_{2A}$ receptor characteristics, described earlier.[101] Therefore, many investigators have studied platelet transporters using either ^{3}H-imipramine or ^{3}H-paroxetine as the radioligand for

measuring the transporters in the platelets of depressed and normal control subjects. It was first reported by Langer et al.[102] and Paul et al.[44] that the binding of ^{3}H-imipramine was significantly decreased in the platelets of drug-free depressed patients compared with normal control subjects. These results have been replicated by several studies. Many investigators have also used ^{3}H-paroxetine for studying 5HT uptake sites, which may be more specific in labeling transporter sites than ^{3}H-imipramine. Though many investigators find decreased 5HT uptake sites in the platelets of depressed patients, several investigators do not. Recently Ellis and Salmond[103] did a meta-analysis of the studies of platelet imipramine binding sites in depressed patients and found that more than 70 studies have compared imipramine binding sites between depressed patients and healthy normal control subjects. Based on the meta-analysis, it appears that there is a highly significant decrease in the B_{max} of 5HTT binding in platelets of depressed subject groups, and this decrease is even greater among those subjects who were free of medication for 4 weeks at the time of the study.

Noradrenergic Markers in Mood Disorders

Norepinephrine and MHPG

To validate the original NE hypothesis of MD, many studies primarily focused on measuring the levels of NE and 3-methoxy-4-hydroxyphenylglycol (MHPG), a metabolic of NE, in urine, plasma, and CSF obtained from patients with MDD and normal control subjects.[26,27,104–112] The major reason for measuring the NE and the MHPG in urine was that this represented the cumulative turnover of NE and MHPG over a 24h period. Also, some studies have suggested that a major portion of MHPG in urine is derived from the brain. Although the exact amount is unclear, some estimate that 20% of MHPG in urine is derived from the brain. Early studies found that the urinary levels of MHPG were lower in bipolar patients during the depressed phase than during periods of euthymia.[113,114] Subsequently, it was reported that MHPG levels were significantly lower in bipolar depressed patients compared with unipolar depressed patients or normal control subjects.[104,108,113] However, some studies have not observed differences between unipolar or bipolar patients and normal control subjects.[115,116]

Similarly plasma NE levels have been reported to be higher in depressed patients than in healthy controls.[106,109,112] The collaborative studies of Secunda et al.[116] found that compared with healthy control subjects depressed patients demonstrated significant elevations in urine levels of both NE and epinephrine (EP), as well as of their metabolites, with the exception of MHPG. Shiah et al.[117] found no differences in plasma MHPG between unipolar and bipolar patients. A collaborative study of the NIMH reported that depressed patients with suicidal behavior showed decreased urinary excretion of MHPG and a lower plasma level of MHPG compared with patients with no suicidal behavior.[116] Brown et al.,[118] observed that suicidal patients with personality disorders had significantly higher CSF levels of NE and higher levels of MHPG compared with non-suicidal patients with personality disorders.

Taken together, these studies indicate a decreased level of MHPG, primarily in bipolar depressed patients, and altered levels of NE and of its metabolites (other than MHPG) in the urine of depressed patients. However, the utility of these findings as markers for either depression or bipolar illness is questionable because of the inconsistency of research the findings.

Adrenergic Receptors in Mood Disorders

There are many lines of evidence that suggest that alterations in adrenergic receptor sensitivity may be associated with MD or schizophrenia. These evidences are derived from the observations that chronic treatment with antidepressants causes the downregulation of α- and β-adrenergic receptors, as well as of 5HT receptors, in the rat brain.[119-124] The effects of chronic treatment with antidepressants on β-adrenergic receptors appear to be very consistent, since almost all antidepressants, with the possible exception of fluoxetine,[125] cause a downregulation of β-adrenergic receptors. These observations suggest that alterations in β-adrenergic receptors may be associated with depressive illness.

Presynpatic α_2-adrenergic receptors play an important role in the release of NE.[126-129] The concentration of NE in the synaptic cleft may regulate the release of NE through α_2-adrenergic receptors by a feedback inhibition. Since α_2-adrenergic receptors regulate the concentrations of NE in the synaptic cleft, they also play an important role in regulating the sensitivity of postsynaptic β-adrenergic receptors.[130] These observations thus suggest that alterations in both α_2-adrenergic and β-adrenergic receptors may be associated with the pathophysiology of depression.

The major strategy used to study β-adrenergic receptors in depression involves the use of leukocytes or lymphocytes. α_2-Adrenergic receptors have been studied by several strategies. These include the determination of the number and the affinity of α_2-adrenergic receptors, as well as of the responsiveness of these receptors, in the platelets. Other strategies include neuroendocrine challenges. Insulin-or clonidine-induced growth hormone (GH) release, as well as yohimbine-induced changes in MHPG and cortisol, have also been used for studying α_2-adrenergic receptor sensitivity. We have studied α_2-adrenergic receptor function in MD and schizophrenia by binding techniques, as well as cyclic AMP studies, in platelets obtained from patients and determined the number and the affinity of the receptors. We studied the responsiveness of the receptors by determining β-adrenergic receptor-stimulated cyclic AMP formation. In this chapter, we describe the results from our studies along with a brief review of the results obtained by other investigators on adrenergic receptor function in MD and schizophrenia.

α_2-Adrenergic Receptor Studies in the Platelets of Patients and Normal Controls

The presence of α_2-adrenergic receptors in human platelets has been demonstrated by binding techniques, as well as cyclic AMP response studies.[131,132] Prostaglandin (PGE)-1-stimulated adenylyl cyclase (AC) activity in human platelets is inhibited by stimulating α_2-adrenergic receptors with NE or EP. α_2-Adrenergic receptors are also responsible for the inhibition of platelet aggregation. Binding studies, as well as cyclic AMP response studies, indicate that the α-adrenergic receptors in human platelets are of the α_2-type and not α_1 in nature.[133,134] Human platelets thus offer a very suitable peripheral source for studying α_2-adrenergic receptor function, which has been studied by (1) determining the number of α_2-adrenergic receptors present in human platelets using binding techniques, (2) determining the response of α_2-adrenergic receptors by measuring NE inhibition of PGE-1-stimulated AC, or (3) studying platelet aggregation.

Platelet α_2-Adrenergic Receptor Binding Studies in Depression

Several radiolabelled ligands have been used for α_2-adrenergic receptor binding studies in platelets obtained from depressed patients and normal controls. These ligands can be classified into three different types: antagonists, such as ^{3}H-yohimbine and ^{3}H-rauwolscine, as well as the ergot antagonist ^{3}H-dihydroergocryptine (DHE); partial agonists, which include ^{3}H-clonidine and ^{3}H-para-aminoclonidine (PAC); and full agonists, such as NE and EP.[132,135] Initial studies focused on antagonist binding using ^{3}H-yohimbine, ^{3}H-rauwolscine, or ^{3}H-DHE. Daiguji et al.[136] studied ^{3}H-yohimbine binding in 11 depressed patients and 9 normal controls. They did not find any significant differences either in B_{max} or the K_D of ^{3}H-yohimbine binding between depressed patients and normal controls. They also had a small number of bipolar depressed patients and did not find any significant differences between unipolar and bipolar depressed patients. Stahl et al.[137] studied 16 drug-free patients with MDD and 29 normal controls using ^{3}H-yohimbine as the ligand, and they too did not observe any significant differences between drug-free depressed patients and normal control subjects. They also did not observe any correlation between the Hamilton Depression Rating Scale (HDRS) and either B_{max} or K_D for ^{3}H-yohimbine binding. ^{3}H-yohimbine binding in depressed patients has also been studied by many other investigators.[138–142] None of them found any significant differences in ^{3}H-yohimbine binding indices between depressed patients and normal controls. Wolfe et al.[142] did not observe any significant differences in the B_{max} of ^{3}H-yohimbine binding between depressed patients and normal controls; however, they observed that the K_D of ^{3}H-yohimbine binding was significantly lower in depressed patients compared with normal control subjects.

Pimoule et al.[143] used ^{3}H-rauwolscine for binding studies but did not observe any significant differences in platelet ^{3}H-rauwolscine binding (B_{max} or K_D) between depressed patients and normal controls.

Several investigators studied α_2-adrenergic receptors in platelets from depressed patients using ^{3}H-DHE.[81,135,144]

Whereas Kafka et al.[81] and Healy et al.[144] observed increased ^{3}H-DHE binding in depressed patients compared with normal controls, Wood and Coppen[145] observed decreased ^{3}H-DHE binding sites in depressed patients compared with normal controls. It should, however, be noted that Kafka et al.[81] and Siever et al.[146] determined ^{3}H-DHE binding at one concentration, whereas Healy et al.[144] and Wood and Coppen[145] determined the B_{max} and the K_D. B_{max} was observed to be increased in depressed patients by Healy et al.[144] but was observed to be decreased in depressed patients by Wood and Coppen.[145]

Platelet α_2-adrenergic receptors in depression have also been studied by using either full or partial agonists for binding. α_2-Adrenergic receptor binding using the partial agonist clonidine was studied by Garcia-Sevilla and colleagues,[147–150] Doyle et al.,[151] Georgotas et al.,[152] and by our group.[153] Castens et al.[154] and Piletz and Halaris[155] studied α_2-adrenergic receptors by using the partial agonist ^{3}H-PAC.

We studied ^{3}H-clonidine binding in platelets obtained from depressed, schizoaffective, bipolar, or schizophrenia patients admitted to the research wards of the Illinois State Psychiatric Institute.[153] The patients went through a drug-free washout period of up to 2 weeks, at which time blood was drawn for determining platelet α_2-adrenergic receptors. Normal controls were nonhospitalized volunteers who were free of any major medical disorders or psychiatric disorders and were not on active medications at the time of the study. ^{3}H-clonidine binding was determined in the platelets of patients and normal controls and B_{max} and K_D of ^{3}H-clonidine binding was correlated with clinical parameters.[153] We observed that B_{max} values of clonidine binding in depressed, schizoaffective, or schizophrenia patients were significantly higher compared with normal control subjects. However, there were no significant differences between bipolar patients and normal control subjects in the B_{max}. To examine if the ^{3}H-clonidine binding values obtained during the drug-free period in the patient population are related to either the baseline HDRS or the Brief Psychiatric Rating Scale (BPRS) scores or to the HDRS or the BPRS change scores after 4–6 weeks of psychoactive drug treatment, we determined the correlations between the B_{max} values and the HDRS and the BPRS scores. We observed a correlation between the baseline ^{3}H-clonidine B_{max} and the BPRS change scores in the schizophrenia patients, which suggests that the baseline α_2-adrenergic receptor number could be a predictor of clinical response in schizophrenia patients. No correlations of B_{max} with the HDRS or the BPRS change scores or the baseline scores were observed in the other diagnostic groups.

[3]H-Clonidine binding in the platelets of depressed patients was also studied by Garcia-Sevilla et al.,[149] who observed are increased B_{max} in the platelets of depressed patients compared with normal controls, which is similar to our results.

Carstens et al.[154] and Piletz and Halaris[155] studied platelet α_2-adrenergic receptors by using the partial agonist ^{3}H-PAC. Whereas Piletz and Halaris[155] observed a significantly higher number of super-high affinity ^{3}H-PAC binding sites in the platelets of depressed patients compared with normal control subjects, Carstens et al.[154] observed a significant decrease in the ^{3}H-PAC binding in depressed patients compared with normal controls.[156] determined ^{3}H-EP (full agonist) binding in platelets obtained from depressed patients and normal controls and observed that the B_{max} was significantly higher in depressed patients than in normal controls. Again, they did not observe any significant differences in the K_D between depressed patients and normal controls.

Although the results of platelet α_2-adrenergic receptor binding studies may appear to be conflicting, in fact, they are quite consistent. The studies that used antagonists for binding studies in platelets, for example, yohimbine or rauwolscine, observed no significant differences either in the B_{max} or the K_D between depressed patients and normal control subjects. Most of the studies that used either partial agonists, for example, clonidine or PAC, or a full agonist, for example, EP, found increased B_{max} in the platelets of depressed patients compared with normal controls.[148,150,153,155,156] Also, Sacchetti et al.[157] studied ^{3}H-clonidine binding in the platelets of depressed patients, and although they did not determine ^{3}H-clonidine binding in the normal controls, the values of B_{max} obtained by them are very similar to those reported by Garcia-Sevilla et al.[149] and Pandey et al.[153] The only discrepancy in the results obtained with agonist binding is that reported by Carstens et al.,[154] who observed a significant decrease in the B_{max} of ^{3}H-PAC binding between depressed patients and normal controls. There may be several reasons for this discrepancy. ^{3}H-clonidine has been shown to label at least two different subpopulations of α_2-adrenergic receptors: a high affinity state, coupled with cell functions, and a low affinity state.[147,148] In those studies that used partial agonists or full agonists for binding, the

differences between depressed patients and normal controls were observed in the high affinity binding sites. It appears that the differences between the results of other investigators and Carstens et al.[154] may be attributed to differences in the methodology and/or the in the patient populations. The reason for the difference in the results obtained by Siever et al.[146] and Healy et al.,[158] who reported increased ^{3}H-DHE binding in depressed patients compared with normal controls, and those of Wood and Coppen,[145] who observed decreased ^{3}H-DHE binding in depressed patients is not clear. One of the factors may be that Siever et al.[146] measured ^{3}H-DHE binding at one concentration and not the B_{max}. Also, the differences observed in the results obtained by using antagonist binding versus those obtained with agonist or partial agonist binding may be related to the subpopulation of α_2-adrenergic receptors by these ligands.

In summary, the described observations suggest that depressive illness may be associated with an increased number of high affinity α_2-adrenergic receptor binding sites in the platelets, and this can be identified by binding studies using a partial or a full agonist as the ligand.

α_2-Adrenergic Receptor Responsiveness in Depression and Schizophrenia

α_2-adrenergic receptors in human platelets have been shown to be linked to the AC system. Stimulation of platelet α_2-adrenergic receptors by agonists, such as NE or EP, causes the inhibition of PGE-1-stimulation of AC in platelet membranes and thus of PGE-l-stimulated cyclic AMP accumulation in intact platelets. PGE1 stimulates AC activity in human platelets by promoting the interaction of its receptors with the stimulatory protein (G_S) and thus activating the catalytic unit of AC, which results in increased formation of cyclic AMP. Occupancy of α_2-adrenergic receptors in human platelets, on the other hand, promotes the interaction of the inhibitory protein (G_i) with the catalytic unit of AC, inhibiting the formation of cyclic AMP. Therefore, the inhibition of cyclic AMP formation by prostaglandins has been used as an index of α_2-adrenergic receptor responsiveness.

This strategy has therefore been employed to examine the biochemical responsiveness of α_2-adrenergic receptors in platelets obtained from psychiatric patients and normal controls. Wang et al.[159] determined the PGE-1-stimulated cyclic AMP accumulation and NE inhibition of PGE-1-stimulated cyclic AMP accumulation in intact platelets obtained from depressed patients and normal controls. They did not observe any significant differences in either of these measures between depressed patients and normal controls, and concluded that α_2-adrenergic receptor responsiveness was not altered in depressed patients. Murphy et al.[160] also observed no significant differences between depressed patients and normal controls in either PGE-1-stimulation or NE inhibition of PGE-1-stimulated cyclic AMP accumulation in intact platelets. However, Siever et al.[146] observed decreased PGE-1-stimulated cyclic AMP accumulation in platelets obtained from 23 depressed patients compared with 53 normal control subjects. They also observed that the inhibition of this response by NE was significantly reduced in depressed patients compared with normal control subjects. Kanof et al.[161] determined PGE-1-stimulated cyclic AMP levels in the platelets of depressed OR schizophrenia patients and normal controls. They observed that the PGE-1-stimulated cyclic AMP response in the platelets of depressed or schizophrenic patients was significantly lower. However, they did not determine the NE inhibition of PGE-1-stimulated cyclic AMP response. More recently, Kanof et al.[162] reported that platelet α_2-receptor sensitivity, as measured by the NE inhibition of PGE-1-stimulated cAMP formation, does not appear to be altered in schizophrenia or depressed patients compared with normal controls.

Platelet α_2-adrenergic receptor responsiveness has also been studied in schizophrenia patients by several other investigators. Initially, Kafka and Van Kammen[163] observed that the PGE-1-stimulated cyclic AMP production in intact platelets from male, but not female, schizophrenia patients was lower than in normal control subjects. They also observed that NE inhibited PGE-1-stimulated cyclic AMP accumulation was lower in male schizophrenia patients compared with male control subjects. Rotrosen et al.[164] observed decreased PGE-1-stimulated cyclic AMP accumulation in schizophrenia patients compared with normal control subjects. Garver et al.[165] studied PGE-1-stimulated AC in platelet membranes obtained from schizophrenia patients and normal controls. They observed that PGE-1-stimulated cyclic AMP accumulation in 20 schizophrenia patients was significantly lower compared with normal controls. They, however, did not observe any significant differences

in sodium fluoride-stimulated AC in platelet membranes, and hence concluded that the receptor mediation of the AC but not the AC itself was defective in schizophrenia patients.

The results thus obtained in studies of the responsiveness of α_2-adrenergic receptors in platelets of depressive or schizophrenia patients again appears to be mixed. Whereas there appears to be a consistent observation that PGE-1-stimulated cyclic AMP accumulation is decreased in schizophrenia patients, there does not appear to be a consistent finding of decreased PGE-1-stimulated cyclic AMP response in depressed patients. The reasons for this are not clear.

α_2-Adrenergic Receptor Studies by Provocative Strategies

α_2-Adrenergic receptor sensitivity and responsiveness in depression and schizophrenia have been studied using provocative challenges. These include determination of GH leads after clonidine administration and determination of plasma levels of MHPG and cortisol after administration of the α_2-adrenergic antagonist yohimbine. Administration of clonidine, a partial α_2-adrenergic receptor agonist, causes an increase in plasma GH levels in humans. This increase can be blocked by the administration of an α_2-adrenergic antagonist. Using these strategies, secretion of human GH in depression or schizophrenia has been studied by the administration of clonidine, and it was reported by Matussek et al.[166], Siever et al.[167] and Checkley et al.[168] that the plasma GH response to clonidine administration in depressed patients is blunted compared with that in normal control subjects, thus indicating decreased α_2-adrenergic receptor responsiveness in depression.

Changes in plasma MHPG, a metabolite of NE, after administration of the α_2-adrenergic agonist clonidine, or the antagonist yohimbine, have also been used as a measure of α_2-adrenergic receptor sensitivity. Henninger et al.[169] studied clonidine-induced MHPG levels in the plasma of depressed patients and normal controls; however, they did not observe any significant differences between the two groups. They also studied plasma MHPG, blood pressure, pulse, subjective mood, and somatic symptoms before and during the administration of yohimbine or placebo. Although they observed that the administration of yohimbine caused a significant decrease in plasma MHPG after 120, 180, and 240 min of yohimbine administration, these changes were not significantly different in 45 depressed patients compared with 20 normal control subjects. These studies thus indicated that there were no significant differences in α_2-adrenergic receptor response between depressed patients and normal controls.

Price et al.[170] determined plasma cortisol levels after the administration of yohimbine. They observed that the mean baseline cortisol level was significantly higher in depressed patients compared with normal controls, which suggests a relative subsensitivity of postsynaptic α_2-adrenergic receptors in depression. They also observed that yohimbine caused a greater cortisol increase in depressed patients compared with normal controls, which suggests decreased α_2-adrenergic receptor responsiveness.

The results of neuroendocrine studies, although encouraging, are again mixed. Studies using clonidine-induced GH release clearly indicate subsensitive postsynaptic α_2-adrenergic receptors. Similar results are obtained by studying yohimbine-lnduced cortisol release, which also indicates subsensitive postsynaptic α_2-adrenergic receptors. However, yohimbine- or clonidine-induced changes in plasma MHPG do not show any differences in α_2-adrenergic receptor response between depressed patients and normal control subjects. Whereas neuroendocrine studies indicate subsensitive α_2-adrenergic receptors in depression, the studies of platelet α_2-adrenergic receptors clearly indicate an increased number in the platelets. Thus, there appears to be dissociation between the neuroendocrine responses of α_2-adrenergic receptors and platelet α_2-adrenergic receptor density. Such dissociation between α_2-adrenergic receptors responses has been observed in several other studies.[169,171]

β-Adrenergic Receptor Studies in the Leukocytes of Patients with Mood Disorders

The observation by several groups of investigators that chronic treatment with almost all antidepressant drugs causes the downregulation of β_2-adrenergic receptors in the rat brain[122,172–174] led to the suggestion that alterations in β_2-adrenergic receptor sensitivity may be associated with the pathophysiology of depressive illness.

Furthermore, neuronal signal transduction is a function of both presynaptic receptors and of the transmitter levels in the synaptic cleft. Postsynaptic receptor sensitivity, which is regulated by the concentrations of amine levels at the synaptic cleft, may thus play an important role in neuronal function. Whereas platelets have been used as a peripheral model for studying α_2-adrenergic receptor function, leukocytes offer an appropriate peripheral source for studying β_2-adrenergic receptor sensitivity.[175] The presence of β_2-adrenergic receptors in human leukocytes has been demonstrated both by cyclic AMP response studies and by ligand binding techniques, and it has been shown that the β-adrenergic receptors in human leukocytes are primarily of the β_2-adrenergic type, as opposed to the predominant β_1 type in the brain. Leukocytes have often been used as a model for studying β_2-adrenergic receptor sensitivity in several pathological conditions, including asthma. Therefore, β_2-adrenergic receptor function and density have been studied in leukocytes of depressed or schizophrenia patients. As mentioned earlier, both cyclic AMP-AC, as well as ligand binding techniques, have been used to study the role of β-adrenergic receptors in psychiatric disorders.

β-Adrenergic Receptor-Mediated AC and Cyclic AMP Studies in Leukocytes of Depressed Patients

Initial studies of β-adrenergic receptors in depression were carried out by determining agonist-stimulated cyclic AMP formation in leukocytes or lymphocytes of depressed patients and normal controls by our group[15,176] and by other investigators.[13,14,17,18,144,177,178] We studied β-adrenergic receptor function in patients with MD or schizophrenia by determining NE- or isoproterenol (ISOP)-stimulated ^{3}H-cyclic AMP accumulation in the intact leukocytes.[15] The subjects of the study were patients admitted to the research wards of the Illinois State Psychiatric Institute who underwent a 2–3-week drug-free washout period, after which blood was drawn from these patients for the studies. Patients were diagnosed according to the Research Diagnostic Criteria (RDC) of Spitzer et al.[179] Normal controls were nonhospitalized volunteers who were free of any major medical or psychiatric disorders and were not on any active medication. We observed that the

NE-stimulated ^{3}H-cyclic AMP accumulation in the leukocytes of depressed patients was significantly lower compared with normal control subjects. NE-stimulated cyclic AMP accumulation in the leukocytes of schizophrenia patients was also significantly lower compared with normal control subjects. However, when we compared the isoproterenol-stimulated cyclic AMP accumulation between patients and controls, we observed that although it was significantly lower in depressed patients compared with normal control subjects, no significant differences were observed in the isoproterenol-stimulated cyclic AMP accumulation between schizophrenia patients and normal controls. When we compared PGE-1-stimulated cyclic AMP accumulation or the basal cyclic AMP accumulation in the depressed patients, schizophrenia patients, and normal controls, we did not observe any significant differences among the groups or between patients and normal controls. This suggested that the differences in the NE and/or isoproterenol-stimulated cyclic AMP accumulation in depressed patients compared with normal controls were primarily related to decreased responsiveness of the β_2-adrenergic receptors.

β_2-Adrenergic receptor responsiveness using AC-cyclic AMP response studies has been examined by several other investigators. Extein et al.[17] reported decreased isoproterenol-stimulated cyclic AMP accumulation in the lymphocytes of depressed patients compared with normal control subjects. Mann and associates[18] studied isoproterenol-stimulated cyclic AMP accumulation in a group of depressed patients, and they observed significantly lower isoproterenol-stimulated cyclic AMP levels in the lymphocytes of patients with endogenous depression compared with normal control subjects. When they divided the depressed patients according to psychomotor manifestations, they observed that the patients with psychomotor agitation had significantly lower isoproterenol-stimulated cyclic AMP accumulation compared with normal controls; the patients with motor retardation; however, had cyclic AMP accumulation that was not significantly different from that of normal control subjects. Ebstein et al.[14,178] determined isoproterenol-, forskolin- or PGE-1-stimulated cyclic AMP accumulation in the lymphocytes of depressed patients and normal controls. They observed significantly decreased isoproterenol-stimulated cyclic AMP levels in the lymphocytes of depressed patients compared with normal controls. Forskolin-stimulated cyclic AMP accumulation was nonsignificantly decreased

in the lymphocytes of depressed patients. They also observed significantly lower isoproterenol-stimulated cyclic AMP accumulation in the depressed patients who did not respond to treatment compared with those who did respond. The PGE-1-stimulated cyclic AMP accumulation was nonsignificantly higher in depressed patients. Whereas these studies reported decreased β-adrenergic receptor-mediated cyclic AMP responsiveness in depressed patients, Klysner et al.[177] observed increased β_2-adrenergic receptor responsiveness in bipolar patients compared with euthymic patients treated with antidepressants. Kanof et al.[180] determined isoproterenol-stimulated cyclic AMP accumulation in depressed patients, and although they did not observe significant differences, the isoproterenol-stimulated cyclic AMP accumulation in depressed patients tended to be lower than that of normal control subjects. These studies, in general, thus indicate a decreased responsiveness of β_2-adrenergic receptors in depression.

β_2-Adrenergic Receptor Binding Studies in Depression

Several radiolabelled ligands have been used for studying β_2-adrenergic receptors in human leukocytes/lymphocytes. Williams et al.[181] studied β_2-adrenergic receptors using ^{3}H-dihydroalprenolol (DHA) and observed that the kinetic and the pharmacologic characteristics of the binding of ^{3}H-DHA to the lymphocyte membrane were similar to those expected of β_2-adrenergic receptors. Since studies of β_2-adrenergic receptors using ^{3}H-DHA required a relatively large amount of blood for complete saturation studies and for determining the B_{max} and K_D, other investigators have used ^{125}I-cyanopindolol (CYP) as a binding ligand for determining β_2-adrenergic receptors. Brodde et al.[182] and O'Hara and Brodde[183] studied the kinetic characteristics of β_2-adrenergic receptors in human lymphocytes using ^{125}I-CYP as the radioligand.

We determined ^{125}I-CYP binding in lymphocytes of normal controls and of depressed patients and schizophrenia patients during their drug-free washout period and observed that the B_{max} of ^{125}I-CYP binding was significantly lower in depressed patients compared with normal controls.[15] In a subsequent study, we determined ^{125}I-CYP binding in 51 normal controls, 27 depressed patients, and 30 schizophrenia patients, and we observed that the B_{max} (fmol/mg protein) in depressed patients (58.05 ± 20 SD) was significantly lower compared with the B_{max} of normal control subjects (75.18 ± 30 SD). However, the B_{max} in schizophrenia patients (63.02 ± 36.6) was not significantly different from that of normal controls. We also observed that the K_D was not significantly different between any of the groups.

β_2-Adrenergic receptor binding in leukocytes/lymphocytes of depressed patients has been studied by several other investigators. Extein et al.[17] determined ³H-DHA binding in the lymphocytes of depressed patients and normal controls, and although they did not determine B_{max} or K_D, they observed a significant decrease in ³H-DHA specific binding in depressed patients. Carstens et al.,[184] who also determined ³H-DHA binding, observed a significant decrease in B_{max} in the lymphocytes of depressed patients compared with normal controls. They also observed positive correlations between the lymphocyte β-adrenergic receptor B_{max} and the Beck Depression Inventory and the HDR scores. Magliozzi et al.[185] determined ¹²⁵I-CYP binding to the lymphocyte membrane of depressed patients and normal controls, and observed a significantly reduced B_{max} in depressed patients compared with normal controls. Mann et al.,[18] who determined ¹²⁵I-CYP binding in intact lymphocytes, did not observe any significant differences in the B_{max} between depressed patients and normal controls. Healy et al.,[158] on the other hand, observed a significant increase in the B_{max} of ³H-DHA binding in depressed patients compared with normal controls. Wright et al.[186] observed a significant decrease in ³H-DHA binding (B_{max}) in the cell lines obtained from family members of bipolar patients compared with normal controls.

The studies of β_2-adrenergic receptors, either by determining the responsiveness of β_2-adrenergic receptors by the cyclic AMP-AC method or by determining the β_2-adrenergic receptor number and affinity using binding techniques, thus indicate decreased β_2-adrenergic receptor number and function in depressive illness. In this context, the findings of decreased β_2-adrenergic receptor-stimulated cyclic AMP formation appear to be more consistent. The discrepancies in the results obtained by the various binding studies are probably related to several factors, such as method of tissue preparation, type of ligand used, and/or patient population. Whereas Mann et al.[18] studied ¹²⁵I-CYP binding in intact lymphocytes, we[15] and Carstens et al.[184] and Magliozzi et al.[185]

studied β_2-adrenergic receptor binding in the leukocyte/lymphocyte membrane of patients. The finding of decreased β_2-adrenergic receptor function and number in patients with depressive illness suggests its potential usefulness as a biological marker for depression.

Signal Transduction Pathways and Their Components

Studies of Signal Transduction Pathways and Their Components in Peripheral Cells: Focus on PI and AC Pathways in Mood Disorders

As discussed earlier, abnormalities of $5HT_{2A}$ receptors, as well as of β_2- and α_1-adrenergic receptors have been observed in leukocytes and platelets of patients with MD. The physiological significance of these altered receptors lies in their ability to activate the signal transduction pathways to which these receptors are linked and eventually to cause a physiological or behavioral response. The two signaling pathways related to MD, which have been studied not only in peripheral cells but also in postmortem brain samples and by animal studies, are the PI and the AC signaling pathways. A major component of these pathways is a receptor, which regulates intracellular mechanisms by interacting with transducing molecules known as G proteins. Activation of these receptors by an agonist induces a conformational change in the receptor that allows it to interact with G proteins. The receptor-G protein interaction results in the formation of second messengers. The second messengers initiate responses further downstream, resulting in the activation of transcription factors and the transcription of target genes. In the PI signaling system, agonist stimulation of G protein-coupled receptors causes the hydrolysis of the substrate phosphatidylinositol 4,5- biphosphate (PIP_2) by the effector phospholipase C (PLC), resulting in the formation of two second messengers, inositol 1,4,5-triphosphate (IP_3) and diacylglycerol (DAG).[187–189] IP_3 stimulates the release of intracellular calcium from the endoplasmic reticulum, and DAG stimulates the enzyme PKC.[190] PKC is one of the major intracellular mediators of signals generated by the stimulation of cell surface receptors. PKC then interacts with and activates many transcription factors, including CREB, which is involved in the transcription of several target genes, such as brain derived neurotrophic factor (BDNF) and neuropeptide Y. Several studies have indicated abnormalities of the PI signaling system in the postmortem brain of patients with unipolar or bipolar disorders. As mentioned before, it was reported that the hydrolysis of PIP_2 by several agonists is altered in the postmortem brain of depressed suicide victims or bipolar patients.[191–193] PI signal transduction determined by 5HT-, thrombin-, or sodium fluoride (NaF)-stimulated IP formation has also been reported to be increased in platelets of depressed patients.[92,194–196] In this study,[196] it was found that the thrombin-stimulated IP_3 formation was significantly increased in the platelets of depressed patients. Other studies have examined various components of the PI signaling system, such as G protein, PLC, IP_3, and PKC in platelets or leukocytes of patients with MD (for review, see Bezchlibnyk and Young[197]).

G Proteins

G proteins consist of three subunits, known as α, β, and γ subunits, which are tightly bound to one another. G proteins are important for producing signals across the membrane. Abnormalities in G proteins have been reported in postmortem brains of bipolar subjects and suicide subjects (for review, see Bezchlibnyk and Young[197]). Young et al.[19] found an increase in $G_{\alpha S}$ levels in the leukocytes of bipolar patients and a decrease in G_{q11} in the platelets of bipolar patients. On the other hand Avissar et al.[198] found an increase in agonist-induced Gpp (NH)p binding, as well as in $G_{\alpha S}$ and $G_{\alpha i}$ levels in mania and a decrease in depression. Mitchell et al.[199] found an increase in $G_{\alpha S}$ levels in the platelets of bipolar patients resulting from treatment. In general, studies of G proteins have found alterations in G proteins in platelets or leukocytes of patients with bipolar disorders or depression, although these studies are limited.

PLC, PKC and MARCKS in Platelets of Patients with Mood Disorders

PLC and PKC are two crucial components of the PI signaling system. Both PLC and PKC have been shown to be involved in a variety of physiological functions, such as the synthesis, release, and reuptake of neurotransmitters, neuronal development, transcription, long-term potentiation (LTP), and behavioral responses.[190] Although the role of PLC in MD has not been fully investigated, some studies have shown abnormalities of PLC in depression and suicide.[192,193,200] Several studies suggest the involvement of PKC in MD, particularly in bipolar disorders.[201–205] This is based on the observation that lithium significantly affects PKC in a number of cell systems, including the central nervous system (CNS).[203,204] Therefore, it has been suggested by some investigators that PKC may be an important target for the action of mood stabilizers, such as lithium and valproate.[206–208] Our lab studied the activities of PLC and PKC, as well as the protein expression of several of the PLC and PKC isozymes, in platelets obtained from bipolar or MDD patients during a drug-free period.[209] We observed significant decreases in PI-PLC activity and in the expression of the PLC δ_1 isozyme in the membrane and the cytosol fractions of platelets obtained from bipolar patients, although no such decrease was observed in PLC β_1 or PLC γ_1. Furthermore, no significant differences were observed either in PLC activity or in the protein expression of any of the PLC isozymes, i.e., PLC β_1, PLC γ_1, or PLC δ_1, in the platelets of patients with major MD. Ebstein et al.[178] did find a decrease in PLC activity in lithium-treated euthymic patients, but they did not study drug-free bipolar patients.

PKC, on the other hand, has been studied by several investigators in the platelets of bipolar or MDD patients. In our recent study,[209] we found that the PKC activity was significantly decreased in the platelets of bipolar but not MDD patients. This decrease was associated with a specific decrease in the protein expression of PKC α, PKC βI, and PKC βII isozymes in the membrane and the cytosol fractions of platelets from bipolar but not MDD patients. However, there was no difference in the protein expression of PKC δ, either in the membrane or the cytosol fraction among bipolar, unipolar, and normal control subjects.

Although activation of PKC causes a variety of physiological responses, the steps involved between the activation of PKC and the subsequent physiological responses are not known. Activation of PKC causes the phosphorylation of a number of membrane-associated protein substrates, such as myristoylated alanine-rich C-kinase substrate (MARCKS), which is a 30 kilo

Dalton (kD) acidic and heat stable protein that is highly expressed in the developing and the adult brain.[210,211] MARCKS contains PKC phosphorylation sites that are in close proximity to the effector domain. It is anchored to membranes by an amino terminal myristic acid moiety and by the electrostatic interaction of the effector domain with the acidic membrane phospholipids phosphatidyl serine and PIP_2. There has been some research that suggests that MARCKS may be involved in the pathophysiology of bipolar illness and possibly suicide.[209,212–216] Our lab has studied MARCKS in the platelets of bipolar and unipolar patients.[209] Because we found decreases in PKC activity and in some PKC isozymes in the platelets of bipolar patients, we examined whether these decreases were associated with changes in MARCKS expression. We found that MARCKS expression levels were significantly increased in the cytosol and the membrane fractions of platelets from bipolar but not from MDD patients compared with normal control subjects. This suggests that an upregulation of MARCKS may be related to decreases in PKC activity and in the protein expression of some of its isozymes.

AC Signaling System in Mood Disorders: Role of α- and β-Adrenergic Receptors

To examine the functional consequences of altered α- and β-adrenergic receptors, the responsiveness of the AC signaling system has been studied in patients with MD. Whereas the $5HT_{2A}$ receptor is linked to the PI signaling system, both α_2- and β-adrenergic receptors are linked to the AC system in either a negative or a positive manner.

PKA in Mood Disorders

The major target of the second messenger cAMP, formed after the activation of the enzyme AC by a receptor agonist, such as isoproterenol, is the complex protein known as PKA. To examine further downstream abnormalities in the AC signaling pathway, the role of PKA has been studied in the postmortem brain of depressed suicide victims and in platelets or fibroblasts obtained from patients with MD.[217–225] PKA is a complex protein composed of two distinct regulatory (R) and two catalytic (C) subunits that form a tetrameric holoenzyme. On the basis of elution patterns on DEAE chromatography, two major forms of PKA have been identified and they are known as Type-1 and Type-2. Both types consist of subunits, known as RIα, RIβ, RIIα, RIIβ, as well as Cα, Cβ, and Cγ. Our studies of postmortem brain samples of suicide victims revealed abnormalities not only of PKA activity but also of regulatory PKA-RIIβ and catalytic Cβ subunits in the postmortem brain of adult suicide victims. Our studies also revealed a decrease in PKA activity associated with a specific decrease in RIα and RIβ in the postmortem brain of teenage suicide victims. Shelton et al.[220] and Manier et al.[221] found a significant decrease in β-adrenergic receptor-stimulated PKA activity in the presence of cAMP in the fibroblasts of depressed patients when compared with normal controls. It was also reported by Shelton et al.[220] that β-adrenergic receptor-stimulated PKA activity is decreased in melancholic depressed patients but not in other depressive subtypes. Akin et al.[20] found decreased PKA activity, which was associated with decreased expression of PKA RIIα, Cα, and Cβ subunits, in the fibroblasts of melancholic depressed subjects compared with nonmelancholic or normal control subjects. These studies suggest a decrease in PKC activity in the peripheral cells of depressed patients, with possible abnormalities of specific subunits of PKA. Manier et al.[222] also studied PKA in the fibroblasts of depressed patients. They observed a decrease in B_{max} of ^{3}H-cAMP binding to the regulatory subunit of PKA in the supernatant fraction of the fibroblasts of patients with major depression, but no change in the K_D value. They also found a significant reduction in the isoproterenol-stimulated phosphorylation of the transcription factor CREB. Perez et al.[225] also observed decreased PKA activity in the platelets of depressed patients. Overall, these studies suggest abnormalities of PKA in patients with depression.

PKA has also been studied in the postmortem brain and the lymphoblasts of bipolar patients. Chang et al.[226] observed altered PKA activity in the postmortem brain of bipolar patients. Karege et al.[227,228] found that PKA activity is significantly increased in the cytosol fraction of cultured lymphoblasts of euthymic bipolar patients compared with normal controls and that cAMP binding is decreased in the cytosol fraction of lymphoblasts of bipolar patients.

BDNF in Mood Disorders

BDNF, a member of the neurotrophin (NT) family of highly basic proteins that includes nerve growth factor (NGF), NT_3, NT_4, and NT_5.[229] It is involved in the maintenance of neuroplasticity in adults and regulates synaptic activity and neurotransmitter synthesis.[229–234] In the adult CNS, neurotrophins are required for the maintenance of neural function, structural integrity of neurons, and neurogenesis.[12,232,235] Structure abnormalities and brain atrophy have been observed in the brains of patients with MD.[236–241] These abnormalities include reductions in the volumes of the PFC and the hippocampus (as determined by MRI), as well as reduced neuronal and glial density in the postmortem brain of patients with MD. It has been suggested that these structural abnormalities in the brain may have resulted from an abnormality of BDNF. Other lines of evidence suggesting the involvement of BDNF in the pathophysiology of MD are derived from the observations that treatment with antidepressants, electroconvulsive shock, or lithium causes an increase in BDNF level in the rat brain.[242–250] Stress is a major risk factor for MD and suicide.[251–253] It has been also observed in animals, that acute or repeated restraint stress causes a rapid decrease in the expression of BDNF.[254,255]

Because of these findings, the neurotrophin hypothesis for depression was postulated by some investigators. Direct evidence for the involvement of BDNF in MD has been obtained by studies of BDNF and of its receptors in the postmortem brain of suicide victims and in the serum of depressed patients. We have observed that the protein and the mRNA expression of BDNF are significantly decreased in the postmortem brain of adult and teenage suicide victims.[256] Karege et al.[257] found that BDNF levels were significantly lower in the serum of depressed patients than in control subjects and were negatively related to the Montgomery Asberg Depression Rating Scale (MADRS) scores. Female patients had lower BDNF levels than male patients. Shimizu et al.[258] determined the levels of serum BDNF in antidepressant-naïve patients with MDD, antidepressant-treated patients with MDD, and normal control subjects. They found that the serum BDNF was significantly lower in antidepressant-naïve MDD patients than in antidepressant-treated patients or in the normal control group. The researchers also found that serum BDNF levels were negatively correlated with the severity of depression. These data suggest that decreased serum BDNF concentration may be a biomarker of patients with MD.

Serum and plasma BDNF levels have also been studied in patients with bipolar disorders, during both depressive and manic episodes. It was reported by Cunha et al.[259] that serum BDNF levels were decreased in manic or depressed bipolar patients compared with euthymic patients or normal control subjects and were negatively correlated with the severity of the manic or the depressive symptoms. Machado-Vieiria et al.[260] found that the plasma BDNF was significantly decreased in drug-free naïve bipolar subjects compared with healthy controls and that the severity of the manic episode was significantly negatively correlated with the plasma BDNF level. Karege et al.[227] did not observe a difference in BDNF mRNA levels in the lymphoblasts between bipolar subjects and normal control subjects.

In a recent study,[261] we determined the protein expression of BDNF in the platelets of patients with pediatric bipolar disorder (PBD) and the mRNA level of BDNF in the lymphocytes of PBD patients and of normal control subjects. We found that the protein expression levels of BDNF in the platelets of PBD patients were significantly decreased compared with normal control subjects. We also found that the mRNA levels of BDNF were significantly decreased in the lymphocytes of PBD patients compared with normal control subjects. Treatment with mood stabilizing drugs increased the mRNA levels of BDNF in the lymphocytes of PBD patients, and their levels were then similar to those of the normal control subjects. Our studies suggest not only that protein and mRNA expression of BDNF are decreased in the platelets or lymphocytes of PBD patients but also that increased BDNF expression may be a marker for treatment response. All the sources of the BDNF in the serum are not unknown, and hence the changes in serum BDNF in patients with MD could be an epi-phenomenon. Platelets are a major source of BDNF in the serum. Platelets bind with BDNF and store it in α granules, releasing BDNF upon activation and in response to several stimuli. It is quite possible that BDNF binds to the platelet's surface, which then results in an internalization of this protein similarly to what occurs with 5HT. BDNF has been shown to cross the blood-brain barrier in either direction. Some studies indicate that quite possibly the major source of the BDNF in serum and platelets is the brain. Further evidence to support this suggestion is derived from a study conducted by Karege et al.,[262] who determined BDNF levels in both the serum and the brain of rats. The BDNF levels underwent similar changes during the development and aging processes, and there was a positive correlation between the serum and the cortical BDNF levels in these animals. Taken together, these studies suggest that the protein expression of BDNF in serum or platelets and the mRNA expression of BDNF in lymphocytes may be a useful biomarker for patients with MD.

Cytokines in Mood Disorders

There are many interactions among the neural, the immune, and the neuroendocrine systems, and this has led many investigators question if the immune system may also be involved in some of the brain-related disorders, such as depression.[263–270] In recent years there have been suggestions that depression, which is one of the major psychiatric disorders known to be related to changes in the neuroendocrine system, may also be related to, or caused by, changes in the immune system.

Cytokines are a diverse group of proteins that can be considered as the hormones of the immune system. These small molecules are secreted by various cells and act as signals between the cells to regulate the immune responses to injury and infection. The responses of cytokines are mediated through cytokine receptors. As is the case with other receptors, specific cytokine receptors respond to the presence of specific cytokines and thus produce their physiological responses. Cytokine receptors are present in soluble forms, as well as being associated with the membranes.

The classification and the nomenclature of the cytokines are complex. The cytokines have been classified into families of IL, TNF, interferons (IFN), chemokines, and colony-stimulating factors. In terms of biological activity, the cytokines have been classified as pro-inflammatory or anti-inflammatory. The pro-inflammatory cytokines, which are involved in the inflammatory process, include IL-1, IL-6, IFN-α, and TNFα. The anti-inflammatory cytokines include IL-4, IL-10, and IL-13. Cytokines have been shown to regulate the growth, the differentiation, and the function of many cell types. Several of the cytokines occur in more than one form, and these are termed as α, β, or γ.

One of the two major lines of evidence suggesting that cytokines play an important role in the pathophysiology of depression is derived from the observation that administration of some cytokines, such as IFN-α, to patients with hepatitis or to cancer patients with melanoma produced a symptom known as "sickness behavior," which is very similar to depression.[271–282] The other major evidence for a role of cytokines in the pathophysiology of depression is derived from the observations that the levels of pro- and/or anti-inflammatory cytokines are altered in the serum[283–298] or the CSF of depressed patients.[299,300] Indirect evidence suggesting a role for cytokines in depression is the observation that stress, which is a major risk factor for depressive illness, alters not only the immune system but also the levels of several cytokines.[300–305]

Several observations suggest that altered immune states may be present in patients suffering from depressive disorders. For example, Kronfol et al.[306] measured the lymphocyte response to mitogens in patients diagnosed with melancholic depression, and these patients were found to have a decreased lymphocyte response compared with controls. These findings of lower lymphocyte responsiveness have been replicated by several studies.[306–310] Several investigators have also observed that natural killer cell activity was decreased in patients with depression compared with normal controls.[306,308,310–312] The other immune function studied in depressed patients, neutrophil activity, was found to be lower among depressed patients than control subjects. These studies do indicate that the immune process may be reduced in patients with depression (for review, see Weisse[298]).

Cytokines have also been shown to interact with the neuroendocrine system, such as the HPA axis,[313,314] as well as some of the neurotransmitter systems, such as 5HT[264,265,272] and NE.[263,266] Abnormalities of the HPA axis[315] 5HT,[23] and of NE[105] have been implicated in the pathophysiology of depression.

Cytokines in Depression

Because of the suggested relationship between abnormal immune function and depression and because of the observation that a reduction in mitogen-stimulated lymphocyte proliferation, as well as reduced natural killer cell activity, has been observed in depressed patients,[287,311,312,316] the levels of cytokines, which are the major mediators of immune function, have been determined primarily in the serum of depressed patients and normal control subjects. A significant number of studies in this area have been conducted by the group of Maes and colleagues.

In an initial study, Maes et al.[287] determined the levels of soluble IL-6 receptor (sIL-6R), sIL-2R, and transferrin receptors (TfR) in the plasma of subjects with major depression. They found that the plasma concentrations of IL-6, sIL-6R, sIL-2R, and TfR were significantly higher in MD patients compared with normal control subjects. They also found positive correlations between the plasma concentrations of IL-6 and sIL-6R, of IL-6 and sIL-2R, of IL-6 and TfR, and also between sIL-2R and TfR. In addition, they observed that chronic treatment with antidepressant drugs did not cause significant changes in these cytokines. In a subsequent study, Sluzewska et al.[294] compared the plasma concentrations of IL-6, sIL-6R, sIL-2R, TfR, C-reactive protein (CPR), and α_1-acid glycoprotein (AGP) between 49 subjects with MDD during an acute phase of the illness and 15 normal control subjects and found that they were all significantly higher in patients with MDD, thus replicating their previous finding in another group of patients in Poland. Maes et al.[290] also determined the level of the IL-1 receptor antagonist (IL-1-rA) in 68 depressed subjects and 22 normal control subjects and found that the serum level of IL-1-rA was significantly higher in depressed patients compared with normal healthy control subjects. In this study they also found that there was a significant and positive relationship between the serum IL-1-rA and severity of illness in 44% of the depressed patients. However, in this study they did not find any significant relationship between serum IL-1-rA concentration and HPA axis activity in these subjects (determined by measuring the 24 hr urinary cortisol and post-dexamethasone cortisol levels). In another study, Maes et al.[288] determined the levels of some cytokines in treatment-resistant depressed patients before and after subchronic treatment with antidepressants. They found that the serum levels of IL-6 and IL-1-rA were significantly higher in patients with MDD compared with normal control subjects. They also found that subchronic treatment with antidepressants had no significant effect on the serum levels of IL-6, IL-1-rA, or IL-6R.

Several other groups have also studied the plasma levels of cytokines in the plasma of depressed patients and normal control subjects. Berk et al.[284] found that the level of IL-6, when detected, was significantly higher in depressed patients compared with normal control subjects. However, they also found that they could not detect IL-6 in 51% of depressed patients and 58% of normal control subjects. However, whenever they could detect IL-6, they found that the level of IL-6 was significantly increased in the serum of depressed patients compared with normal control subjects.

In another study, the Maes' group[286] found that the serum levels of IL-6, IL-10, and IL-1-rA were much higher in patients with MDD compared with normal control subjects, although these results were not statistically significant, and that there was no significant effect of antidepressant treatment on the serum levels of these cytokines.

Another group of investigators who have done significant work on the role of cytokines in depression are Anisman and Merali in Canada. In one study they determined not only the levels of cytokines but also HPA axis function in the same patients.[283] In this study they examined if the alterations in cytokines observed in depression are related to the neuro-vegetative symptom profile or to the chronicity of the illness. They found that the level of IL-1-β was increased in dysthymic patients and was highly correlated with the age of onset and the duration of the illness. They also found that IL-2 production was decreased in all of the depressed groups, which included depression, atypical depression, dysthymic depression, and atypical dysthymia groups. Kim et al.[285] determined the plasma levels of IL-12 in 102 psychiatric patients consisting of 34 with MDD, 25 bipolar patients, and 43 schizophrenia patients, as well as 85 normal control subjects. They found that the level of IL-12 in patients with major depression was significantly higher than that of the control group, whereas no differences were found in the bipolar or the schizophrenia groups compared with normal control subjects. Eight weeks of treatment caused a significant decrease in the level of IL-12 in these patients.

TNF-α has important effects on behavioral, endocrine, and immune parameters in the rat,[287,300,317,318] and therefore the levels of TNF-α have also been determined in the serum of depressed patients. In a recent study, Tuglu et al.[297] determined the levels of TNF-α and CRP in the serum of 26 patients with MDD and 17 normal control subjects. They found that the level of TNF-α was significantly higher in MDD patients compared with normal control subjects. When these patients were treated with antidepressants, their levels of TNF-α decreased and were very similar to those observed in normal control subjects. These results provide strong evidence that not only is the level of the cytokine

TNF-α decreased in depressed patients but also that treatment with antidepressants normalizes the level of TNF-α.

The levels of cytokines in the plasma have also been determined in late-life depression in a study of elderly depressed patients by Thomas et al.[296] They found that the levels of plasma IL-1β were significantly higher in elderly depressed patients compared with normal control subjects, and that the higher levels correlated with current depression severity. Trzonkowski et al.[319] studied pro-inflammatory cytokines in depressed elderly subjects and found elevated levels of TNF-α and IL-6, as well as of cortisol, and a decreased level of ACTH, and insufficient production of IL-10 in depressed patients compared with normal control subjects.

Although most of these studies were performed in the plasma of depressed patients, Carpenter et al.[299] determined the levels of cytokines in the CSF of depressed patients and control subjects. They determined the level of IL-6 in the CSF of 18 subjects with MDD and 26 normal control subjects but did not find any significant differences.

In summary, these studies strongly suggest that, in general, the levels of several cytokines, which include IL-2, IL-6, as well as IL-12, and of their soluble receptors are increased in the plasma of depressed patients compared with normal control subjects. It has also been observed that mitogen-elicited production of the pro-inflammatory cytokines IL-1β, IL-6, and TNF-α is increased in the serum of depressed patients.

An important question is whether these increased levels of cytokines observed in depressed patients are normalized by treatment with antidepressants. As described earlier, several studies determined the levels of cytokines before and after treatment with antidepressant drugs. Whereas it was found that levels of IL-1β and IL-6 normalize after treatment with antidepressant medication,[320,321] treatment with antidepressants did not cause changes in the increased production of soluble IL-2 receptors, soluble IL-6 receptors, or IL-6 receptors in MDD.[289] It has also been reported that the levels of IL-6 and of the anti-inflammatory cytokine IL-10 and of IL-1-rA were also moderately increased in depressed patients compared with normal control subjects; however, the increased levels of these cytokines were not affected by treatment with antidepressants.

Although studies of cytokines are limited, most do suggest an increase in the levels of cytokines in the serum of depressed patients. One study that compared cytokines in the CSF of depressed patients and normal control subjects did not find any differences between them, and hence it is unclear if the levels of cytokines in the brain are different between depressed patients and normal control subjects. Studies need to be carried out in the postmortem brain from depressed or depressed suicide subjects to further examine and clarify the role of cytokines in depression.

In order for a diagnostic biomarker to be useful, certain criteria need to be met. These criteria include the following (see Chapter 1 of this book, Ritsner and Gottesman):

1. The biomarker should reflect some basic pathophysiological process, and detect a fundamental feature of the disease with high sensitivity and specificity.
2. The biomarker should be specific for the disease compared with related disorders.
3. The biomarker should not reflect symptomatology of disorder.
4. The biomarker can be measured repeatedly over time and should be reproducible.
5. The biomarker should be measured in noninvasive and easy-to-perform tests that can be done at the bedside or in the outpatient setting.
6. The biomarker should not cause harm to the individual being assessed.
7. The biomarker should be reliable in many testing environments/labs.
8. The biomarker should be cost effective.

In addition to the above criteria, an ideal biomarker also must have a high sensitivity and specificity in order to have a reasonably predictive value. It has been extremely difficult to find a biomarker which meets all these criteria, but some of the biomarkers have good potential. The major limitation has been that some of these biomarkers are not specific for a particular disease. For example, it may be specific for mood disorders, but not for the major depressive disorders or bipolar disorder. Serum BDNF could be a good biomarker for depression, as it has been found to be low in serum of depressed patients and it also increases after treatment with antidepressants. However, later it was found that it is also low in serum of bipolar patients and is increased after treatment with lithium. Platelet 5HT$_{2A}$ receptors, on the other hand, appear to be quite specific to suicidal behavior, but not to a particular disease (i.e., depression or bipolar disorders). However, it appears to be slightly more elevated in these disorders compared with normal controls. The search for potential biomarkers needs to be continued, whether they are trait markers or state markers. Even if any of the biomarkers do not completely satisfy the above requirements for the biomarkers, they may still be quite useful.

Conclusions and Future Directions

The diagnosis as well as the therapeutic response in patients with psychiatric disorders is primarily based on clinical assessments. This often provides false/positive or false/negative results. A search for a simple, easily accessible biomarker is therefore necessary to complement the clinical assessments for both diagnostic as well as prognostic purposes. A biomarker is an indicator of a normal or pathologic process or pharmacologic response to a therapeutic intervention. Biomarkers may be very useful in early identification of illness (e.g., depression or bipolar illness), or in identification of symptoms (e.g., suicidal behavior) not only for early intervention but also for prevention of suicide, which is a major risk factor in these illnesses. Biomarkers can therefore be used as a diagnostic tool for the identification of individuals with certain illnesses, in this case bipolar illness or depression. They can also be used as indicators of disease progression and/or prognosis, and for monitoring and predicting clinical response to a particular therapeutic intervention.

Because of the inaccessibility of the human brain, peripheral biomarkers (e.g., in serum, blood cells, saliva or urine) more accessible than brain or the CSF may be very useful. Abnormalities of the serotonergic and noradrenergic system in peripheral tissues have been studied in great detail, as reviewed in this chapter. Many of these results have been quite encouraging but many of the studies were not only conflicting but produced inconsistent results. Nonetheless, these studies do indicate some of the potentially useful biomarkers for these illnesses, such as the use of platelet $5HT_{2A}$ receptor for depression and suicidal behavior, platelet serotonin transporter sites for depression, PKC for bipolar illness. The inconsistent results may be related to the heterogeneity of these illnesses or to other methodological issues (e.g., studies performed during a drug-free period or those performed during treatment, the choice of biochemical or biological methods). Another advantage of these biomarkers is that eventually

they may also help in identifying more homogeneous populations, based either on therapeutic response, on the sub-diagnosis or on particular behaviors such as suicide or impulsive aggressive behavior. Further work on these biomarkers in combination with genetic studies may prove to be much more useful.

References

1. Hirschfeld, R.M. and M.M. Weissman, Risk factors for major depression and bipolar disorder, in Neuropharmacology – the Fifth Generation of Progress, K.L. Davis et al., Editors. 2002, Lippincott Williams & Wilkins: Philadelphia, pp. 1017–25.
2. Kessler, R.C., K.A. McGonagle, S. Zhao, et al. Lifetime and 12-month prevalence of DSM-III-R psychiatric disorders in the United States. Results from the National Comorbidity Survey. Arch Gen Psychiatry 1994; 51:8–19.
3. Reducing Suicide, A National Imperative. Committee on Pathophysiology and Prevention of Adolescent and Adult Suicide, Board on Neuroscience and Behavioral Health, Institute of Medicine of the National Academies, S.K. Goldsmith et al., Editors, Washington, DC: The National Academies Press; 2002.
4. Botsis, A.F., C.R. Soldatos, and C.N. Stefanis, Suicide. Biopsychosocial Approaches. Amsterdam: Elsevier; 1997.
5. Moscicki, E.K., P. O'Carroll, D.S. Rae, B.Z., et al. Suicide attempts in the Epidemiologic Catchment Area Study. Yale J Biol Med 1988; 61:259–68.
6. National Center for Health Statistics: Advance report of final mortality statistics. 1994, NCSH Monthly Vital Statistics Report 1992; 40 (Suppl 2).
7. Bunney, W.E., Jr. and J.M. Davis. Norepinephrine in depressive reactions. A review. Arch Gen Psychiatry 1965; 13:483–94.
8. Schildkraut, J.J. The catecholamine hypothesis of affective disorders: a review of supporting evidence. Am J Psychiatry 1965; 122:509–22.
9. Coppen, A. The biochemistry of affective disorders. Br J Psychiatry 1967; 113:1237–64.
10. Lapin, I.P. and G.F. Oxenkrug. Intensification of the central serotoninergic processes as a possible determinant of the thymoleptic effect. Lancet 1969; 1:132–6.
11. van Praag, H.M. Toward a biochemical classification of depression. Adv Biochem Psychopharmacol 1974; 11:357–68.
12. Duman, R.S. Synaptic plasticity and mood disorders. Mol Psychiatry 2002; 7 (Suppl 1):S 29–34.
13. Halper, J.P., R.P. Brown, J.A. Sweeney, J.H. Kocsis, A. Peters, and J.J. Mann. Blunted beta-adrenergic responsivity of peripheral blood mononuclear cells in endogenous depression. Isoproterenol dose-response studies. Arch Gen Psychiatry 1988; 45:241–4.
14. Ebstein, R.P., B. Lerer, B. Shapira, Z. Shemesh, D.G. Moscovich, and S. Kindler. Cyclic AMP second-messenger signal amplification in depression. Br J Psychiatry 1988; 152:665–9.

15. Pandey, G.N., M.W. Dysken, D.L. Garver, and J.M. Davis. Beta-adrenergic receptor function in affective illness. Am J Psychiatry 1979; 136:675–8.

16. Ebstein, R.P., D. Moscovich, S. Zeevi, Z. Amiri, and B. Lerer. Effect of lithium in vitro and after chronic treatment on human platelet adenylate cyclase activity: postreceptor modification of second messenger signal amplification. Psychiatry Res 1987; 21:221–8.

17. Extein, I., J. Tallman, C.C. Smith, and F.K. Goodwin. Changes in lymphocyte beta-adrenergic receptors in depression and mania. Psychiatry Res 1979; 1:191–7.

18. Mann, J.J., R.P. Brown, J.P. Halper, et al. Reduced sensitivity of lymphocyte beta-adrenergic receptors in patients with endogenous depression and psychomotor agitation. N Engl J Med 1985; 313:715–20.

19. Young, L.T., P.P. Li, S.J. Kish, and J.J. Warsh. Cerebral cortex beta-adrenoceptor binding in bipolar affective disorder. J Affect Disord 1994; 30:89–92.

20. Akin, D., D.H. Manier, E. Sanders-Bush, and R.C. Shelton. Signal transduction abnormalities in melancholic depression. Int J Neuropsychopharmacol 2005; 8:5–16.

21. Freis, E.D. Mental depression in hypertensive patients treated for long periods with large doses of reserpine. N Engl J Med 1954; 251:1006–8.

22. Harris, T.H. Depression induced by Rauwolfia compounds. Am J Psychiatry 1957; 113:950.

23. Maes, M. and H.Y. Meltzer, The serotonin hypothesis of major depression, in Psychopharmacology: The Fourth Generation of Progress, F.E. Bloom and D.J. Kupfer, Editors. 1995, Raven: New York, pp. 933–44.

24. Pandey, G.N. and Y. Dwivedi, The serotonergic system, in Psychopharmacology: Treatment of Psychiatric Disorders, J. Ananth, Editor. 1999, Jaypee Brothers Medical Publishers: New Delhi, India, pp. 11–30.

25. Pandey, G.N. and Y. Dwivedi. Monoamine receptors and signal transduction mechanisms in suicide. Curr Psychiatr Rev 2006; 2:51–75.

26. Kopin, I.J. Catecholamine metabolism: basic aspects and clinical significance. Pharmacol Rev 1985; 37:333–64.

27. Schatzberg, A.F., J.A. Samson, K.L. Bloomingdale, et al. Toward a biochemical classification of depressive disorders. X. Urinary catecholamines, their metabolites, and D-type scores in subgroups of depressive disorders. Arch Gen Psychiatry 1989; 46:260–8.

28. Asberg, M., L. Bertilsson, B. Martensson, G.P. et al. CSF monoamine metabolites in melancholia. Acta Psychiatr Scand 1984; 69:201–19.

29. Asberg, M., L. Traskman, and P. Thoren. 5-HIAA in the cerebrospinal fluid. A biochemical suicide predictor? Arch Gen Psychiatry 1976; 33:1193–7.

30. Agren, H. Symptom patterns in unipolar and bipolar depression correlating with monoamine metabolites in the cerebrospinal fluid: II. Suicide. Psychiatry Res 1980; 3:225–36.

31. Meltzer, H.Y. and M.T. Lowy, The serotonin hypothesis of depression, in Psychopharmacology: The Third Generation of Progress, H.Y. Meltzer, Editor. 1987, Raven: New York.

32. Nordstrom, P., M. Samuelsson, M. Asberg, et al. CSF 5-HIAA predicts suicide risk after attempted suicide. Suicide Life Threat Behav 1994; 24:1–9.

33. Roy, A., D. Pickar, M. Linnoila, et al. Cerebrospinal fluid monoamine and monoamine metabolite concentrations in melancholia. Psychiatry Res 1985; 15:281–92.

34. Van Praag, H.M. Depression, suicide and the metabolism of serotonin in the brain. J Affect Disord 1982; 4:275–90.

35. van Praag, H.M. CSF 5-HIAA and suicide in non-depressed schizophrenics. Lancet 1983; 2:977–8.

36. Spreux-Varoquaux, O., J.C. Alvarez, I. Berlin, et al. Differential abnormalities in plasma 5-HIAA and platelet serotonin concentrations in violent suicide attempters: relationships with impulsivity and depression. Life Sci 2001; 69:647–57.

37. Lidberg, L., H. Belfrage, L. Bertilsson, et al. Suicide attempts and impulse control disorder are related to low cerebrospinal fluid 5-HIAA in mentally disordered violent offenders. Acta Psychiatr Scand 2000; 101:395–402.

38. Tuomisto, J. and E. Tukiainen. Decreased uptake of 5-hydroxytryptamine in blood platelets from depressed patients. Nature 1976; 262:596–8.

39. Meltzer, H.Y., R.C. Arora, R. Baber, and B.J. Tricou. Serotonin uptake in blood platelets of psychiatric patients. Arch Gen Psychiatry 1981; 38:1322–6.

40. Rausch, J.L., D.S. Janowsky, S.C. Risch, and L.Y. Huey. A kinetic analysis and replication of decreased platelet serotonin uptake in depressed patients. Psychiatry Res 1986; 19:105–12.

41. Muck-Seler, D., N. Pivac, M. Mustapic, et al. Platelet serotonin and plasma prolactin and cortisol in healthy, depressed and schizophrenic women. Psychiatry Res 2004; 127:217–6.

42. Oxenkrug, G.F. The content and uptake of 5-HT by blood platelets in depressive patients. J Neural Transm 1979; 45:285–9.

43. Franke, L., H.J. Schewe, B. Muller, et al. Serotonergic platelet variables in unmedicated patients suffering from major depression and healthy subjects: relationship between 5HT content and 5HT uptake. Life Sci 2000; 67:301–5.

44. Paul, S.M., M. Rehavi, P. Skolnick, et al. Depressed patients have decreased binding of tritiated imipramine to platelet serotonin "transporter". Arch Gen Psychiatry 1981; 38:1315–7.

45. Owens, M.J. and C.B. Nemeroff. Role of serotonin in the pathophysiology of depression: focus on the serotonin transporter. Clin Chem 1994; 40:288–95.

46. Asberg, M., P. Thoren, L. Traskman, et al. "Serotonin depression" – a biochemical subgroup within the affective disorders? Science 1976; 191:478–80.

47. Hou, C., F. Jia, Y. Liu, and L. Li. CSF serotonin, 5-hydroxyindolacetic acid and neuropeptide Y levels in severe major depressive disorder. Brain Res 2006; 1095:154–8.

48. Reddy, P.L., S. Khanna, M.N. Subhash, et al. CSF amine metabolites in depression. Biol Psychiatry 1992; 31:112–8.

49. Stanley, M. and J.J. Mann. Increased serotonin-2 binding sites in frontal cortex of suicide victims. Lancet 1983; 1:214–6.

50. Pandey, G.N., S.C. Pandey, Y. Dwivedi, R.P. Sharma, P.G. Janicak, and J.M. Davis. Platelet serotonin-2A receptors: a potential biological marker for suicidal behavior. Am J Psychiatry 1995; 152:850–5.

51. Humphrey, P.P., P. Hartig, and D. Hoyer. A proposed new nomenclature for 5-HT receptors. Trends Pharmacol Sci 1993; 14:233–6.

52. Teitler, M. and K. Herrick-Davis. Multiple serotonin receptor subtypes: molecular cloning and functional expression. Crit Rev Neurobiol 1994; 8:175–88.

53. Hoyer, D. and G. Martin. 5-HT receptor classification and nomenclature: towards a harmonization with the human genome. Neuropharmacology 1997; 36:419–28.

54. Barnes, N.M. and T. Sharp. A review of central 5-HT receptors and their function. Neuropharmacology 1999; 38:1083–152.

55. Conn, P.J. and E. Sanders-Bush. Regulation of serotonin-stimulated phosphoinositide hydrolysis: relation to the serotonin 5-HT-2 binding site. J Neurosci 1986; 6:3669–75.

56. Kusumi, I., T. Koyama, and I. Yamashita. Serotonin-stimulated Ca2+ response is increased in the blood platelets of depressed patients. Biol Psychiatry 1991; 30:310–2.

57. Kusumi, I., T. Koyama, and I. Yamashita. Effect of various factors on serotonin-induced Ca2+ response in human platelets. Life Sci 1991; 48:2405–12.

58. Elliott, J.M. and A. Kent. Comparison of [125I]iodolysergic acid diethylamide binding in human frontal cortex and platelet tissue. J Neurochem 1989; 53:191–6.

59. Andres, A.H., M.L. Rao, S. Ostrowitzki, and W. Entzian. Human brain cortex and platelet serotonin2 receptor binding properties and their regulation by endogenous serotonin. Life Sci 1993; 52:313–21.

60. Biegon, A., A. Weizman, L. Karp, et al. Serotonin 5-HT2 receptor binding on blood platelets–a peripheral marker for depression? Life Sci 1987; 41:2485–92.

61. Pandey, G.N., S.C. Pandey, P.G. Janicak, R.C. Marks, and J.M. Davis. Platelet serotonin-2 receptor binding sites in depression and suicide. Biol Psychiatry 1990; 28:215–22.

62. Hrdina, P.D., D. Bakish, J. Chudzik, A. Ravindran, and Y.D. Lapierre. Serotonergic markers in platelets of patients with major depression: upregulation of 5-HT2 receptors. J Psychiatry Neurosci 1995; 20:11–9.

63. Ostrowitzki, S., M.L. Rao, J. Redei, and A.H. Andres. Concurrence of cortex and platelet serotonin2 receptor binding characteristics in the individual and the putative regulation by serotonin. J Neural Transm Gen Sect 1993; 93:27–35.

64. Arora, R.C. and H.Y. Meltzer. Increased serotonin2 (5-HT2) receptor binding as measured by 3H-lysergic acid diethylamide (^{3}H-LSD) in the blood platelets of depressed patients. Life Sci 1989; 44:725–34.

65. Biegon, A., N. Essar, M. Israeli, et al. Serotonin 5-HT2 receptor binding on blood platelets as a state dependent marker in major affective disorder. Psychopharmacology (Berl) 1990; 102:73–5.

66. Mendelson, S.D. The current status of the platelet 5-HT(2A) receptor in depression. J Affect Disord 2000; 57:13–24.

67. Butler, J. and B.E. Leonard. The platelet serotonergic system in depression and following sertraline treatment. Int Clin Psychopharmacol 1988; 3:343–7.

68. Hrdina, P.D., D. Bakish, A. Ravindran, et al. Platelet serotonergic indices in major depression: up-regulation of 5-HT2A receptors unchanged by antidepressant treatment. Psychiatry Res 1997; 66:73–85.

69. Rao, M.L., B. Hawellek, A. Papassotiropoulos, A. Deister, and C. Frahnert. Upregulation of the platelet Serotonin2A receptor and low blood serotonin in suicidal psychiatric patients. Neuropsychobiology 1998; 38:84–9.

70. Sheline, Y.I., M.E. Bardgett, J.L. Jackson, et al. Platelet serotonin markers and depressive symptomatology. Biol Psychiatry 1995; 37:442–7.

71. Rosel, P., B. Arranz, J. Vallejo, et al. Altered [3H]imipramine and 5-HT2 but not [3H]paroxetine binding sites in platelets from depressed patients. J Affect Disord 1999; 52:225–33.

72. Cowen, P.J., E.M. Charig, S. Fraser, and J.M. Elliott. Platelet 5-HT receptor binding during depressive illness and tricyclic antidepressant treatment. J Affect Disord 1987; 13:45–50.

73. Bakish, D., P. Cavazzoni, J. Chudzik, et al. Effects of selective serotonin reuptake inhibitors on platelet serotonin parameters in major depressive disorder. Biol Psychiatry 1997; 41:184–90.

74. Mann, J.J., P.A. McBride, R.P. Brown, et al. Relationship between central and peripheral serotonin indexes in depressed and suicidal psychiatric inpatients. Arch Gen Psychiatry 1992; 49:442–6.

75. Sheline, Y.I., K.J. Black, M.E. Bardgett, and J.G. Csernansky. Platelet binding characteristics distinguish placebo responders from nonresponders in depression. Neuropsychopharmacology 1995; 12:315–22.

76. Biegon, A., A. Grinspoon, B. Blumenfeld, et al. Increased serotonin 5-HT2 receptor binding on blood platelets of suicidal men. Psychopharmacology (Berl) 1990; 100:165–7.

77. McBride, P.A., R.P. Brown, M. DeMeo, et al. The relationship of platelet 5-HT2 receptor indices to major depressive disorder, personality traits, and suicidal behavior. Biol Psychiatry 1994; 35:295–308.

78. Crow, T.J., A.J. Cross, S.J. Cooper, et al. Neurotransmitter receptors and monoamine metabolites in the brains of patients with Alzheimer-type dementia and depression, and suicides. Neuropharmacology 1984; 23:1561–9.

79. Owen, F., D.R. Chambers, S.J. Cooper, et al. Serotonergic mechanisms in brains of suicide victims. Brain Res 1986; 362:185–8.

80. Arranz, B., A. Eriksson, E. Mellerup, et al. Brain 5-HT1A, 5-HT1D, and 5-HT2 receptors in suicide victims. Biol Psychiatry 1994; 35:457–63.

81. Kafka, M.S., L.J. Siever, J.I. Nurnberger, et al. Platelet alpha-adrenergic receptor function in affective disorders and schizophrenia. Psychopharmacol Bull 1985; 21:599–602.

82. Joyce, J.N., A. Shane, N. Lexow, A. Winokur, M.F. Casanova, and J.E. Kleinman. Serotonin uptake sites and serotonin receptors are altered in the limbic system of schizophrenics. Neuropsychopharmacology 1993; 8:315–36.

83. Stockmeier, C.A., G.E. Dilley, L.A. Shapiro, et al. Serotonin receptors in suicide victims with major depression. Neuropsychopharmacology 1997; 16:162–73.

84. Lowther, S., F. De Paermentier, M.R. Crompton, et al. Brain 5-HT2 receptors in suicide victims: violence of death, depression and effects of antidepressant treatment. Brain Res 1994; 642:281–9.

85. Rosel, P., B. Arranz, L. San, et al. Altered 5-HT(2A) binding sites and second messenger inositol trisphosphate (IP(3)) levels in hippocampus but not in frontal cortex from depressed suicide victims. Psychiatry Res 2000; 99:173–81.

86. Arora, R.C. and H.Y. Meltzer. 3H-imipramine binding in the frontal cortex of suicides. Psychiatry Res 1989; 30:125–35.

87. Laruelle, M., A. Abi-Dargham, M.F. Casanova, et al. Selective abnormalities of prefrontal serotonergic receptors in schizophrenia. A postmortem study. Arch Gen Psychiatry 1993; 50:810–8.

88. Hrdina, P.D., E. Demeter, T.B. Vu, P. Sotonyi, and M. Palkovits. 5-HT uptake sites and 5-HT2 receptors in brain of antidepressant-free suicide victims/depressives: increase in 5-HT2 sites in cortex and amygdala. Brain Res 1993; 614:37–44.

89. Arango, V., P. Ernsberger, P.M. Marzuk, et al. Autoradiographic demonstration of increased serotonin 5-HT2 and beta-adrenergic receptor binding sites in the brain of suicide victims. Arch Gen Psychiatry 1990; 47:1038–47.

90. Mann, J.J., M. Stanley, P.A. McBride, and B.S. McEwen. Increased serotonin2 and beta-adrenergic receptor binding in the frontal cortices of suicide victims. Arch Gen Psychiatry 1986; 43:954–9.

91. Pandey, G.N., Y. Dwivedi, H.S. Rizavi, et al. Higher expression of serotonin 5-HT(2A) receptors in the postmortem brains of teenage suicide victims. Am J Psychiatry 2002; 159:419–29.

92. Mikuni, M., I. Kusumi, A. Kagaya, et al. Increased 5-HT-2 receptor function as measured by serotonin-stimulated phosphoinositide hydrolysis in platelets of depressed patients. Prog Neuropsychopharmacol Biol Psychiatry 1991; 15:49–61.

93. Meltzer, H.Y., B. Umberkoman-Wiita, A. Robertson, B.J. Tricou, M. Lowy, and R. Perline. Effect of 5-hydroxytryptophan on serum cortisol levels in major affective disorders. I. Enhanced response in depression and mania. Arch Gen Psychiatry 1984; 41:366–74.

94. Meltzer, H.Y., R. Perline, B.J. Tricou, et al. Effect of 5-hydroxytryptophan on serum cortisol levels in major affective disorders. II. Relation to suicide, psychosis, and depressive symptoms. Arch Gen Psychiatry 1984; 41:379–87.

95. Meltzer, H.Y., M. Lowy, A. Robertson, P. Goodnick, and R. Perline. Effect of 5-hydroxytryptophan on serum cortisol levels in major affective disorders. III. Effect of antidepressants and lithium carbonate. Arch Gen Psychiatry 1984; 41:391–7.

96. Meltzer, H.Y. Serotonergic function in the affective disorders: the effect of antidepressants and lithium on the 5-hydroxytryptophan-induced increase in serum cortisol. Ann N Y Acad Sci 1984; 430:115–37.

97. Coccaro, E.F., M.E. Berman, R.J. Kavoussi, and R.L. Hauger. Relationship of prolactin response to d-fenfluramine to behavioral and questionnaire assessments of aggression in personality-disordered men. Biol Psychiatry 1996; 40:157–64.

98. Malone, K.M., E.M. Corbitt, S. Li, and J.J. Mann. Prolactin response to fenfluramine and suicide attempt lethality in major depression. Br J Psychiatry 1996; 168:324–9.

99. O'Keane, V. and T.G. Dinan. Prolactin and cortisol responses to d-fenfluramine in major depression: evidence for diminished responsivity of central serotonergic function. Am J Psychiatry 1991; 148:1009–15.

100. Cleare, A.J., R.M. Murray, and V. O'Keane. Reduced prolactin and cortisol responses to d-fenfluramine in depressed compared to healthy matched control subjects. Neuropsychopharmacology 1996; 14:349–54.

101. Cook, E.H., Jr., K.E. Fletcher, M. Wainwright, et al. Primary structure of the human platelet serotonin 5-HT2A receptor: identify with frontal cortex serotonin 5-HT2A receptor. J Neurochem 1994; 63:465–9.

102. Langer, S.Z., E. Zarifian, M. Briley, R. Raisman, and D. Sechter. High-affinity binding of 3H-imipramine in brain and platelets and its relevance to the biochemistry of affective disorders. Life Sci 1981; 29:211–20.

103. Ellis, P.M. and C. Salmond. Is platelet imipramine binding reduced in depression? A meta-analysis. Biol Psychiatry 1994; 36:292–9.

104. Goodwin, F.K. and R.M. Post, Studies of amine metabolites in affective illness and in schizophrenia: a comparative analysis., in Biology of the Major Psychosis, D.X. Freedman, Editor. 1975, Raven: New York, pp. 299–332.

105. Schatzberg, A.F. and J.J. Schildkraut, Recent Studies on Norepinephrine systems in mood disorders, in Psychopharmacology: The Fourth Generation of Progress, F.E. Bloom and D.J. Kupfer, Editors. 1995, Raven: New York, pp. 911–20.

106. Barnes, R.F., R.C. Veith, S. Borson, et al. High levels of plasma catecholamines in dexamethasone-resistant depressed patients. Am J Psychiatry 1983; 140:1623–5.

107. Elsworth, J.D., D.E. Redmond, Jr., and R.H. Roth. Plasma and cerebrospinal fluid 3-methoxy-4-hydroxyphenylethylene glycol (MHPG) as indices of brain norepinephrine metabolism in primates. Brain Res 1982; 235:115–24.

108. Jimerson, D.C., J.I. Nurnberger, Jr., R.M. Post, E.S. Gershon, and I.J. Kopin. Plasma MHPG in rapid cyclers and healthy twins. Arch Gen Psychiatry 1981; 38:1287–90.

109. Jimerson, D.C., T.R. Insel, V.I. Reus, and I.J. Kopin. Increased plasma MHPG in dexamethasone-resistant depressed patients. Arch Gen Psychiatry 1983; 40:173–6.

110. Leckman, J.F. and J.W. Maas, Plasma MHPG: relationship to brain noradrenergic system and emerging clinical applications., in Neurobiology of Mood Disorders, R.M. Post et al., Editors. 1984, Williams & Wilkins: Baltimore, MD, pp. 529–38.

111. Muscettola, G., W.Z. Potter, D. Pickar, and F.K. Goodwin. Urinary 3-methoxy-4-hydroxyphenylglycol and major affective disorders. A replication and new findings. Arch Gen Psychiatry 1984; 41:337–42.

112. Roy, A., D.C. Jimerson, and D. Pickar. Plasma MHPG in depressive disorders and relationship to the dexamethasone suppression test. Am J Psychiatry 1986; 143:846–51.

113. Jones, F.D., J.W. Maas, H. Dekirmenjian, and J.A. Fawcett. Urinary catecholamine metabolites during behavioral changes in a patient with manic-depressive cycles. Science 1973; 179:300–2.

114. Bond, P.A., F.A. Jenner, and G.A. Sampson. Daily variations of the urine content of 3-methoxy-4-hydroxyphenylglycol in two manic-depressive patients. Psychol Med 1972; 2:81–5.

115. Agren, H. Depressive symptom patterns and urinary MHPG excretion. Psychiatry Res 1982; 6:185–96.

116. Secunda, S.K., C.K. Cross, S. Koslow, M.M. Katz, J.H. Kocsis, and J.W. Maas. Studies of amine metabolites in depressed patients. Relationship to suicidal behavior. Ann N Y Acad Sci 1986; 487:231–42.

117. Shiah, I.S., H.C. Ko, J.F. Lee, and R.B. Lu. Platelet 5-HT and plasma MHPG levels in patients with bipolar I and bipolar II depressions and normal controls. J Affect Disord 1999; 52:101–10.

118. Brown, G.L., F.K. Goodwin, J.C. Ballenger, et al. Aggression in humans correlates with cerebrospinal fluid amine metabolites. Psychiatry Res 1979; 1:131–9.

119. Sulser, F. New Perspectives on the Mode of Action of Antidepressant Drugs. TIPS 1979; 92–94.

120. Pandey, G.N., B. Brown, and J.M. Davis. Effect of Treatment with Some Atypical Antidepressants on 3H-DHA Binding in the Rat Brain. Drug Dev Res 1985; 251–59.

121. Pandey, G.N. and J.M. Davis. Treatment with antidepressants and down regulation of beta-adrenergic receptors. Drug Dev Res 1985; 393–406.

122. Green, A.R. Evolving concepts on the interactions between antidepressant treatments and monoamine neurotransmitters. Neuropharmacology 1987; 26:815–22.

123. Peroutka, S.J. and S.H. Snyder. Long-term antidepressant treatment decreases spiroperidol-labeled serotonin receptor binding. Science 1980; 210:88–90.

124. Smith, C.B., J.A. Garcia-Sevilla, and P.J. Hollingsworth. alpha 2-Adrenoreceptors in rat brain are decreased after long-term tricyclic antidepressant drug treatment. Brain Res 1981; 210:413–8.

125. Mishra, R., A. Janowsky, and F. Sulser. Subsensitivity of the norepinephrine receptor-coupled adenylate cyclase system in brain: effects of nisoxetine versus fluoxetine. Eur J Pharmacol 1979; 60:379–82.

126. Langer, S.Z. Presynaptic regulation of the release of catecholamines. Pharmacol Rev 1980; 32:337–62.

127. Starke, K. Regulation of noradrenaline release by presynaptic receptor systems. Rev Physiol Biochem Pharmacol 1977; 77:1–124.

128. Starke, K. Presynaptic receptors. Annu Rev Pharmacol Toxicol 1981; 21:7–30.

129. Westfall, T.C. Local regulation of adrenergic neurotransmission. Physiol Rev 1977; 57:659–728.

130. Schweitzer, J.W., R. Schwartz, and A.J. Friedhoff. Intact presynaptic terminals required for beta-adrenergic receptor regulation by desipramine. J Neurochem 1979; 33:377–9.

131. Elliott, J.M. Platelet receptor binding studies in affective disorders. J Affect Disord 1984; 6:219–39.

132. Piletz, J.E., D.S. Schubert, and A. Halaris. Evaluation of studies on platelet alpha 2 adrenoreceptors in depressive illness. Life Sci 1986; 39:1589–616.

133. Grant, J.A. and M.C. Scrutton. Novel alpha2-adrenoreceptors primarily responsible for inducing human platelet aggregation. Nature 1979; 277:659–61.

134. Jakobs, K.H., W. Saur, and G. Schultz. Metal and metal-ATP interactions with human platelet adenylate cyclase: effects of alpha adrenergic inhibition. Mol Pharmacol 1978; 14:1073–8.

135. Kafka, M.S. and S.M. Paul. Platelet alpha 2-adrenergic receptors in depression. Arch Gen Psychiatry 1986; 43:91–5.

136. Daiguji, M., H.Y. Meltzer, C. Tong, D.C. U'Prichard, M. Young, and H. Kravitz. alpha 2-Adrenergic receptors in platelet membranes of depressed patients: no change in number of 3H-yohimbine affinity. Life Sci 1981; 29:2059–64.

137. Stahl, S.M., P.M. Lemoine, R.D. Ciaranello, and P.A. Berger. Platelet alpha 2-adrenergic receptor sensitivity in major depressive disorder. Psychiatry Res 1983; 10:157–64.

138. Braddock, L., P.J. Cowen, J.M. Elliott, S. Fraser, and K. Stump. Binding of yohimbine and imipramine to platelets in depressive illness. Psychol Med 1986; 16:765–73.

139. Lenox, R.H., J. Ellis, D.A. VanRiper, et al., Platelet alpha2-adrenergic receptor activity in clinical studies of depressions, in Frontiers in Neuropsychiatric Research, E. Usdin, M. Goldstein, and A. Friedhoff, Editors. 1983, MacMillan: New York, p. 331.

140. Smith, C.B., P.J. Hollingsworth, J.A. Garcia-Sevilla, and A.P. Zis. Platelet alpha 2 adrenoreceptors are decreased in number after antidepressant therapy. Prog Neuropsychopharmacol Biol Psychiatry 1983; 7:241–7.

141. Campbell, I.C., R.M. McKernan, S.A. Checkley, I.B. Glass, C. Thompson, and E. Shur. Characterization of platelet alpha 2 adrenoceptors and measurement in control and depressed subjects. Psychiatry Res 1985; 14:17–31.

142. Wolfe, N., B.M. Cohen, and A.J. Gelenberg. Alpha 2-adrenergic receptors in platelet membranes of depressed patients: increased affinity for 3H-yohimbine. Psychiatry Res 1987; 20:107–16.

143. Pimoule, C., M.S. Briley, C. Gay, et al. 3H-Rauwolscine binding in platelets from depressed patients and healthy volunteers. Psychopharmacology (Berl) 1983; 79:308–12.

144. Healy, D., P.A. Carney, and B.E. Leonard. Monoamine-related markers of depression: changes following treatment. J Psychiatr Res 1982; 17:251–60.

145. Wood, K. and A. Coppen. Peripheral alpha-adrenergic activity in the affective disorders. Adv Biochem Psychopharmacol 1982; 32:13–9.

146. Siever, L.J., M.S. Kafka, S. Targum, and C.R. Lake. Platelet alpha-adrenergic binding and biochemical responsiveness in depressed patients and controls. Psychiatry Res 1984; 11:287–302.

147. Garcia-Sevilla, J.A., J. Guimon, P. Garcia-Vallejo, and M.J. Fuster. Biochemical and functional evidence of supersensitive platelet alpha 2-adrenoceptors in major affective disorder. Effect of long-term lithium carbonate treatment. Arch Gen Psychiatry 1986; 43:51–7.

148. Garcia-Sevilla, J.A. and M.J. Fuster. Labelling of human platelet alpha 2-adrenoceptors with the full agonist [3H](-) adrenaline. Eur J Pharmacol 1986; 124:31–41.

149. Garcia-Sevilla, J.A., P.J. Hollingsworth, and C.B. Smnith. Alpha 2-adrenoreceptors on human platelets: selective labelling by [3H]clonidine and [3H]yohimbine and competitive inhibition by antidepressant drugs. Eur J Pharmacol 1981; 74:329–41.

150. Garcia-Sevilla, J.A., A.P. Zis, P.J. Hollingsworth, J.F. Greden, and C.B. Smith. Platelet alpha 2-adrenergic receptors in major depressive disorder. Binding of tritiated clonidine before and after tricyclic antidepressant drug treatment. Arch Gen Psychiatry 1981; 38:1327–33.

151. Doyle, M.C., A.J. George, A.V. Ravindran, and R. Philpott. Platelet alpha 2-adrenoreceptor binding in elderly depressed patients. Am J Psychiatry 1985; 142:1489–90.

152. Georgotas, A., J. Schweitzer, R.E. McCue, M. Armour, and A.J. Friedhoff. Clinical and treatment effects on 3H-clonidine and 3H-imipramine binding in elderly depressed patients. Life Sci 1987; 40:2137–43.

153. Pandey, G.N., P.G. Janicak, J.I. Javaid, and J.M. Davis. Increased 3H-clonidine binding in the platelets of patients with depressive and schizophrenic disorders. Psychiatry Res 1989; 28:73–88.

154. Carstens, M.E., A.H. Engelbrecht, V.A. Russell, et al. Alpha 2-adrenoceptor levels on platelets of patients with major depressive disorders. Psychiatry Res 1986; 18:321–31.

155. Piletz, J.E. and A. Halaris. Super high affinity 3H-para-aminoclonidine binding to platelet adrenoceptors in depression. Prog Neuropsychopharmacol Biol Psychiatry 1988; 12:541–53.

156. Garcia-Sevilla, J.A., C. Udina, M.J. Fuster, E. Alvarez, and M. Casas. Enhanced binding of [3H] (-) adrenaline to platelets of depressed patients with melancholia: effect of long-term clomipramine treatment. Acta Psychiatr Scand 1987; 75:150–7.

157. Sacchetti, E., G. Conte, A. Pennati, A. Vita, A. Alciati, and C.L. Cazzullo. Platelet alpha 2-adrenoceptors in major depression: relationship with urinary 4-hydroxy-3-methoxyphenylglycol and age at onset. J Psychiatr Res 1985; 19:579–86.

158. Healy, D., P.A. Carney, A. O'Halloran, and B.E. Leonard. Peripheral adrenoceptors and serotonin receptors in depression. Changes associated with response to treatment with trazodone or amitriptyline. J Affect Disord 1985; 9:285–96.

159. Wang, Y.C., G.N. Pandey, J. Mendels, and A. Frazer. Platelet adenylate cyclase responses in depression: implications for a receptor defect. Psychopharmacologia 1974; 36:291–300.

160. Murphy, D.L., C. Donnelly, and J. Moskowitz. Catecholamine receptor function in depressed patients. Am J Psychiatry 1974; 131:1389–91.

161. Kanof, P.D., C. Johns, M. Davidson, L.J. Siever, E.F. Coccaro, and K.L. Davis. Prostaglandin receptor sensitivity in psychiatric disorders. Arch Gen Psychiatry 1986; 43:987–93.

162. Kanof, P.D., C.A. Johns, M. Davidson, L.J. Siever, E.F. Coccaro, and K.L. Davis. Platelet alpha 2-adrenergic receptor function in psychiatric disorders. Psychiatry Res 1988; 23:11–22.

163. Kafka, M.S. and D.P. van Kammen. alpha-Adrenergic receptor function in schizophrenia. Receptor number, cyclic adenosine monophosphate production, adenylate cyclase activity, and effect of drugs. Arch Gen Psychiatry 1983; 40:264–70.

164. Rotrosen, J., A.D. Miller, D. Mandio, L.J. Traficante, and S. Gershon. Prostaglandins, platelets, and schizophrenia. Arch Gen Psychiatry 1980; 37:1047–54.

165. Garver, D.L., C. Johnson, and D.R. Kanter. Schizophrenia and reduced cyclic AMP production: evidence for the role of receptor-linked events. Life Sci 1982; 31:1987–92.

166. Matussek, N., M. Ackenheil, H. Hippius, et al. Effect of clonidine on growth hormone release in psychiatric patients and controls. Psychiatry Res 1980; 2:25–36.

167. Siever, L.J., T.W. Uhde, and D.L. Murphy. Possible subsensitization of alpha 2-adrenergic receptors by chronic monoamine oxidase inhibitor treatment in psychiatric patients. Psychiatry Res 1982; 6:293–302.

168. Checkley, S.A., A.P. Slade, and E. Shur. Growth hormone and other responses to clonidine in patients with endogenous depression. Br J Psychiatry 1981; 138:51–5.

169. Heninger, G.R., D.S. Charney, and L.H. Price. alpha 2-Adrenergic receptor sensitivity in depression. The plasma MHPG, behavioral, and cardiovascular responses to yohimbine. Arch Gen Psychiatry 1988; 45:718–26.

170. Price, L.H., D.S. Charney, A.L. Rubin, and G.R. Heninger. Alpha 2-adrenergic receptor function in depression. The cortisol response to yohimbine. Arch Gen Psychiatry 1986; 43:849–58.

171. Brodde, O.E., M. Anlauf, J. Arroyo, R. Wagner, F. Weber, and K.D. Buck. Hypersensitivity of adrenergic receptors and blood-pressure response to oral yohimbine in orthostatic hypotension. N Engl J Med 1983; 308:1033–4.

172. Sulser, F. New perspectives on the mode of action of antidepressant drugs. TIPS 1979; 92–94.

173. Pandey, G.N. and J.M. Davis. Treatment with antidepressants and down-regulation of beta-adrenergic receptors. Drug Dev Res 1983; 13:393–406.

174. Pandey, G.N., B.D. Brown, and J.M. Davis. Effect of treatment with some atypical antidepressants on ^{3}H-DHA binding in rat brain. Drug Dev Res 1985; 5:251–59.

175. Stahl, S.M. Peripheral models for the study of neurotransmitter receptors in man. Psychopharmacol Bull 1985; 21:663–71.

176. Pandey, G.N., P.G. Janicak, and J.M. Davis. Decreased beta-adrenergic receptors in the leukocytes of depressed patients. Psychiatry Res 1987; 22:265–73.

177. Klysner, R., A. Geisler, and R. Rosenberg. Enhanced histamine- and beta-adrenoceptor-mediated cyclic AMP formation in leukocytes from patients with endogenous depression. J Affect Disord 1987; 13:227–32.

178. Ebstein, R.P., B. Lerer, E.R. Bennett, et al. Lithium modulation of second messenger signal amplification in man: inhibition of phosphatidylinositol-specific phospholipase C and adenylate cyclase activity. Psychiatry Res 1988; 24:45–52.

179. Spitzer, R.L., J. Endicott, and E. Robins. Research diagnostic criteria: rationale and reliability. Arch Gen Psychiatry 1978; 35:773–82.

180. Kanof, P.D., E.F. Coccaro, C.A. Johns, M. Davidson, L.J. Siever, and K.L. Davis. Cyclic-AMP production by polymorphonuclear leukocytes in psychiatric disorders. Biol Psychiatry 1989; 25:413–20.

181. Williams, L.T., R. Snyderman, and R.J. Lefkowitz. Identification of beta-adrenergic receptors in human lymphocytes by (-) (3H) alprenolol binding. J Clin Invest 1976; 57:149–55.

182. Brodde, O.E., G. Engel, D. Hoyer, K.D. Bock, and F. Weber. The beta-adrenergic receptor in human lymphocytes: subclassification by the use of a new radio-ligand (+/−)-125 Iodocyanopindolol. Life Sci 1981; 29:2189–98.

183. O'Hara, N. and O.E. Brodde. Identical binding properties of (+/−)- and (−)-125Iodocyanopindolol to beta 2-adrenoceptors in intact human lymphocytes. Arch Int Pharmacodyn Ther 1984; 272:24–39.

184. Carstens, M.E., A.H. Engelbrecht, V.A. Russell, et al. Beta-adrenoceptors on lymphocytes of patients with major depressive disorder. Psychiatry Res 1987; 20:239–48.

185. Magliozzi, J.R., D. Gietzen, R.J. Maddock, et al. Lymphocyte beta-adrenoreceptor density in patients with unipolar depression and normal controls. Biol Psychiatry 1989; 26:15–25.

186. Wright, A.F., D.N. Crichton, J.B. Loudon, J.E. Morten, and C.M. Steel. Beta-adrenoceptor binding defects in cell lines from families with manic-depressive disorder. Ann Hum Genet 1984; 48:201–14.

187. Abdel-Latif, A.A. Calcium-mobilizing receptors, polyphosphoinositides, and the generation of second messengers. Pharmacol Rev 1986; 38:227–72.

188. Berridge, M.J. and R.F. Irvine. Inositol phosphates and cell signalling. Nature 1989; 341:197–205.

189. Nishizuka, Y. The molecular heterogeneity of protein kinase C and its implications for cellular regulation. Nature 1988; 334:661–5.

190. Nishizuka, Y. Intracellular signaling by hydrolysis of phospholipids and activation of protein kinase C. Science 1992; 258:607–14.

191. Jope, R.S., L. Song, P.P. Li, et al. The phosphoinositide signal transduction system is impaired in bipolar affective disorder brain. J Neurochem 1996; 66:2402–9.

192. Pacheco, M.A. and R.S. Jope. Phosphoinositide signaling in human brain. Prog Neurobiol 1996; 50:255–73.

193. Pacheco, M.A., C. Stockmeier, H.Y. Meltzer, J.C. Overholser, G.E. Dilley, and R.S. Jope. Alterations in phosphoinositide signaling and G-protein levels in depressed suicide brain. Brain Res 1996; 723:37–45.

194. Karege, F., P. Bovier, W. Rudolph, and J.M. Gaillard. Platelet phosphoinositide signaling system: an overstimulated pathway in depression. Biol Psychiatry 1996; 39:697–702.

195. Pandey, G.N., S.C. Pandey, and J.M. Davis. Peripheral adrenergic receptors in affective illness and schizophrenia. Pharmacol Toxicol 1990; 66 (Suppl 3):13–36.

196. Pandey, G.N., X. Ren, S.C. Pandey, Y. Dwivedi, R. Sharma, and P.G. Janicak. Hyperactive phosphoinositide signaling pathway in platelets of depressed patients: effect of desipramine treatment. Psychiatry Res 2001; 105:23–32.

197. Bezchlibnyk, Y. and L.T. Young. The neurobiology of bipolar disorder: focus on signal transduction pathways and the regulation of gene expression. Can J Psychiatry 2002; 47:135–48.

198. Avissar, S., Y. Nechamkin, L. Barki-Harrington, G. Roitman, and G. Schreiber. Differential G protein measures in mononuclear leukocytes of patients with bipolar mood disorder are state dependent. J Affect Disord 1997; 43:85–93.

199. Mitchell, P.B., H.K. Manji, G. Chen, et al. High levels of Gs alpha in platelets of euthymic patients with bipolar affective disorder. Am J Psychiatry 1997; 154:218–23.

200. Pandey, G.N., Y. Dwivedi, S.C. Pandey, et al. Low phosphoinositide-specific phospholipase C activity and expression of phospholipase C beta1 protein in the prefrontal cortex of teenage suicide subjects. Am J Psychiatry 1999; 156:1895–901.

201. Friedman, E., W. Hoau Yan, D. Levinson, T.A. Connell, and H. Singh. Altered platelet protein kinase C activity in bipolar affective disorder, manic episode. Biol Psychiatry 1993; 33:520–5.

202. Wang, H.Y., P. Markowitz, D. Levinson, A.S. Undie, and E. Friedman. Increased membrane-associated protein kinase C activity and translocation in blood platelets from bipolar affective disorder patients. J Psychiatr Res 1999; 33:171–9.

203. Manji, H.K., R. Etcheberrigaray, G. Chen, and J.L. Olds. Lithium decreases membrane-associated protein kinase C in hippocampus: selectivity for the alpha isozyme. J Neurochem 1993; 61:2303–10.

204. Manji, H.K., W.Z. Potter, and R.H. Lenox. Signal transduction pathways. Molecular targets for lithium's actions. Arch Gen Psychiatry 1995; 52:531–43.

205. Manji, H.K. and R.H. Lenox. Ziskind-Somerfeld Research Award. Protein kinase C signaling in the brain: molecular transduction of mood stabilization in the treatment of manic-depressive illness. Biol Psychiatry 1999; 46:1328–51.

206. Lenox, R.H. and H.K. Manji, Lithium, in American Psychiatric Press Textbook of Psychopharmacology, 2nd ed., A.F. Schatzberg and C.B. Nemeroff, Editors. 1998, American Psychiatric Press: Washington, DC, pp. 379–429.

207. Ikonomov, O.C. and H.K. Manji. Molecular mechanisms underlying mood stabilization in manic-depressive illness: the phenotype challenge. Am J Psychiatry 1999; 156:1506–14.

208. Jope, R.S. Anti-bipolar therapy: mechanism of action of lithium. Mol Psychiatry 1999; 4:117–28.

209. Pandey, G.N., Y. Dwivedi, J. SridharaRao, X. Ren, P.G. Janicak, and R. Sharma. Protein kinase C and phospholipase C activity and expression of their specific isozymes is decreased and expression of MARCKS is increased in platelets of bipolar but not in unipolar patients. Neuropsychopharmacology 2002; 26:216–28.

210. Aderem, A. The MARCKS brothers: a family of protein kinase C substrates. Cell 1992; 71:713–6.

211. Blackshear, P.J. The MARCKS family of cellular protein kinase C substrates. J Biol Chem 1993; 268:1501–4.

212. Watson, D.G. and R.H. Lenox. Chronic lithium-induced down-regulation of MARCKS in immortalized hippocampal cells: potentiation by muscarinic receptor activation. J Neurochem 1996; 67:767–77.

213. Watson, D.G., B.H. Wainer, and R.H. Lenox. Phorbol ester- and retinoic acid-induced regulation of the protein kinase C substrate MARCKS in immortalized hippocampal cells. J Neurochem 1994; 63:1666–74.

214. McNamara, R.K., T.M. Hyde, J.E. Kleinman, and R.H. Lenox. Expression of the myristoylated alanine-rich C kinase substrate (MARCKS) and MARCKS-related protein (MRP) in the prefrontal cortex and hippocampus of suicide victims. J Clin Psychiatry 1999; 60 (Suppl 2):21–6; discussion 40–1, 113–6.

215. Wang, L., D.G. Watson, and R.H. Lenox. Myristoylation alters retinoic acid-induced down-regulation of MARCKS in immortalized hippocampal cells. Biochem Biophys Res Commun 2000; 276:183–8.

216. Pandey, G.N., Y. Dwivedi, X. Ren, et al. Altered expression and phosphorylation of myristoylated alanine-rich C kinase substrate (MARCKS) in postmortem brain of suicide victims with or without depression. J Psychiatr Res 2003; 37:421–32.

217. Pandey, G.N. and Y. Dwivedi. Focus on protein kinase A and protein kinase C, critical components of signal transduction system, in mood disorders and suicide. Int J Neuropsychopharmacol 2005; 8:1–4.

218. Dwivedi, Y., R.R. Conley, R.C. Roberts, C.A. Tamminga, and G.N. Pandey. [(3)H]cAMP binding sites and protein kinase a activity in the prefrontal cortex of suicide victims. Am J Psychiatry 2002; 159:66–73.

219. Dwivedi, Y., H.S. Rizavi, P.K. Shukla, et al. Protein kinase A in postmortem brain of depressed suicide victims: altered expression of specific regulatory and catalytic subunits. Biol Psychiatry 2004; 55:234–43.

220. Shelton, R.C., D.H. Mainer, and F. Sulser. cAMP-dependent protein kinase activity in major depression. Am J Psychiatry 1996; 153:1037–42.

221. Manier, D.H., A. Eiring, R.C. Shelton, and F. Sulser. Beta-adrenoceptor-linked protein kinase A (PKA) activity in human fibroblasts from normal subjects and from patients with major depression. Neuropsychopharmacology 1996; 15:555–61.

222. Manier, D.H., R.C. Shelton, T.C. Ellis, C.S. Peterson, A. Eiring, and F. Sulser. Human fibroblasts as a relevant model to study signal transduction in affective disorders. J Affect Disord 2000; 61:51–8.

223. Rahman, S., P.P. Li, L.T. Young, O. Kofman, S.J. Kish, and J.J. Warsh. Reduced [3H]cyclic AMP binding in postmortem brain from subjects with bipolar affective disorder. J Neurochem 1997; 68:297–304.

224. Fields, A., P.P. Li, S.J. Kish, and J.J. Warsh. Increased cyclic AMP-dependent protein kinase activity in postmortem brain from patients with bipolar affective disorder. J Neurochem 1999; 73:1704–10.

225. Perez, J., D. Tardito, G. Racagni, E. Smeraldi, and R. Zanardi. Protein kinase A and Rap1 levels in platelets of untreated patients with major depression. Mol Psychiatry 2001; 6:44–9.

226. Chang, A., P.P. Li, and J.J. Warsh. cAMP-Dependent protein kinase (PKA) subunit mRNA levels in postmortem brain from patients with bipolar affective disorder (BD). Brain Res Mol Brain Res 2003; 116:27–37.

227. Karege, F., M. Schwald, P. Papadimitriou, C. Lachausse, and M. Cisse. The cAMP-dependent protein kinase A and brain-derived neurotrophic factor expression in lymphoblast cells of bipolar affective disorder. J Affect Disord 2004; 79:187–92.

228. Karege, F., C. Lambercy, M. Schwald, T. Steimer, and M. Cisse. Differential changes of cAMP-dependent protein kinase activity and 3H-cAMP binding sites in rat hippocampus during maturation and aging. Neurosci Lett 2001; 315:89–92.

229. Huang, E.J. and L.F. Reichardt. Neurotrophins: roles in neuronal development and function. Annu Rev Neurosci 2001; 24:677–736.

230. Altar, C.A., N. Cai, T. Bliven, et al. Anterograde transport of brain-derived neurotrophic factor and its role in the brain. Nature 1997; 389:856–60.

231. Bartrup, J.T., J.M. Moorman, and N.R. Newberry. BDNF enhances neuronal growth and synaptic activity in hippocampal cell cultures. Neuroreport 1997; 8:3791–4.

232. Kang, H. and E.M. Schuman. Long-lasting neurotrophin-induced enhancement of synaptic transmission in the adult hippocampus. Science 1995; 267:1658–62.

233. Schinder, A.F. and M. Poo. The neurotrophin hypothesis for synaptic plasticity. Trends Neurosci 2000; 23:639–45.

234. Thoenen, H. Neurotrophins and neuronal plasticity. Science 1995; 270:593–8.

235. Duman, R.S., S. Nakagawa, and J. Malberg. Regulation of adult neurogenesis by antidepressant treatment. Neuropsychopharmacology 2001; 25:836–44.

236. Leibenluft, E., B.A. Rich, D.T. Vinton, et al. Neural circuitry engaged during unsuccessful motor inhibition in pediatric bipolar disorder. Am J Psychiatry 2007; 164:52–60.

237. Pavuluri, M.N., M.M. O'Connor, E. Harral, and J.A. Sweeney. Affective neural circuitry during facial emotion processing in pediatric bipolar disorder. Biol Psychiatry 2007; 62:158–67.

238. Pearlson, G.D. Structural and functional brain changes in bipolar disorder: a selective review. Schizophr Res 1999; 39:133–40; discussion 162.

239. Pearlson, G.D. and A.E. Veroff. Computerised tomographic scan changes in manic-depressive illness. Lancet 1981; 2:470.

240. Rajkowska, G. Cell pathology in mood disorders. Semin Clin Neuropsychiatry 2002; 7:281–92.

241. Rajkowska, G. Cell pathology in bipolar disorder. Bipolar Disord 2002; 4:105–16.

242. Altar, C.A., R.E. Whitehead, R. Chen, G. Wortwein, and T.M. Madsen. Effects of electroconvulsive seizures and antidepressant drugs on brain-derived neurotrophic factor protein in rat brain. Biol Psychiatry 2003; 54:703–9.

243. Chen, B., D. Dowlatshahi, G.M. MacQueen, J.F. Wang, and L.T. Young. Increased hippocampal BDNF immunoreactivity in subjects treated with antidepressant medication. Biol Psychiatry 2001; 50:260–5.

244. Duman, R.S. and V.A. Vaidya. Molecular and cellular actions of chronic electroconvulsive seizures. J ECT 1998; 14:181–93.

245. Fukumoto, T., S. Morinobu, Y. Okamoto, A. Kagaya, and S. Yamawaki. Chronic lithium treatment increases the expression of brain-derived neurotrophic factor in the rat brain. Psychopharmacology (Berl) 2001; 158:100–6.

246. Gonul, A.S., F. Akdeniz, F. Taneli, O. Donat, C. Eker, and S. Vahip. Effect of treatment on serum brain-derived neurotrophic factor levels in depressed patients. Eur Arch Psychiatry Clin Neurosci 2005; 255:381–6.

247. Hashimoto, R., N. Takei, K. Shimazu, L. Christ, B. Lu, and D.M. Chuang. Lithium induces brain-derived neurotrophic factor and activates TrkB in rodent cortical neurons: an essential step for neuroprotection against glutamate excitotoxicity. Neuropharmacology 2002; 43:1173–9.

248. Karege, F., G. Vaudan, M. Schwald, N. Perroud, and R. La Harpe. Neurotrophin levels in postmortem brains of suicide victims and the effects of antemortem diagnosis and psychotropic drugs. Brain Research. Mol Brain Res 2005; 136:29–37.

249. Nibuya, M., S. Morinobu, and R.S. Duman. Regulation of BDNF and trkB mRNA in rat brain by chronic electroconvulsive seizure and antidepressant drug treatments. J Neurosci 1995; 15:7539–47.

250. Shimizu, E., K. Hashimoto, N. Okamura, et al. Alterations of serum levels of brain-derived neurotrophic factor (BDNF) in depressed patients with or without antidepressants. Biol Psychiatry 2003; 54:70–5.

251. Malone, K.M., G.L. Haas, J.A. Sweeney, and J.J. Mann. Major depression and the risk of attempted suicide. J Affect Disord 1995; 34:173–85.

252. Paykel, E.S. Life stress, depression and attempted suicide. J Hum Stress 1976; 2:3–12.

253. Westrin, A. Stress system alterations and mood disorders in suicidal patients. A review. Biomed Pharmacother 2000; 54:142–5.

254. Smith, M.A., S. Makino, R. Kvetnansky, and R.M. Post. Stress and glucocorticoids affect the expression of brain-derived neurotrophic factor and neurotrophin-3 mRNAs in the hippocampus. J Neurosci 1995; 15:1768–77.

255. Ueyama, T., Y. Kawai, K. Nemoto, M. Sekimoto, S. Tone, and E. Senba. Immobilization stress reduced the expression of neurotrophins and their receptors in the rat brain. Neurosci Res 1997; 28:103–10.

256. Dwivedi, Y., H.S. Rizavi, R.R. Conley, R.C. Roberts, C.A. Tamminga, and G.N. Pandey. Altered gene expression of brain-derived neurotrophic factor and receptor tyrosine

kinase B in postmortem brain of suicide subjects. Arch Gen Psychiatry 2003; 60:804–15.

257. Karege, F., G. Perret, G. Bondolfi, M. Schwald, G. Bertschy, and J.M. Aubry. Decreased serum brain-derived neurotrophic factor levels in major depressed patients. Psychiatry Res 2002; 109:143–8.

258. Shimizu, E., K. Hashimoto, N. Okamura, et al. Alterations of serum levels of brain-derived neurotrophic factor (BDNF) in depressed patients with or without antidepressants. Biol Psychiatry 2003; 54:70–5.

259. Cunha, A.B., B.N. Frey, A.C. Andreazza, et al. Serum brain-derived neurotrophic factor is decreased in bipolar disorder during depressive and manic episodes. Neurosci Lett 2006; 398:215–9.

260. Machado-Vieira, R., M.O. Dietrich, R. Leke, et al. Decreased plasma brain derived neurotrophic factor levels in unmedicated bipolar patients during manic episode. Biol Psychiatry 2007; 61:142–4.

261. Pandey, G.N., H.S. Rizavi, Y. Dwivedi, and M.N. Pavuluri. Brain-derived neurotrophic factor gene expression in pediatric bipolar disorder: effects of treatment and clinical response. J Am Acad Child Adolesc Psychiatry 2008; 47:1077–85.

262. Karege, F., M. Schwald, and M. Cisse. Postnatal developmental profile of brain-derived neurotrophic factor in rat brain and platelets. Neurosci Lett 2002; 328:261–4.

263. Kabiersch, A., A. del Rey, C.G. Honegger, and H.O. Besedovsky. Interleukin-1 induces changes in norepinephrine metabolism in the rat brain. Brain Behav Immun 1988; 2:267–74.

264. Morikawa, O., N. Sakai, H. Obara, and N. Saito. Effects of interferon-alpha, interferon-gamma and cAMP on the transcriptional regulation of the serotonin transporter. Eur J Pharmacol 1998; 349:317–24.

265. Myint, A.M. and Y.K. Kim. Cytokine-serotonin interaction through IDO: a neurodegeneration hypothesis of depression. Med Hypotheses 2003; 61:519–25.

266. Zalcman, S., J.M. Green-Johnson, L. Murray, et al. Cytokine-specific central monoamine alterations induced by interleukin-1, -2 and -6. Brain Res 1994; 643:40–9.

267. Anisman, H., L. Kokkinidis, and Z. Merali. Further evidence for the depressive effects of cytokines: anhedonia and neurochemical changes. Brain Behav Immun 2002; 16:544–56.

268. Hopkins, S.J. and N.J. Rothwell. Cytokines and the nervous system. I: expression and recognition. Trends Neurosci 1995; 18:83–8.

269. Kronfol, Z. and D.G. Remick. Cytokines and the brain: implications for clinical psychiatry. Am J Psychiatry 2000; 157:683–94.

270. Muller, N. and M. Ackenheil. Psychoneuroimmunology and the cytokine action in the CNS: implications for psychiatric disorders. Prog Neuropsychopharmacol Biol Psychiatry 1998; 22:1–33.

271. Bonaccorso, S., V. Marino, M. Biondi, F. Grimaldi, F. Ippoliti, and M. Maes. Depression induced by treatment with interferon-alpha in patients affected by hepatitis C virus. J Affect Disord 2002; 72:237–41.

272. Bonaccorso, S., V. Marino, A. Puzella, et al. Increased depressive ratings in patients with hepatitis C receiving interferon-alpha-based immunotherapy are related to

interferon-alpha-induced changes in the serotonergic system. J Clin Psychopharmacol 2002; 22:86–90.

273. Bonaccorso, S., A. Puzella, V. Marino, et al. Immunotherapy with interferon-alpha in patients affected by chronic hepatitis C induces an intercorrelated stimulation of the cytokine network and an increase in depressive and anxiety symptoms. Psychiatry Res 2001; 105:45–55.

274. Collier, J. and R. Chapman. Combination therapy with interferon-alpha and ribavirin for hepatitis C: practical treatment issues. BioDrugs 2001; 15:225–38.

275. Hunt, C.M., J.A. Dominitz, B.P. Bute, B. Waters, U. Blasi, and D.M. Williams. Effect of interferon-alpha treatment of chronic hepatitis C on health-related quality of life. Dig Dis Sci 1997; 42:2482–6.

276. Loftis, J.M. and P. Hauser. The phenomenology and treatment of interferon-induced depression. J Affect Disord 2004; 82:175–90.

277. Malaguarnera, M., I. Di Fazio, S. Restuccia, G. Pistone, L. Ferlito, and L. Rampello. Interferon alpha-induced depression in chronic hepatitis C patients: comparison between different types of interferon alpha. Neuropsychobiology 1998; 37:93–7.

278. Pariante, C.M., M.G. Orru, A. Baita, M.G. Farci, and B. Carpiniello. Treatment with interferon-alpha in patients with chronic hepatitis and mood or anxiety disorders. Lancet 1999; 354:131–2.

279. Pavol, M.A., C.A. Meyers, J.L. Rexer, A.D. Valentine, P.J. Mattis, and M. Talpaz. Pattern of neurobehavioral deficits associated with interferon alfa therapy for leukemia. Neurology 1995; 45:947–50.

280. Renault, P.F., J.H. Hoofnagle, Y. Park, et al. Psychiatric complications of long-term interferon alfa therapy. Arch Intern Med 1987; 147:1577–80.

281. Valentine, A.D., C.A. Meyers, M.A. Kling, E. Richelson, and P. Hauser. Mood and cognitive side effects of interferon-alpha therapy. Semin Oncol 1998; 25:39–47.

282. Yirmiya, R., J. Weidenfeld, Y. Pollak, et al. Cytokines, "depression due to a general medical condition," and antidepressant drugs. Adv Exp Med Biol 1999; 461:283–316.

283. Anisman, H., A.V. Ravindran, J. Griffiths, and Z. Merali. Endocrine and cytokine correlates of major depression and dysthymia with typical or atypical features. Mol Psychiatry 1999; 4:182–8.

284. Berk, M., A.A. Wadee, R.H. Kuschke, and A. O'Neill-Kerr. Acute phase proteins in major depression. J Psychosom Res 1997; 43:529–34.

285. Kim, Y.K., I.B. Suh, H. Kim, et al. The plasma levels of interleukin-12 in schizophrenia, major depression, and bipolar mania: effects of psychotropic drugs. Mol Psychiatry 2002; 7:1107–14.

286. Kubera, M., G. Kenis, E. Bosmans, et al. Plasma levels of interleukin-6, interleukin-10, and interleukin-1 receptor antagonist in depression: comparison between the acute state and after remission. Pol J Pharmacol 2000; 52:237–41.

287. Maes, M. Evidence for an immune response in major depression: a review and hypothesis. Prog Neuropsychopharmacol Biol Psychiatry 1995; 19:11–38.

288. Maes, M., E. Bosmans, R. De Jongh, G. Kenis, E. Vandoolaeghe, and H. Neels. Increased serum IL-6 and IL-1 receptor antagonist concentrations in major depression

and treatment resistant depression. Cytokine 1997; 9:853–8.

289. Maes, M., H.Y. Meltzer, E. Bosmans, et al. Increased plasma concentrations of interleukin-6, soluble interleukin-6, soluble interleukin-2 and transferrin receptor in major depression. J Affect Disord 1995; 34:301–9.

290. Maes, M., E. Vandoolaeghe, R. Ranjan, E. Bosmans, R. Bergmans, and R. Desnyder. Increased serum interleukin-1-receptor-antagonist concentrations in major depression. J Affect Disord 1995; 36:29–36.

291. Musselman, D.L., A.H. Miller, M.R. Porter, et al. Higher than normal plasma interleukin-6 concentrations in cancer patients with depression: preliminary findings. Am J Psychiatry 2001; 158:1252–7.

292. Nassberger, L. and L. Traskman-Bendz. Increased soluble interleukin-2 receptor concentrations in suicide attempters. Acta Psychiatr Scand 1993; 88:48–52.

293. O'Brien, S.M., L.V. Scott, and T.G. Dinan. Cytokines: abnormalities in major depression and implications for pharmacological treatment. Hum Psychopharmacol 2004; 19:397–403.

294. Sluzewska, A., J. Rybakowski, E. Bosmans, et al. Indicators of immune activation in major depression. Psychiatry Res 1996; 64:161–7.

295. Song, C., T. Dinan, and B.E. Leonard. Changes in immunoglobulin, complement and acute phase protein levels in the depressed patients and normal controls. J Affect Disord 1994; 30:283–8.

296. Thomas, A.J., S. Davis, C. Morris, E. Jackson, R. Harrison, and J.T. O'Brien. Increase in interleukin-1beta in late-life depression. Am J Psychiatry 2005; 162:175–7.

297. Tuglu, C., S.H. Kara, O. Caliyurt, E. Vardar, and E. Abay. Increased serum tumor necrosis factor-alpha levels and treatment response in major depressive disorder. Psychopharmacology (Berl) 2003; 170:429–33.

298. Weisse, C.S. Depression and immunocompetence: a review of the literature. Psychol Bull 1992; 111:475–89.

299. Carpenter, L.L., G.R. Heninger, R.T. Malison, A.R. Tyrka, and L.H. Price. Cerebrospinal fluid interleukin (IL)-6 in unipolar major depression. J Affect Disord 2004; 79:285–9.

300. Minami, M., Y. Kuraishi, T. Yamaguchi, S. Nakai, Y. Hirai, and M. Satoh. Immobilization stress induces interleukin-1 beta mRNA in the rat hypothalamus. Neurosci Lett 1991; 123:254–6.

301. Anisman, H. and Z. Merali. Cytokines, stress, and depressive illness. Brain Behav Immun 2002; 16:513–24.

302. Connor, T.J. and B.E. Leonard. Depression, stress and immunological activation: the role of cytokines in depressive disorders. Life Sci 1998; 62:583–606.

303. Leonard, B.E. and C. Song. Stress, depression, and the role of cytokines. Adv Exp Med Biol 1999; 461:251–65.

304. Merali, Z., S. Lacosta, and H. Anisman. Effects of interleukin-1beta and mild stress on alterations of norepinephrine, dopamine and serotonin neurotransmission: a regional microdialysis study. Brain Res 1997; 761:225–35.

305. Tilders, F.J. and E.D. Schmidt. Cross-sensitization between immune and non-immune stressors. A role in the etiology of depression? Adv Exp Med Biol 1999; 461:179–97.

306. Kronfol, Z., J. Silva, Jr., J. Greden, S. Dembinski, R. Gardner, and B. Carroll. Impaired lymphocyte function in depressive illness. Life Sci 1983; 33:241–7.

307. Calabrese, J.R., R.G. Skwerer, B. Barna, et al. Depression, immunocompetence, and prostaglandins of the E series. Psychiatry Res 1986; 17:41–7.

308. Maes, M., E. Bosmans, E. Suy, C. Vandervorst, C. DeJonckheere, and J. Raus. Depression-related disturbances in mitogen-induced lymphocyte responses and interleukin-1 beta and soluble interleukin-2 receptor production. Acta Psychiatr Scand 1991; 84:379–86.

309. Herberman, R.B. Effect of alpha-interferons on immune function. Semin Oncol 1997; 24:S 9–78–S 9–80.

310. Irwin, M. and J.C. Gillin. Impaired natural killer cell activity among depressed patients. Psychiatry Res 1987; 20:181–2.

311. Herbert, T.B. and S. Cohen. Depression and immunity: a meta-analytic review. Psychol Bull 1993; 113:472–86.

312. Irwin, M. Immune correlates of depression. Adv Exp Med Biol 1999; 461:1–24.

313. Capuron, L., C.L. Raison, D.L. Musselman, D.H. Lawson, C.B. Nemeroff, and A.H. Miller. Association of exaggerated HPA axis response to the initial injection of interferon-alpha with development of depression during interferon-alpha therapy. Am J Psychiatry 2003; 160:1342–5.

314. Maes, M., S. Scharpe, H.Y. Meltzer, et al. Relationships between interleukin-6 activity, acute phase proteins, and function of the hypothalamic-pituitary-adrenal axis in severe depression. Psychiatry Res 1993; 49:11–27.

315. Carroll, B.J. The dexamethasone suppression test for melancholia. Br J Psychiatry 1982; 140:292–304.

316. Maes, M. Major depression and activation of the inflammatory response system. Adv Exp Med Biol 1999; 461:25–46.

317. Connor, T.J., C. Song, B.E. Leonard, Z. Merali, and H. Anisman. An assessment of the effects of central interleukin-1beta, -2, -6, and tumor necrosis factor-alpha administration on some behavioural, neurochemical, endocrine and immune parameters in the rat. Neuroscience 1998; 84:923–33.

318. Yirmiya, R. Endotoxin produces a depressive-like episode in rats. Brain Res 1996; 711:163–74.

319. Trzonkowski, P., J. Mysliwska, B. Godlewska, et al. Immune consequences of the spontaneous pro-inflammatory status in depressed elderly patients. Brain Behav Immun 2004; 18:135–48.

320. Frommberger, U.H., J. Bauer, P. Haselbauer, A. Fraulin, D. Riemann, and M. Berger. Interleukin-6-(IL-6) plasma levels in depression and schizophrenia: comparison between the acute state and after remission. Eur Arch Psychiatry Clin Neurosci 1997; 247:228–33.

321. Sluzewska, A., J.K. Rybakowski, M. Laciak, A. Mackiewicz, M. Sobieska, and K. Wiktorowicz. Interleukin-6 serum levels in depressed patients before and after treatment with fluoxetine. Ann N Y Acad Sci 1995; 762:474–6.

Chapter 37
The Diagnosis of Alcoholism Through the Identification of Biochemical Markers in Hair

Nadia De Giovanni

Abstract Alcoholic beverages and the heavy problems linked to their abuse have been familiar in human societies since the beginning of recorded history. Alcoholism is a social, economic and medical question, that involve a wide population in almost all ethnic groups, and the evidence of alcohol abuse is often very difficult mainly for the capability of abusers to keep secret their trouble. Hence the clinician needs various information to make the exact diagnosis, including the acquisition of the complete history of the patient and the investigation of clinical signs. Today, the study of such a topic could be markedly improved by the systematic use of laboratory tests, such as blood ethanol, serum gamma-glutamyl transferase (γ-GT), the mean corpuscular volume of erythrocytes (MCV) and the carbohydrate-deficient transferrin (CDT), currently the most specific marker of alcohol abuse.

In the last years some minor ethanol metabolites in hair matrix (in particular ethyl glucuronide and fatty acid ethyl esters) have been studied, for the unique ability of hair to serve as a long-term storage of xenobiotics with respect to the temporal appearance in blood. Over the last 20 years in fact, hair testing has gained increasing attention for the retrospective investigation of chronic drug abuse because of the window of drug detection is dramatically extended to weeks and months. The chance to detect minor ethanol metabolites in hair have been proposed in the early 2000 and ethyl glucuronide and fatty acid ethyl esters seem to satisfy the prerequisites requested by the alcoholism diagnosis.

Ethyl glucuronide (EtG) is a non-volatile, water-soluble, direct metabolite of ethanol. It has received much recent attention as a sensitive and specific biological marker of alcoholism. Formed in the liver via conjugation of ethanol with activated glucuronate, EtG remains detectable in serum, plasma, and hair for days after ethanol abuse. The use of this marker detected in hair alone and complementary with other biological state markers is expected to lead to significant improvement in treatment outcome, therapy efficacy and cost reduction.

Fatty acid ethyl esters (FAEE) are products of the non-oxidative ethanol metabolism, which are known to be detectable in blood about 24 h after the last alcohol intake. After deposition in hair they should be suitable long-term markers of chronically elevated alcohol consumption. It was shown by some investigations that FAEE are also present in sebum, that there is no strong difference in their concentrations between pubic, chest and scalp hair, and that they are detectable in hair segments after a 2 months period of abstinence. From these experiences it follows that the measurement of FAEE concentrations in hair is a useful way for a retrospective detection of alcohol abuse, able to discriminate between heavy drinkers and teetotallers. Moreover the use of FAEE into neonatal hair to objectively identify children exposed to alcohol in utero may be a helpful approach to diagnose foetal alcohol spectrum disorder.

Keywords Alcoholism · biochemical markers · hair · ethyl glucuronide · fatty acid ethyl esters

Abbreviations ARBDs: Alcohol-related birth defects; ARNDs: Alcohol-related neurodevelopment disorders; BAC: Blood alcohol concentration; CDT: Carbohydrate-deficient transferrin; CE: Cocaethylene; C-FAEE: Sum of concentrations of ethyl myristate, palmitate, oleate, stearate; EtG: Ethyl glucuronide; EtS: Ethyl sulphate;

N. De Giovanni
Istituto di Medicina Legale, Università Cattolica del
Sacro Cuore, Roma, Italy

FAEE: Fatty acid ethyl esters; FAS: Foetal Alcohol Syndrome; γ-GT: γ-Glutamyl transferase; MCV: Erythrocyte Mean Corpuscular Volume; mg: Milligrams; ng: Nanograms; PEth: Phosphatidylethanol; SOHT: Society of Hair Testing; SQ: Squalene.

Introduction

The misuse of alcohol represents one of the leading causes of preventable death, illness and injury in many societies throughout the world, in almost every ethnic group.

The "World Health Organization" has estimated that a high percentage of population drinks alcoholic beverages worldwide, and most of abusers suffer alcohol use disorders.[1]

Despite the wide ranging health implications associated with alcohol abuse, its consumption continues to be socially accepted in most countries. Alcohol can therefore be considered the most widely consumed licit drug in the world and its consumption can be associated with a variety of adverse health and social consequences related to its intoxicating, toxic and dependence-producing properties.[2] Furthermore we must worry about the emerging trend of more harmful patterns in drinking, especially among young people.[3,4]

Adverse effects of alcohol have been demonstrated for many disorders, including liver cirrhosis, mental illness, several types of cancer, pancreatitis, gastrointestinal and cardiovascular diseases, immunological disorders, lung, skeletal and muscular diseases, reproductive disorders[5,6] and pre-natal harm,[7] including an increased risk of prematurity, low birth weight and damage to the foetus among pregnant women.[8,9]

Alcohol use is also strongly related to social consequences such as drink driving injuries and fatalities, aggressive behaviour, family's disruption and reduced job productivity.[10,11] The risk of social harms is generally dose dependent, hence the higher the level of alcohol consumption, the frequency of drinking and the frequency and volume of episodic heavy drinking, the more serious is the crime or injury. Men are more likely than women to drink at all and to drink more when they do, with the gap greater for riskier behaviour.

Considerable lives and money could be saved if one could detect early stages of lapsing/relapsing behaviour in addicted persons (e.g., in safety-sensitive workplaces) and could disclose harmful drinking in social drinkers.

These preliminary remarks suggest that alcoholism is one of the most frequent addictions that present particular interest in forensic and clinical medicine. Unfortunately, the diagnosis of alcoholism is still difficult for the clinician because alcoholics can easily hide their problem when they approach the doctor. At the moment, it could be performed using various information such as the acquisition of accurate and complete history of the patient, the use of questionnaires and the clinical signs of alcoholism.[12,13] Objective laboratory evidence of heavy drinking is needed to help the clinicians to raise the possible issue of alcohol use as the real cause of symptoms. The possibility of detecting the use of ethanol for a relatively long period of time after intake could be also useful for monitoring abstinence in a number of cases of clinical and forensic interest such as withdrawal treatment, driving license reissue/renewal, minor adoption.[14]

The identification of biochemical markers able to give a definite response is hence an issue in the diagnosis of chronic heavy alcohol use, of early stage alcohol abuse and dependence, as well as in preventing/diagnosing/treating chronic diseases with an alcohol related origin.

The search of such indicators has been a primary target for many years with important implications both clinical and forensic[14–16]; unfortunately today a unique substance able to give such information does not exist yet. However researches are in continuous evolution and scientists are pressed to explore for biological markers useful for the objective measure of alcoholism.

Ethanol Metabolism

Alcoholic beverages contain ethanol, a small neuropsychotropic molecule with depressing action on the Central Nervous System. Being a small molecule, weakly charged, it easily crosses cell membranes and readily reaches the equilibrium between blood and tissues.

After alcoholic beverages intake, ethanol is mainly metabolized by a two stage oxidation process that in the liver converts a proportion of the ethanol (about 90–95%) to acetaldehyde then catabolized on to acetyl coenzyme-A which is in turn oxidized through the Krebs cycle (Fig. 37.1). The remaining portion is excreted by kidney (0.5–2%), lung (1.6–6%) and skin

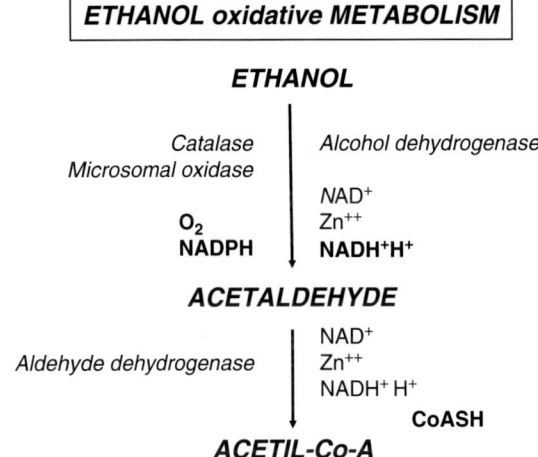

Fig. 37.1 Ethanol oxidative metabolism

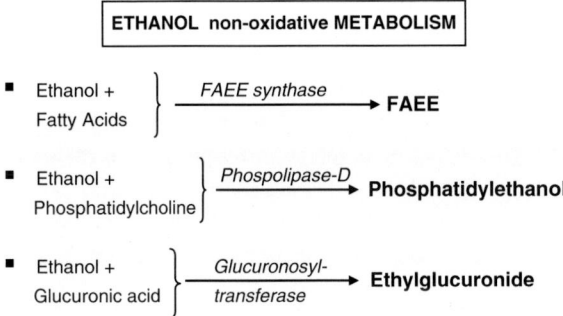

Fig. 37.2 Ethanol non-oxidative metabolism

(0–0.5%). A small amount undergoes to a minor metabolism that produces various substances by the reaction of ethanol with endogenous compounds (Fig. 37.2).

Acute and Chronic Intoxications

Alcohol intake can cause two kinds of intoxications, acute and chronic. Acute intoxication happens with a high amount of alcohol intake in a brief period. This pathology is easily diagnosed also with the support of ethanol detection in blood, which amount is strictly related to the impairment of the subject.[17]

Chronic intoxication vice versa could be very difficult to pick out, because alcoholics are able to hide their trouble often drinking in secret, so neither the family nor the friends have idea of their problem. The diagnosis of chronic alcohol abuse can consequently be problematic because people can long abstain drinking before to present himself to the doctor.

Chronic alcoholism, which is strictly associated with hepatic, pancreatic, and myocardial diseases, can be considered one of the major health problems in the world. In addition to the chronic diseases that may develop in those who drink large amounts of alcohol over a number of years, alcohol use is also associated with an increased risk of acute health conditions, such as injuries, including traffic accidents.[18]

The frequency and volume of episodic heavy drinking are of particular importance in increasing the risk of injuries and certain cardiovascular diseases (coronary heart disease and stroke). Alcohol hence places a significant burden on several aspects of human life.

Considering the large dimension of alcohol-associated problems, the diagnosis of excessive alcohol consumption is an important task from the medical point of view.

Alcohol Abuse During Pregnancy

Although many women give up alcohol when pregnant, a significant number continue to drink, and some continue to drink at harmful levels. Hence an important people's category is constituted by women which alcohol use during pregnancy can lead to serious foetal malformations.[19] Only a few neonates exposed in utero to a high maternal use of alcohol, show typical changes of "Foetal Alcohol Syndrome" and information obtained by mothers are often not credible; consequently it is frequently impossible to perform a correct diagnosis.[20,21]

Foetal alcohol syndrome (FAS), alcohol-related birth defects (ARBDs), and alcohol-related neurodevelopment disorders (ARNDs) in neonates are the more frequent consequences of maternal alcohol consumption during pregnancy.[22,23] Facial characteristics are generally associated with FAS, while ARBDs and ARNDs are more difficult to diagnose. Moreover foetal exposure to alcohol can cause central nervous system dysfunction, pre- and post-natal growth problems, cardiac defects in neonates, attention deficit disorders and mental retardation in older children.

Up to now, diagnosis of foetal alcohol effect has depended largely on maternal interview.[24] Today clinical tests are becoming more widely used, and the analysis could not only be performed on biological samples collected by the mothers, but also using meconium[23,25,26] and neonates hair.[22,27] These biological samples are already a well-established approach to investigate in-utero drug exposure, and they are now becoming a promising reference to detect mother use/abuse of alcohol.[28]

Obviously, foetal hair growth, hair cycle and the steady substance exchange with amniotic fluid before birth must be taken into account when performing neonate's hair analysis.

Objective Diagnosis of Alcoholism

In clinical as well as in forensic practice, biological markers of high sensitivity and specificity capable of monitoring alcohol consumption of abusers in treatment for alcohol dependence, are required.[14–16] The markers routinely used today cannot be considered satisfactory in respect of these parameters. They are usually enzymatic or haematological parameters with sensitivity and specificity not always satisfactory, for the possible contamination by different pathologies not related to alcoholism, by nutritional factors, metabolic disorders.[29] Furthermore, they can be influenced by age, gender and a variety of substances besides non-alcohol-associated diseases and do not cover fully the time axis for alcohol intake.[30] Therefore, current biological markers remain suboptimal with regard to sensitivity and specificity for monitoring recent alcohol consumption in various settings.

Difficulties linked to the diagnosis of alcohol abuse could be exceeded with the knowledge of an absolute marker able to give a definite response. Such a marker doesn't exist yet, but in the last years various indicators have been studied that could help in the diagnosis of use/abuse of alcohol and in alcoholism related problems.

Biological Markers

Biomarkers of alcohol intake can be short-term or long-term markers, depending on the compounds identified in different biological samples.[14,16]

The measure of ethanol itself in blood, urine or expired air could be considered a short-term marker of ethanol intake because of its rapid metabolism and elimination from the body fluids; this particular measure doesn't distinguish between chronic and acute use, and the time window is confined to just a few hours.

On the contrary, the detection of chronic alcohol abuse requires long-term markers, able to identify alcoholism with high specificity.

Long-term markers of alcoholism can be classified in two main groups constituted by indirect and direct markers.

Indirect markers involve all the pathological changes in metabolism and biochemistry of alcoholics, such as the increase of some clinical parameters, liver enzyme activity gamma-glutamyl transferase (γ-GT) and some other hepatic enzymes, the increase of erythrocyte mean corpuscular volume (MCV) or the carbohydrate deficient transferrin (CDT).[29,31] Of the known markers of chronic heavy drinking, CDT is the most widely studied and used.[32,33] Nevertheless, a significant percentage of heavy drinkers test negative for CDT, resulting in a highly specific but moderately sensitive tool. All clinical parameters currently used to diagnose the alcoholic abuse today are indirect markers.

In spite of the enormous advantages of these indirect markers, we must take into account that anomalous levels of such parameters have not been enough specificity, because they can also originate from hepatic diseases or other pathological reasons. Most of these compounds can be measured in serum or blood.

Direct markers contain all the carbon atoms of ethanol and include ethanol itself, minor alcohol metabolites, follow-up products of acetaldehyde.

In particular ethanol metabolites show the level of alcohol use associated to biochemical changes in the body and they can be successfully used as markers. Some of these compounds can be considered short-term or long-term indicators depending on the biological samples used. In fact, as for drugs, ethanol metabolites could be incorporated in hair, biological matrix useful to diagnose past use/abuse.[34]

Hair Matrix

Hair is a unique biological matrix that shows a number of advantages over traditional media, like urine and blood. First of all the wide time window of drug detection, dramatically extended up to weeks, months or even years, in relation to hair length, allows the evaluation of a long-term history of use/abuse. Hence, hair analysis gives information regarding a time window earlier and longer than blood and urine, so becoming an important tool in forensic field when it is necessary to confirm or to exclude drug addiction.[35–37]

Hair is composed of protein (keratin essentially, 65–95%), water (15–35%), lipids (1–9%) and minerals (<1%). The shaft consists of keratinized cells glued by cell membrane complex that together form three concentric structures: cuticle, cortex and medulla. Hair originates from hair follicle located 3–5 mm below the skin surface, which is surrounded by a rich capillary system that provides the metabolic material necessary to hair growth (Fig. 37.3). The follicle is closely associated with glands (sebaceous and apocrine) and each hair belongs to a sebaceous gland with the duct leading to the upper part of the root to ensure that the mature hair is bathed in sebum for 2–3 days prior to reaching the skin surface.

Hair shaft grows in cycles that are probably the most rapid of all human tissues. The cycles cover three stages. The active growing stage called "anagen" lasts about 4–8 years (<6 months for non-head hair); the transition stage called "catagen" has a life of a few weeks; the resting stage "telogen" goes on 4–6 months. The individual length of hair depends on stage duration and growth rate, which is approximately 0.22–0.52 mm/day (or 0.6–1.42 cm/month) for head hair; moreover growth rate is determined by hair type and anatomical location. In fact there are significant differences both in the proportions anagen/telogen hair and in growth rate from various anatomical sites.

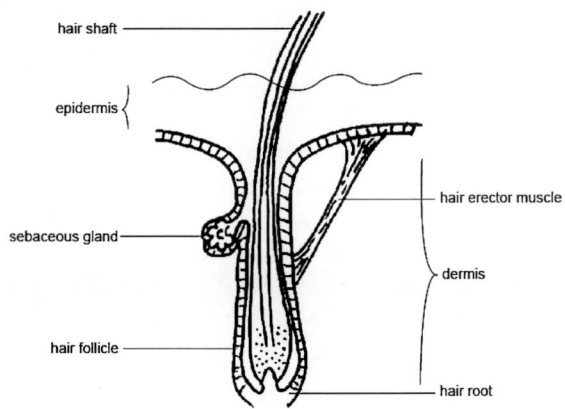

Fig. 37.3 Schematic representation of human hair shaft

These parameters also depend on race, sex, age and state of health. At any one time, approximately 85% of adult scalp hair is in the growing phase (anagen) with the remaining 15% in the resting phase (telogen). The significant consequence of cyclic growth is hair age heterogeneity respect to the distance from the skin.

Hair samples are particularly useful to prove chronic exposure to drugs or other poisons. Therefore, they can be used as diagnostic tool for clinical detection of drug abuse, for gestational drug exposure in neonates and for elucidation of other chronic poisonings; its use is also growing in pre-employment screening, in forensic sciences and for doping control.[38] Hair collection is non-invasive and easy to perform and may be achieved under close supervision. The usefulness of the latter is evident in forensic situations where it is necessary to prevent adulteration or substitution.

Historically the use of hair matrix for the detection of xenobiotics can be traced back to the early 1900 with the analysis of heavy metals; the first determination of arsenic in hair of an exhumed body was published in 1858.[39] Since then the possible use of hair in the search of heavy metals has been suggested.

The opportunity to detect drugs of abuse on the contrary, dates to the end of the seventieth. In 1979 Baumgartner et al.[40] published their experience over the determination of opiates in hair, demonstrating the chance to use hair analysis as a tool to identify the previous use of such compounds. Hence, since 1980 researchers began to test hair matrix for drugs contributing to progress in this field.[41]

Today the analysis of xenobiotics in hair is a routine task in forensic laboratories,[42] because the scientific community (Society of Hair Testing, founded in 1995 – SOHT)[43] expressed favourable consensus about the role of "drug testing" in hair and made some recommendations about specimen collection, specimen handling procedures, criteria for obtaining positive results, metabolites to be detected and metabolite-to-parent drug ratios.[44,45]. Moreover it has been demonstrated that the concentration along the hair shaft differed and corresponded with the time course of drug intake, hence allowing re-enacting the abuse history.

The importance of pre-analytical steps in hair analysis is often underestimated and collection procedures have not been standardized yet, even if the SOHT published the recommendations.[44] In most published studies, samples are obtained from random locations on the scalp, also if the best area to collect hair is at the back of the head, called the "*vertex posterior*".[46] Compared with other regions of the head, this area has less variability in the hair growth rate, the number of hairs in the growing phase is more constant and the hair is less subjected to age- and sex-related influences. The effect of heterogeneity of hair growth is that hair cannot be collected singularly, but it is necessary to take a lock to ensure the samples of hair in anagen phase.[46]

The proximal end of the sample (the nearest to the skin) must be clearly marked when the shaft is longer than 2–3 cm. It would be possible in this case to perform the segmentation of hair, hence allowing reconstructing the history of abuse. Hair samples should be stored under dry and dark conditions at room temperature. The simplest way is in an appropriately labelled paper envelope. Storage in plastic bags should be avoided because of contamination by softeners (i.e., plasticizers) and since the plastic can potentially extract lipophilic substances from hair.

Using proper conditions, most drugs or metabolites in hair are very stable and can be detected after years of storage.

At the end of the twentieth century, biochemical markers of alcoholism were firstly detected in hair of alcoholics, and just as for drugs in hair, the detection of minor ethanol metabolites in this matrix may help the objective diagnosis of alcohol use/abuse.[34]

Incorporation of Drugs

The mechanisms of drug incorporation and retention into hair have not been completely clarified and many factors are likely to affect the concentration of drugs in hair, so complicating the interpretation of results. It is generally proposed that xenobiotics can enter into hair by at least three stages: from blood during hair formation, from sweat and sebum and from external environment.[46] To explain the incorporation of exogenous substances in hair some theoretical models have been proposed.[47,48]

The first incorporation model hypothesizes an active or passive diffusion from blood capillaries into growing cells on the basis of the follicle. When the hair grows this area keratinizes and trap the substances becoming impervious to the external surroundings. This period would correspond to a phase of drug exposure of about 3 days.

Since hair biology is very complex, the incorporation of drugs cannot be completely explained by the first model, hence an alternative representation was proposed, supported by experimental data.[46,47] The new theory envisages a combination of the previous hypothesis and the diffusion from body secretions (sweat, sebaceous glands) during the development of the shaft, indicating that drugs enter hair by various mechanisms in a variety of locations, times and sources. Substances in fact can be incorporated also from deep skin compartments during hair shaft formation. The deposition by diffusion from sweat or sebum secretions into the completed hair shaft is an important alternative mechanism.

A third hypothesis estimate the previous together with the possible environmental pollution after the growth of the shaft. In fact, substances can be deposited from the surroundings, and this is particularly important when it is necessary to distinguish active intake of drugs from passive consumption. This model rules out the existence of inaccessible areas to the environment.

Drug transport through bio-membranes is also influenced by some other parameters that also affect their absorption from cells in hair growing follicles. In particular molecular weight and chemical structure of the substance, its chemical properties, micro-environment, the nature of bio-membranes, haematic flow, the bond with plasmatic proteins, lipidic solubility of drugs, the ratio between ionized and non-ionized substances, as well as the physical/physiological characteristics of the individual, strongly influence which mechanism will dominate.[48–50]

From the structural point of view, three main factors influence the drug incorporation: the melanin content of hair, the lipophilicity and the basicity of the substance itself.[36,37]

In vivo and in vitro binding studies on drug–melanin interaction [47] have demonstrated differences in drug concentrations for various substances from pigmented and non-pigmented hair fibres. The concentration of basic drugs in pigmented hair could be about tenfold higher than non-pigmented hair.

Lipophilicity of substances is the second important element that controls drug absorption; in fact lipophilic organic molecules can easily penetrate membranes and spread according to the concentration gradient in matrix cells. On the contrary, for hydrophilic molecules or organic ions of medium molecular mass, membranes form an impermeable barrier.

Basicity of substances is also important, resulting extremely low the concentration of acidic compounds in hair due to their weak incorporation rate into the matrix.

Both effects, pH and binding to melanin, lead to the accumulation of lipophilic and basic drugs in matrix cells with clear preference for pigmented hair.

Segmental Analysis

When the length of the hair is greater than 2 – 3 cm, it is possible to divide the shaft into segments, each corresponding to a period life, hence tracing the abuse back to time.[51] The hair is cut into segments of about 1, 2 or 3 cm, which corresponds approximately to about 1, 2 or 3 months' growth. As a consequence of these statements, hair analysis provides important information with respect to time course of an individual's drug use.

Prerequisites of a correct diagnosis are very careful sampling, accurate segmentation and subsequent decontamination.[44] To ensure the correct interpretation of results, hair must be cut as close as possible to the scalp and particular care is also required to ensure that each segment retain the position they originally had beside one another.

Cosmetic Treatment and External Contamination

Drug concentration declines dramatically after cosmetic treatment including bleaching, dyeing and permanent waving.[37,46] Moreover UV or water exposure, soil or weathering have negative outcome on drug incorporation. On the other hand some treatments can also produce modifications on the keratinized tissues allowing permeability to the environment.

Such effects have to be taken into account when interpreting drug abuse analyses in hair samples, being necessary to consider both the reduction due to the cosmetic treatment, and the permeability of the treated shaft to external contamination. The last could bring to false positive hair testing results due to passive exposure to drugs in the environment. This is particularly important with alcoholics, being ethanol a volatile molecule that can easily penetrate the hair shaft.

To distinguish between actual consumption and contamination there exist a number of approaches like decontamination using various solvents, the measurement of metabolite to parent drug concentration ratios, the use of cut-off levels setting the limits for passive endogenous drug exposure and the reproducibility of results (including segmental analysis) using a newly collected hair specimen.[46,52,53] However, a definite differentiation between active use of drugs and passive exposition is not possible.

Hair from Other Body Areas

When scalp hair is not available, other types of hair (pubic, arm or axillary hair, etc.) can be suggested as an alternative source for drug detection. A comparison of drug concentrations between scalp hair and hair from other sites of the body has been described in several papers.[54–56] In fact additional information can be obtained from alternative sources of hair, for example in subjects with very short or no scalp hair. These alternative samples are traditionally subject to less exposure to cosmetic and environmental influence such as light and weather; they represent another growth time period because of the different growth cycle and they show differences in pigmentation. Segmental analysis is not useful due to a high proportion of telogen hair.

Higher drug concentrations have been found for a number of drugs in pubic, axillary, arm, chest or thigh hair versus scalp hair. In general, these differences could be explained by increased incorporation from sweat or sebum during the longer telogen stage, and they are in the order of the inter-individual variations of scalp hair concentrations. Therefore, apart from a different time period they represent, drug concentrations in hair from other body sites can be interpreted in a similar fashion as scalp hair.

Forensic Employ of Hair

Hair analysis could be acceptable in forensic applications if some rules are respected, but the results must always be critically discussed according to some evaluation, first of all the chance of passive exposure, then the use of suitable cut-off values.[44,53] Moreover drug incorporation in hair depends on various different factors such as growth speedy, ethnic group, anatomic origin, drug metabolism, bioavailability, age, sex, inter-individual variability. Even so, it is however possible to extrapolate a relation between drug concentration in hair and the amount taken, and between the time of the drug use and the distribution of the metabolite along the hair shaft. A precise estimation of the time of consuming however is not possible, only month space.[46] The analysis of keratinous matrix could hence be considered effective in determining numerous drugs of abuse, and today it is a routinely practise in forensic toxicology.

Alcoholism Biomarkers in Hair

Since alcoholic beverages are legal in many countries, a reliable alcohol test able to differentiate social drinkers from abusers is needed. Hence, the analysis of suitable ethanol markers in hair would be an advantageous tool for chronic alcohol abuse control because of the wide diagnostic window allowed by this specimen and the possibility of segmental investigation useful to reconstruct the history of abuse.

Although alcohol is the most frequently abused substance, it has played only a minor role in hair analysis because of its volatility and not durably incorporation. In fact, just like water and many other solvents, ethanol itself can pass through the hair shaft.[34] The majority of the ethanol absorbed from the surroundings is rapidly replaced by water from bond sites, either during the washing procedures or for the interaction of air humidity, so decreasing ethanol amount in hair. On the contrary alkaline hydrolysis can bring to the production of ethanol from compounds normally present in hair, leading to an increase of ethanol. Therefore ethanol physically absorbed into hair does not represent the true measure of alcohol ingested and cannot be used as biochemical marker of alcohol abuse. This is the main reason of the marginal attention obtained by experts.

As ethanol itself is not a suitable compound for hair analysis, research has focused on other alcohol markers, in fact after alcohol is ingested it is metabolized by various pathways to a number of secondary products, particularly fatty acid ethyl esters, ethyl glucuronide, ethyl sulphate and cocaethylene.

At the end of the nineties some researcher conjectured the deposition of these minor ethanol metabolites in hair[57,58] just like drugs and in the last years literature reports have been increased enormously.

As already extensively explained, hair has the unique ability to serve as a long-term storage of foreign substances with respect to the temporal appearance in blood. Over the last 20 years, hair testing has gained increasing attention and recognition for the retrospective investigation of chronic drug abuse as well as intentional or unintentional poisoning. In the same way the detection of minor ethanol metabolites in this matrix could help the objective diagnosis of alcohol use/abuse.

Two main groups of alcohol abuse biomarkers in hair matrix can be discerned, indirect and direct markers.

Indirect markers, that arise from pathological changes in biochemistry and metabolism of alcoholics, are routinely employed in biological fluids but it is really difficult to detect them in hair matrix. However, in 1978 the possibility to obtain indication of alcoholism situation from the protein status of hair collected from alcoholics was investigated,[59] measuring the root diameter and the volume of the hair. The research showed the correlation of these parameters with protein content demonstrating to be a sensitive indicator of protein deficiency. Unfortunately these findings are not unequivocally linked to alcohol abuse, hence their usefulness is limited.

Chronic alcoholics frequently show associated malnutrition, and both ethanol and malnutrition exert profound changes on zinc and copper metabolism.[60] The amount of copper in hair was significantly related to the amount of ethanol consumed, whereas hair zinc was found to be higher in consumers of distilled beverages. No relation between hair zinc and copper and nutritional status, kind of diet consumed, style of life, and liver cirrhosis was observed. Consequently, levels of zinc and copper in hair could be related only with alcohol intake. More recently,[61] the amount of trace element in hair was again correlated to the prior alcohol history: zinc and copper concentrations were found significantly higher in alcoholics rather than the control group, and copper concentration could be related to the amount of alcohol used. The study of these markers however did not proceed maybe because it was considered not much specific for this pathology.

Direct markers that can be detected in hair include the entrapping of molecular ethanol itself (not useful to detect chronic alcoholism as already discussed), the follow-up products of acetaldehyde and the deposition of minor metabolites of ethanol carrying the C_2H_5-group, including ethyl glucuronide, fatty acid ethyl esters, phosphatidylethanol, cocaethylene and other ethyl esters such as ethyl sulfate.

Acetaldehyde Adducts to Hair Protein

Acetaldehyde is the first metabolic product of ethanol; it possesses an aldehyde group characterized by the ability to react with intracellular proteins to form stable and non-stable complexes (Fig. 37.4). These complexes are named "adducts" and induce cellular damages because they alter biological properties of proteins.

Acetaldehyde is able to form stable adducts with haemoglobin in erythrocytes, with serous proteins but also with hair protein, particularly keratin. Nevertheless the use of such protein adducts as alcoholism markers, is still limited[62] and it needs experimental studies not yet performed.

Minor Ethanol Metabolites

Minor ethanol metabolites containing the intact group $-C_2H_5$ can be used as biochemical markers of alcoholism, because directly related to alcohol consumption. In fact the finding of these molecules in biological medium can be evidence of alcoholic use, contributing to the objective diagnosis of alcohol use/abuse. As already demonstrated for drugs of abuse, these molecules can be incorporated in hair remaining irreversibly trapped in the keratinised matrix.

Phosphatidylethanol

Up to now no literature data about the determination of phosphatidylethanol (PEth) in hair are reported, even if the compound could theoretically be incorporated in the keratinised matrix.

PEth is a group of phospholipids formed in cellular membranes by the catalytic action of the enzyme phospholipase-D on phosphatidylcholine[63] only in the presence of ethanol. In fact ethanol and other primary alcohols with short chain can interfere in the hydrolysis as substrate alternative to the water. Its degradation speedy is low respect to its production, and for its high specificity and slow elimination, it was proposed as biomarker of alcohol abuse. A single alcohol intake is not sufficient to the production of phosphatidylethanol, hence respect to other markers, it can be identified in biological samples only after various intakes, but it remain recognizable for a longer time.[64,65] PEth could be considered a promising long-term indicator of alcohol abuse, being detectable in blood up to 14 days after sobriety. In comparison to the traditional alcohol markers, PEth has the advantage of being a direct derivative of ethanol which makes it significantly more alcohol specific.

Cocaethylene

Cocaethylene (CE) is a minor active metabolite of cocaine that can be produced in the human body concurrently with alcohol intake[66] (Fig. 37.5). This ethyl homologue of cocaine keeps the same psycho activity of the parent drug, and can be yield exclusively during the simultaneous intake of cocaine and ethanol. Cocaethylene can in fact be formed from cocaine and ethanol during the re-esterification in the cells of liver

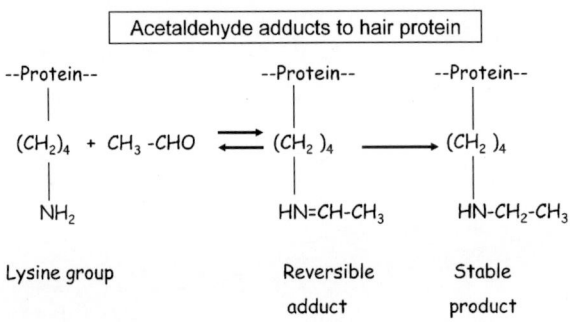

Fig. 37.4 Reaction scheme of acetaldehyde adducts to hair protein

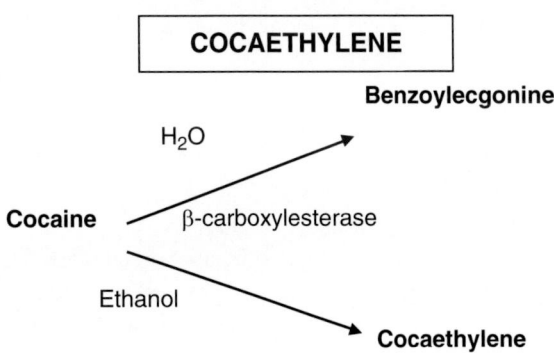

Fig. 37.5 Synthesis of cocaethylene

and kidney of alcoholics. The reaction is catalyzed by the hepatic carboxyltransferase, which is the same non-specific esterase that lead to the synthesis of benzoylecgonine (the main cocaine metabolite) when there is no ethanol.[67]

In the presence of substantial blood ethanol concentration, approximately 17% of intravenous cocaine is transformed to CE over a wide range of doses. For these reasons it could be considered a specific marker of alcoholism. As expected, CE has been detected in the hair of ethanol users by various authors,[68,69] nevertheless, many factors are supposed to affect the amount of the two substances incorporated in the hair matrix, such as the subject's habits in ethanol and cocaine use, genetic variability in the metabolism of both substances, and the different chemical and physical properties of CE. Its use is obviously restricted to a population of simultaneous abusers of cocaine and alcohol.

Ethyl Sulphate

Ethyl sulphate (EtS), another direct ethanol metabolite, appears to offer potential as a biomarker for recent alcohol consumption.[70,71] Although its window of assessment is similar to that of ethyl glucuronide (EtG) (extensively described afterwards), there are differences between the two markers in their pathways for formation and degradation. Studies have been recently performed to assess the excretion of EtS compared to ethylglucuronide and ethanol in drinking experiments with healthy volunteers; moreover some authors tried to elucidate the possibility of using the two metabolites for monitoring abstinence in substance use disorder patients during rehabilitation treatment.[72]

Because EtG and EtS are formed via different pathways and they could be considered a proof of ethanol consumption, they might be used conjointly thereby increasing sensitivity.

Fatty Acid Ethyl Esters

In the last few years, two ethanol metabolites – ethyl glucuronide and fatty acid ethyl esters – have been mainly investigated in hair samples for their ability to be incorporated into this biological matrix.

FAEE is the acronym of "Fatty Acid Ethyl Esters", minor metabolites of ethanol that represent a group of non polar compounds easily hydrolyzed by alkaline medium.

Their synthesis in vivo takes place by the esterification of ethanol with free fatty acids, and trans-esterification of triglycerides, lipoproteins or phospholipids (Fig. 37.6). The reaction occurs under the action of the specific enzyme "FAEE synthase" both cytosolic and microsomal. However, it has been proven that some non-specific enzymes, such as carboxylesterases, lipoprotein lipases, carboxyl ester lipases, may have the same action.[73] Fatty acid ethyl esters are hence products of the non-oxidative ethanol metabolism, and they are known to be formed and detectable in blood and almost all human tissues after alcohol consumption, so being useful biochemical markers of ethanol intake.

Acetaldehyde was firstly proposed as the cause of the organic damage in alcoholics, but later it has been demonstrated that its production in pancreas, heavily damaged by alcohol abuse, is scarce.[74] The correlation of fatty acid ethyl esters with alcohol-induced damage to the liver and pancreas have been demonstrated studying autoptical cases,[75,76] in fact FAEE and FAEE synthase can be detected mainly in the organs more frequently damaged by ethanol abuse, that is pancreas and liver.

FAEE are therefore mediators of organic damage consequence of ethanol intake.

The recovery of ethyl stearate, palmitate, oleate, linoleate and arachidonate in biological fluids, as products of non-oxidative metabolism of ethanol, has been firstly depicted by Lange et al.[77] and successively

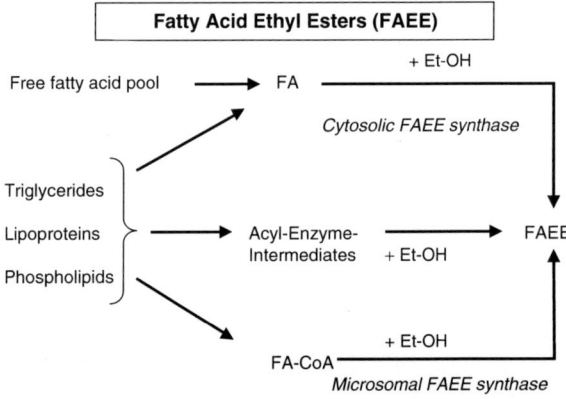

Fig. 37.6 Production of Fatty Acid Ethyl Esters

deeply investigated by Laposata et al.[73,78,79] Published data demonstrated that the amount or type of FAEE in traditional biological fluids can be used to differentiate a chronic alcoholic from an episodic heavy drinker at or near peak blood ethanol concentrations and approximately 24 h after the last alcohol intake. Thus, FAEE can be considered short-term markers of ethanol intake in blood and tissues, useful in distinguishing chronic alcoholics from social drinkers.

Forensic pathology can also benefit from the detection of FAEE in biological specimens; in fact post-mortem studies[80,81] evaluated these final products consequent to the ingestion of alcoholic beverages, in human liver and adipose tissue. Results showed that they could be considered as long-term metabolites and their deposit in adipose tissue can lead to an accurate diagnosis even when ethanol has been completely eliminated by the organism. Hence they are post-mortem markers of pre-mortem ethanol intake.

Because it could be useful to locate an alcoholic also various days after last consumption, the searches of different matrices able to give retrospective information have been performed. Recently it was demonstrated that these compounds can also be incorporated from sebum into hair where they can be used as long-term markers for excessive alcohol consumption.

FAEE in Hair

Some years ago Pragst et al.[57] reported the characterization of FAEE in hair, demonstrating that these compounds should be considered suitable long-term markers of chronically elevated alcohol consumption after their deposition in hair. Of the various esters identified, ethyl myristate, palmitate, stearate and oleate have been chosen as markers of ethanol intake. The study referred concentrations of FAEE in social drinkers (30–60 g ethanol intake per weeks) about one order of magnitude smaller than alcoholics, in particular ethyl palmitate was found at concentrations lower than 0.40 ng/mg, always under the value obtained from alcoholics. Negative results or traces of ethyl palmitate were observed in teetotallers. Hair collected from alcoholics showed higher values of ethyl palmitate and oleate respect to ethyl stearate and myristate. Only traces of other esters were identified in some of the examined subjects.

In another study on the contrary, ethyl palmitate and myristate were the main compounds encountered, while only a negligible amount of ethyl oleate was noted.[82] Quantitative determinations showed very low amounts of FAEE (expressed as the sum of ethyl myristate and palmitate) in teetotallers (less than 0.24 ng/mg), FAEE ranging from 0.12 to 0.98 ng/mg in social drinkers, values scattered in a wide range (0.06–6.29 ng/mg) in alcoholics.

As many other exogenous substances, FAEE can be deposited in hair not only from blood, but also through other routes. It was however established that the main way of incorporation compels FAEE synthesis from ethanol in the sebaceous glands,[83] excretion with sebum in the superior region of the root, distribution on the hair surface with sebum, and finally diffusion into the hair matrix. This finding can cause some confusion when FAEE are used as marker of excessive alcohol consumption, because of inter-individual differences on the activity of the sebum glands. In fact, sebum production is monitored by hormones, and it depends on age, sex and it is subdued to individual differences.

FAEE can also be detected in other body hair such as pubic, chest, beard without significant differences in their concentrations when compared with scalp hair.[84] For the demonstration of such a statement, samples collected from scalp, pubis, axilla, beard and body hair of teetotallers, moderate social drinkers and fatalities were analysed for the presence of FAEE. The sum of the concentrations of ethyl myristate, ethyl palmitate, ethyl oleate and ethyl stearate (C-FAEE) in the hair samples was compared with information about the drinking behaviour of the individuals. Although there were large differences in the esters concentrations in hair from different sites in the same individual, cases of chronic excessive alcohol consumption were characterized by C-FAEE > 1.0 ng/mg in almost all samples, testifying the reliability of these specimens.

To characterize the history of drug abuse, segmentation can be performed cutting hair in small pieces by the side of the length. Moving from this assumption, ester distribution along the hair shaft should give information about the history of alcohol abuse; in fact the lipophilic character of ethyl esters suggests their excretion from sebaceous glands and successive distribution on the surface of the hair.

A segmental hair analysis to detect alcoholism history[83] established that esters concentrations increased from proximal to distal with no agreement between the self-reported drinking histories and FAEE concentrations along the hair length. Moreover it was demonstrated that they remain detectable in hair segments after a 2 months period of abstinence. These observations could be explainable by the mechanism of deposition through the sebum that compels an increase in concentration from proximal to distal region.

The knowledge that sebum contains about 10–20% of squalene (SQ) and that its amount depends on various inter-individual factors, had suggested the possibility to correct the measure of FAEE. For this purpose some studies were addressed to improve the analysis, adjusting the amount of sebum layered on the hair surface dividing the sum of the esters by the quantity of squalene.[85] A sensitive and reliable method for the determination of squalene in addition to FAEE from the same hair extracts was then developed.[85] The results showed that squalene enables a control of the lipid content of hair and a correction of C-FAEE in cases with deviations from the usual lipid content in a similar way as creatinine in urine. However, the relative concentration C-FAEE/C-SQ cannot completely replace the absolute concentration C-FAEE, and both data should regularly be used for an improved interpretation with respect to alcohol abuse.

Cosmetic Evaluation and External Contamination

It is well known that the amount of exogenous substances in hair decreases after some cosmetic treatment, and this drop in concentration depends both on the kind of the treatment and on the damage produced. On the contrary permanent wave or bleaching seems to promote the increase of absorption of drugs, from sweat, sebum or external sources.

The influence of hair care and hair cosmetics in FAEE detection has been investigated[86] in order to establish the possible elimination of FAEE by hair treatments. The study performed on hair care products, demonstrated that commonly used cosmetics, such as colouring, bleaching, perms, do not have any significant effects in FAEE concentration even if traces of the four ethyl esters were detected in all the products tested. On the contrary false positive results have been obtained with hair daily submitted to treatment with lotions, deodorants, hair spray containing high concentration of ethanol. As an explanation, it is assumed that FAEE are formed in the sebum glands also after regular topical application of products with higher ethanol content.

To assess the influence of ethanol contamination, an in vitro experiment was performed,[87] leaving hair in an atmosphere saturated with ethanol vapours for 15 days. The spontaneous production of FAEE was demonstrated by analyzing hair day by day. Although the experiment was managed in a stressed way and could not represent real life, it focused the attention of researchers on the problem of hair contamination that can occur, for example, with ethanol-containing cosmetics. Therefore, care in interpretation must be taken into account, especially with such a volatile molecule, carefully examining the anamnesis of the patient.

In conclusion FAEE concentration in hair seems to be influenced from various factors. Frequent washing lead to the removal of sebaceous layer before the deposition, hence a decrease of FAEE can be expected. Cosmetic treatments involving chemical hydrolysis, oxidation of esters or high basic pH, bring to a reduction of the FAEE. The daily use of cosmetic products containing the esters implies on the contrary their deposition on the surface of the shaft, so increasing the concentrations of FAEE.

In spite of all these variables, only the latter shows a significant effect of adulteration, but this problem could be overcome for example using hair collected from other body regions.

FAEE in Newborn

As already discussed, maternal alcohol consumption during pregnancy may lead to various diseases in newborns. The analysis of FAEE in meconium can be today considered useful for the objective detection of foetal exposure; in fact it has been proven that such compounds don't pass placenta, hence their presence in meconium represents a valid estimation of ethanol circulating in the liver.[28]

FAEE had been the first studied neonatal biologic marker in hair for babies at risk for alcohol-related birth defects.[22,88] The determination of FAEE in newborn hair may revolutionize current methods used to diagnose FASD, allowing intervention and treatment at stages where the adverse effects of alcohol can still be mitigated.

To be able to use this test, it is necessary to establish baseline FAEE levels in infants born to non-alcoholic women because ethanol occurs in the baby naturally even without drinking. Literature refers about a study performed on scalp hair of infants born to non-alcoholic women[89] that measured levels of FAEE detected in almost all hair samples. In this way baseline levels of FAEE in neonatal hair of non-alcoholic mothers have been established. Mild infrequent maternal drinking does not elevate these baseline levels. These data may help avoid false-positive determinations when assessing infants born to problem drinking mothers. Results of this study suggest that FAEE in neonatal hair may be useful biomarkers in identifying in utero alcohol exposure and may facilitate the early diagnosis and treatment of FASD.

Ethyl Glucuronide

Ethyl glucuronide (ethyl-beta-D-6-glucosiduronic acid, EtG) (Fig. 37.7), a recent and promising marker of alcohol use, is the product of a minor metabolic transformation of ethanol accomplished by the mitochondrial UDP-glucuronosyl transferase. It is a phase-II metabolite mainly formed in the liver by the conjugation reaction with glucuronic acid; it accounts for the 0.02–0.06% of the ingested ethanol.

From the chemical point of view, EtG is a water-soluble, non-volatile, stable, slightly acidic compound with a molecular weight of 222 g/mol. Its production occurs shortly after consumption of ethanol and in serum it is found to peak 2–3.5 h after ethanol has reached its maximum, and to stay detectable up to 8 h after ethanol elimination. In urine EtG has been found starting from 1 h after intake and persisting up to 80 h. In these biological fluids, it closes the gap between short-term markers for alcohol consumption, such as ethanol, and long-term markers for alcohol misuse usually employed such as γ-GT, MCV and CDT. In hair it has been demonstrated to be a true long-term marker of alcohol abuse.[58,90–92]

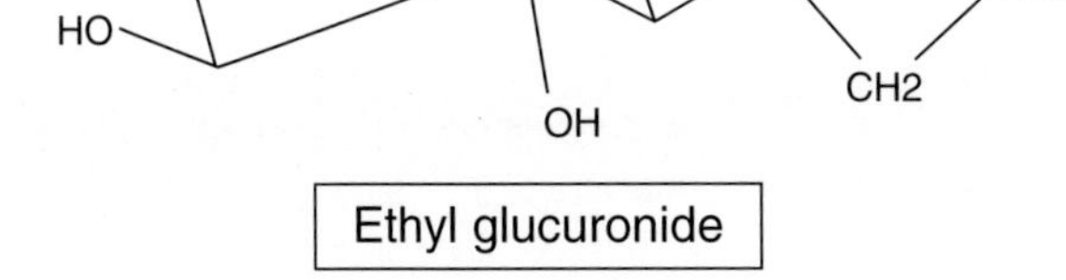

Being a unique metabolite of ethanol, it has received much attention as a sensitive and specific biological marker of ethanol consumption, because it can offer an extended window for assessment of drinking status.

EtG was firstly isolated in 1952 by Kamil[93] from rabbit urine, and in 1967 it was identified in human urine[94]; since then various authors confirmed its presence in serum and urine from alcoholics,[95,96] others studied its pharmacokinetic properties and its feasibility as marker of the use/abuse of alcohol.[92,97]

In 1993, Sachs first[98] highlighted the chance to detect EtG in hair, and this hypothesis was later demonstrated.[99] The molecule has never been identified in teetotallers, occasional drinkers showed very low concentrations (lower than 1–5 ng/mg), while social drinkers exhibited amount ranging from 0 to 40 ng/mg.

Being EtG a carboxylic acid extremely hydrophilic and polar, on the base of the knowledge of drugs incorporation into keratinic matrix, it can be considered an unusual compound to be found in hair. Lipophilicity is in fact one of the factors that affect the bond with drugs during the growth of the hair shaft. In spite of its characteristics, EtG incorporation in hair has been demonstrated, even though the incorporation mechanism wasn't explained yet.[90,91,100]

The experts didn't always find EtG in hair from alcoholics, hence they observed that the non-recovery of EtG is not a definite prove of abstention of habitual intake of alcoholic beverages, but its identification in hair is indicative of high probability of alcohol abuse.[100,101] They suggested that EtG can be removed from hair during the wash for its high polarity, and the use of cosmetic treatments may influence the incorporation of the molecule in hair just as drugs of abuse.[100]

The development of new analytical procedures[58,91,97] with high sensitivity allowed the detection of EtG at very low concentrations, demonstrating the chance to use it as a biochemical marker of alcoholism, even if it continue to be not always identified in hair of alcoholics. As years go by, analytical techniques improve and extremely sophisticated methods are discovered,[101–103] allowing the determination of very low amounts of EtG.

It has been demonstrated that its incorporation in hair is not influenced by the presence or absence of melanin, and unlike many other substances, the EtG determination

Ethyl glucuronide

Fig. 37.7 Structure of Ethyl glucuronide

in hair has not to take into account the hair colour for the correct interpretation of hair testing results.[104]

In some specific situations, head hair can be missing, and therefore, alternative anatomical locations of hair are of interest. A study that examined paired hair specimens (head hair and pubic hair) from social drinkers[105] demonstrated that surprisingly, EtG was identified at high concentrations in pubic hair, in the range 12–1,370 pg/mg. This finding, therefore, established that it is not possible to document the drinking status of a subject by simply switching from head hair to pubic hair.

After the best analytical determination of EtG was established, the major controversial issue was the implementation of a cut-off allowing the discrimination between teetotallers and social drinkers and between social and heavy drinkers. Pragst and Yegles[106] have proposed that an EtG concentration higher than 25 pg/mg corresponds to a chronic alcohol abuser. EtG concentrations below the limit of quantization may indicate weak social drinkers or teetotallers, but do not completely exclude alcohol abuse. On the other hand, a positive result with a cut-off at a high value corresponds more likely to a subject having an alcoholic behaviour with a high level of certainty.

In spite of this evolution, no correlation between the absolute amount of ethanol intake and EtG concentration in hair has been demonstrated, and the exact incorporation modality in the hair structure is not known yet, hence more investigations are needed. Politi et al.[101] explored this field, finding EtG concentration of up to 434.7 pg/mg in the hair of alcoholics while no EtG was detected in the hair of teetotallers, however they could not prove the correlation.

Several reasons may contribute to the lack of correlation between alcohol consumption and EtG concentration in hair. On the one hand, EtG may be washed out due to its polarity during normal body care or hair washing. In addition, information on alcohol consumption is based at least in part on personal statements and, therefore, may not be sufficiently reliable to allow a detailed interpretation on this point. Nevertheless, in accordance with previous studies, the current results corroborate the observation that the absence of EtG in hair does not indicate a person to be abstaining from alcohol.[101] However, alcohol consumption can be assumed if EtG is detectable in hair samples. In fact, although increased analytical sensitivity has generally decreased the number of false negative results, positive EtG results have also been found in hair from social drinkers.

Combined Markers

In various studies EtG was compared with other possible bio-markers such as phosphatidylethanol, cocaethylene, ethyl sulphate and/or fatty acid ethyl esters.

The complementary use of EtG together with other upcoming markers of alcohol consumption like phosphatidylethanol[107] should lead to an improvement in treatment outcome, quality of life and cost reduction. Evaluation of liver histology and anamnestic evidence of alcohol abuse in a group of autoptical cases were taken in consideration for the interpretation. Measurable levels of EtG were present in 49 of the 70 autopsy cases whereas PEth was present in 36. The findings suggest that measurements of EtG in hair may provide improved diagnostic information on alcohol abuse, due to a long retrospective time-window for detection and stability of EtG in hair in the decaying cadaver. It is still open the problem involving false negative results, in fact an EtG level below the cut-off does not completely exclude previous alcohol abuse.

The correlation with cocaethylene,[108] in the hair of cocaine users demonstrated no quantitative correlation between EtG and CE. Nevertheless, many factors are supposed to affect the amount of the two substances incorporated in the hair matrix, such as the subject's habits in ethanol and cocaine use, genetic variability in the metabolism of both substances, and the different chemical and physical properties of EtG and CE. According to these data, EtG appears to be a more sensitive and specific marker of non-moderate alcohol users than CE.

Ethyl glucuronide (EtG) and ethyl sulphate (EtS) are two ethanol metabolites that can be detected in serum up to 8 h after ethanol elimination. Their presence is therefore indicative of recent ethanol consumption in case of delayed sampling after an event (e.g. car crash). The comparison between the two molecules has already been discussed.[72]

Finally the combination of the two specific long-term markers EtG and FAEE in hair can give a high contribution on the diagnosis of excessive alcohol consumption, even though no significant correlation between the two concentrations in the positive cases was demonstrated yet.

The results of recent papers[109,110,111] confirm that by using a cut-off value of C-FAEE > 1 ng/mg and/or a positive EtG result in hair, excessive alcohol consumption can be identified using hair analysis.

Segmental analysis of some of the specimens did not reveal the same distribution for EtG compared to FAEE in hair, and no chronological accordance compared to the self-reported alcohol consumption could be observed for both parameters.[110]

These different results of both methods could be explained in terms of differences between EtG and FAEE in mechanism of formation and incorporation into hair and elimination from hair. The measurement and combined interpretation of FAEE and EtG in hair could be very helpful in many forensic and clinical cases.

Conclusions and Future Directions

The ability of alcoholics to mask their difficulties with alcohol implies much confusion for the clinicians, who therefore need objective instrumentations to make the diagnosis. The use of alcohol in pregnancy is also an underestimated issue, often causing foetus malformations. The use of questionnaires is unsatisfactory and the need of biomarkers is advisable. Moreover the use of long-term markers is a useful tool in numerous other settings, including safety sensitive work contexts where use is dangerous or in other settings where alcohol use may be risky (e.g. driving, work-place or monitoring physicians or other professionals who are in recovery and working) or for resolving forensic questions.

Many biochemical markers have been studied for this purpose in blood, urine or other biological samples. In addition to traditional markers, a combination of direct ethanol metabolites can be useful in the expert assessment of diagnose alcoholism. A careful individual interpretation of the results for the different markers, however, is an absolute necessity.

The measurement of long-term biochemical markers is becoming crucial for the objective diagnosis of alcoholism. For this purpose particular strategies have to be embraced, like the choice of suitable biological matrices able to allow a long-term detection. Hair could be considered a reliable matrix for its wider detection time frame respect to traditional specimens.

Improved analytical technology has resulted in improved sensitivity and accuracy thus providing better scientific understanding and test interpretation[112]. These advances will further promote the use of hair analysis as a useful and objective tool of evidence. Introduction of automation techniques and improved methods for sample preparation will enable application of hair analysis for large-scale testing and more general use.

Particular interest in forensic field has been obtained by the detection of minor metabolites of ethanol in hair.

FAEE concentrations in hair could be considered a useful way for a retrospective detection of alcohol abuse, and the sum of FAEE is considered as a potentially valuable marker of chronic intake of high quantities of ethanol. It is necessary to keep in mind however that the concentration of FAEE in hair seems to be affected by some factors. First of all FAEE concentration may decrease due to frequent wash that can remove the sebaceous layer before their incorporation. A decrease due to chemical hydrolysis of esters can also occur (in particular the oxidation of unsaturated esters, or the high pH). An increase of FAEE concentration can also be speculated due to two main reasons, the deposition from cosmetic products already containing the ethyl esters, and their local synthesis from the cosmetics containing ethanol, hence environmental pollution. Only the last event showed significant effect,[86] but this problem could be probably overrun using hair collected from other body area, like for example pubic hair.

Even if more investigations are advisable to establish the real usefulness of these markers, FAEE concentration in hair could be probably considered a true biomarker of chronic intake of ethanol.

Due to its specific time-frame of detection and its high sensitivity and specificity, EtG is a promising marker for alcohol consumption and for relapse control that enables the therapist to intervene at an early stage of relapsing behaviour.

Growing interest was observed for this marker during the last years and the many technical efforts performed to improve sensitivity made the limit of detection decrease from about 2 ng/mg hair[113] to values close to 2 pg/mg hair.[110]

Finally, a promising approach to diagnosing alcohol misuse could be the use of a combination of biochemical markers such as FAEE and EtG in hair to increase sensitivity and specificity.

In conclusion the researches carried out so far indicate that these minor ethanol metabolites may provide a powerful tool in the diagnosis of chronic alcohol intake. However, further extended studies are needed to fully understand the influence of biological variables (including differences in hair growth and mechanisms of drug incorporation) and to improve correlations between concentration of the marker and extent of ethanol intake.

References

1. World Health Organization (www.who.int/topics/alcohol_drinking/en/)
2. Gmel G, Rehm J. Harmful alcohol use. Alcohol Res Health 2003;27:52–62. Review.
3. Bellis MA, Hughes K, Morleo M, et al. Predictors of risky alcohol consumption in schoolchildren and their implications for preventing alcohol-related harm. Subst Abuse Treat Prev Policy 2007;2:15.
4. Rivara FP, Relyea-Chew A, Wang J, et al. Drinking behaviours in young adults: the potential role of designated driver and safe ride home programs. Inj Prev 2007;13:168–172.
5. Lieber CS, Medical disorders of alcoholism. N Engl J Med 1995; 333:1058–1065.
6. Room R, Babor T, Rehm J. Alcohol and public health. Lancet 2005; 365:519–530
7. Chang G, Goetz MA, Wilkins-Haug L. A brief intervention for prenatal alcohol use: an in-depth look. J Subst Abuse Treat 2000;18:365–369.
8. Maier SE, West JR. Drinking patterns and alcohol-related birth defects. Alcohol Res Health 2001;25:168–174.
9. Allebeck P, Olsen J. Alcohol and fetal damage. Alcohol Clin Exp Res 1998;22:329S–332S.
10. Stockwell T, Chikritzhs T, Brinkman S. The role of social and health statistics in measuring harm from alcohol. J Subst Abuse 2000;12:139–154.
11. Huckle T, Pledger M, Casswell S Trends in alcohol-related harms and offences in a liberalized alcohol environment. Addiction 2006;101:232–240.
12. Neumann T, Spies C. Use of biomarkers for alcohol use disorders in clinical practice. Addiction 2003;98 (Suppl 2):81–91.
13. Aalto M, Seppä K. Use of laboratory markers and the audit questionnaire by primary care physicians to detect alcohol abuse by patients. Alcohol Alcohol 2005;40:520–523.
14. Niemelä O. Biomarkers in alcoholism. Clin Chim Acta 2007;377:39–49
15. Sillanaukee P. Laboratory markers of alcohol abuse. Alcohol Alcohol 1996,31:613–616.
16. Hannuksela ML, Liisanantti MK, Nissinen AE, et al. Biochemical markers of alcoholism. Clin Chem Lab Med 2007;45:953–961.
17. Kraut JA, Kurtz I. Toxic alcohol ingestions: clinical features, diagnosis, and management. Clin J Am Soc Nephrol 2008;3:208–225.
18. Khiabani HZ, Opdal MS, Mørland J. Blood alcohol concentrations in apprehended drivers of cars and boats suspected to be impaired by the police. Traffic Inj Prev 2008;9:31–36.
19. Morse BA, Hutchins E. Reducing complications from alcohol use during pregnancy through screening. J Am Med Women Assoc 2000;55:225–227.
20. Caprara DL, Nash K, Greenbaum R, et al. Novel approaches to the diagnosis of fetal alcohol spectrum disorder. Neurosci Biobehav Rev 2007;31:254–260.
21. Bearer CF, Markers to detect drinking during pregnancy. Alcohol Res Health 2001;25:210–218.
22. Caprara DL, Klein J, Koren G. Diagnosis of fetal alcohol spectrum disorder (FASD): fatty acid ethyl esters and neonatal hair analysis. Ann Ist Super Sanita 2006;42:39–45.
23. Chan D, Klein J, Karaskov T, et al. Fetal exposure to alcohol as evidenced by fatty acid ethyl esters in meconium in the absence of maternal drinking history in pregnancy. Ther Drug Monit 2004;26:474–481.
24. Wurst FM, Kelso E, Weinmann W, et al. Measurement of direct ethanol metabolites suggests higher rate of alcohol use among pregnant women than found with the AUDIT – a pilot study in a population-based sample of Swedish women. Am J Obstet Gynecol 2008;198:407.e1–5.
25. Moore C, Jones J, Lewis D, et al. Prevalence of fatty acid ethyl esters in meconium specimens. Clin Chem 2003;49:133–136.
26. Ostrea EM Jr. Testing for exposure to illicit drugs and other agents in the neonate: a review of laboratory methods and the role of meconium analysis. Curr Probl Pediatr 1999; 29:37–56.
27. Caprara DL, Brien JF, Iqbal U, et al. A Guinea pig model for the identification of in utero alcohol exposure using fatty acid ethyl esters in neonatal hair. Pediatr Res 2005; 58:1158–1163.
28. Cook JD. Biochemical markers of alcohol use in pregnant women: review. Clin. Biochem 2003;36: 9–19.
29. Rinck D, Frieling H, Freitag A, et al. Combinations of carbohydrate-deficient transferrin, mean corpuscular erythrocyte volume, gamma-glutamyltransferase, homocysteine and folate increase the significance of biological markers in alcohol dependent patients. Drug Alcohol Depend 2007 15;89:60–65.
30. Das SK, Dhanya L, Vasudevan DM. Biomarkers of alcoholism: an updated review. Scand J Clin Lab Invest 2008; 68:81–92.
31. Hock B, Schwarz M, Domke I, et al. Validity of carbohydrate-deficient transferrin (%CDT), gamma-glutamyltransferase (gamma-GT) and mean corpuscular erythrocyte volume (MCV) as biomarkers for chronic alcohol abuse: a study in patients with alcohol dependence and liver disorders of non-alcoholic and alcoholic origin. Addiction 2005; 100:1477–1486.
32. Jeppsson JO, Arndt T, Schellenberg F, et al. Toward standardization of carbohydrate-deficient transferrin (CDT) measurements: I. Analyte definition and proposal of a candidate reference method. Clin Chem Lab Med 2007;45:558–562.
33. Bortolotti F, De Paoli G, Tagliaro F. Carbohydrate-deficient transferrin (CDT) as a marker of alcohol abuse: a critical review of the literature 2001–2005. J Chromatogr B Analyt Technol Biomed Life Sci 2006;841:96–109.
34. Pragst F, Spiegel K, Sporkert F, et al. Are there possibilities for the detection of chronically elevated alcohol consumption by hair analysis? A report about the state of investigation. Forensic Sci Int 2000;107:201–223.

35. Villain M, Cirimele V, Kintz P. Hair analysis in toxicology. Clin Chem Lab Med 2004;42:1265–1272.

36. Kintz P, Villain M, Cirimele V. Hair analysis for drug detection. Ther Drug Monit 2006;28:442–446.

37. Pragst F, Balikova MA. State of the art in hair analysis for detection of drug and alcohol abuse. Clin Chim Acta 2006;370:17–49.

38. Huestis MA. Judicial acceptance of hair tests for substances of abuse in the US courts: scientific, forensic and ethical aspects. Ther. Drug Monit 1996;18: 456–459.

39. Camper JL. Praktisches Handbuch der Gerichtlichen Medizin (2 vols). A. Hirschwald, Berlin, 1857–1858.

40. Baumgartner AM, Jones PF, Baumgartner WA, et al. Radioimmunoassay of hair for determining opiate abuse histories. J Nucl Med 1979;20:748–752.

41. Sachs H. History of hair analysis. Forensic Sci Int 1997;84:7–16.

42. Kintz P. Drug testing in hair, 1st ed. CRC Press, Boca Raton, FL, 1996.

43. www.soht.org

44. Society of Hair Testing. Statement of the Society of Hair Testing concerning the examination of drugs in human hair. Forensic Sci Int 1997;84:3–6.

45. Bost RO. Consensus opinion summarizing the current applicability of hair analysis to testing for drugs of abuse. SOFT ToxTalk 1990;14.

46. Wennig R. Potential problems with the interpretation of hair analysis results Forensic Sci Int 2000;107:5–12.

47. Cone EJ. Mechanism of drug incorporation into hair. Ther Drug Monit 1996;148:438–443.

48. Balíková M. Hair analysis for drugs of abuse. Plausibility of interpretation. Biomed Pap Med Fac Univ Palacky Olomouc Czech Repub 2005;149:199–207.

49. Wada M, Nakashima K. Hair analysis: an excellent tool for confirmation of drug abuse. Anal Bioanal Chem 2006;385:413–415.

50. Musshoff F, Madea B. New trends in hair analysis and scientific demands on validation and technical notes. Forensic Sci Int 2007;165:204–215.

51. Clauwaert KM, Van Bocxlaer JF, Lambert WE, et al. Segmental analysis for cocaine and metabolites by HPLC in hair of suspected drug overdose cases. Forensic Sci Int 2000;110:157–166.

52. Tsanaclis L, Wicks JF. Differentiation between drug use and environmental contamination when testing for drugs in hair. Forensic Sci Int 2008;176:19–22.

53. Kintz P, Mangin P. What constitutes a positive result in hair analysis: proposal for the establishment of cut-off values. Forensic Sci Int 1995;70:3–11.

54. Moeller MR. Hair analysis as evidence in forensic cases. Ther Drug Monit 1996;18:444–449.

55. Drummer OH. Postmortem toxicology of drugs of abuse. Forensic Sci Int 2004;142:101–113.

56. Lehrmann E, Afanador ZR, Deep-Soboslay A, et al. Postmortem diagnosis and toxicological validation of illicit substance use. Addict Biol 2008;13:105–117.

57. Pragst F, Auwaerter V, Sporkert F, Spiegel K. Analysis of fatty acid ethyl esters in hair as possible markers of chronically elevated alcohol consumption by headspace solid-phase microextraction (HS-SPME) and gas chromatography-mass spectrometry (GC-MS). Forensic Sci Int 2001;121:76–88.

58. Skopp G, Schmitt G, Pötsch L, et al. Ethyl glucuronide in human hair. Alcohol Alcohol 2000;35:283–285.

59. Bregar RR, Gordon M, Whitney EN. Hair root diameter measurement as an indicator of protein deficiency in non-hospitalized alcoholics. Am J Clin Nutr 1978;31:230–236.

60. Rodriguez-Moreno F, Gonzalez-Reimers E, Santolaria-Fernandez F, et al. Zinc, copper, manganese, and iron in chronic alcoholic liver disease. Alcohol 1997;14:39–44.

61. González-Reimers E, Alemán-Valls MR, Barroso-Guerrero F, et al. Hair zinc and copper in chronic alcoholics. Biol Trace Elem Res 2002;85:269–275.

62. Watson RR, Solkoff D, Wang JY, et al. Detection of ethanol consumption by ELISA assay measurement of acetaldehyde adducts in murine hair. Alcohol 1998;16:279–284.

63. Gustavsson L, Alling C. Formation of phosphatidylethanol in rat brain by phospholipase D. Biochem Biophys Res Commun 1987;142:958–963.

64. Varga A, Hansson P, Lundqvist C, et al. Phosphatidylethanol in blood as a marker of ethanol consumption in healthy volunteers: comparison with other markers. Alcohol Clin Exp Res 1998;22:1832–1837.

65. Hartmann S, Aradottir S, Graf M, et al. Phosphatidylethanol as a sensitive and specific biomarker: comparison with gamma-glutamyl transpeptidase, mean corpuscular volume and carbohydrate-deficient transferrin. Addict Biol 2007;12:81–84.

66. Landry MJ. An overview of cocaethylene, an alcohol-derived, psychoactive cocaine metabolite. J Psychoactive Drugs 1992;24:273–276.

67. Dean RA, Christian CD, Sample RH, et al. Human liver cocaine esterase: ethanol-mediated formation of cocaethylene. FASEB J 1991;5:2735–2739.

68. Bermejo AM, López P, Alvarez I, et al. Solid-phase microextraction for the determination of cocaine and cocaethylene in human hair by gas chromatography-mass spectrometry. Forensic Sci Int 2006;156:2–8.

69. de Toledo FC, Yonamine M, de Moraes Moreau RL, et al. Determination of cocaine, benzoylecgonine and cocaethylene in human hair by solid-phase microextraction and gas chromatography-mass spectrometry. J Chromatogr B Analyt Technol Biomed Life Sci 2003;798:361–365.

70. Dresen S, Weinmann W, Wurst FM. Forensic confirmatory analysis of ethyl sulphate – a new marker for alcohol consumption – by liquid-chromatography/electrospray ionization/tandem mass spectrometry. J Am Soc Mass Spectrom 2004;15:1644–1648.

71. Helander A, Beck O. Ethyl sulfate: a metabolite of ethanol in humans and a potential biomarker of acute alcohol intake. J Anal Toxicol 2005;29:270–274.

72. Morini L, Politi L, Zucchella A, et al. Ethyl glucuronide and ethyl sulphate determination in serum by liquid chromatography-electrospray tandem mass spectrometry. Clin Chim Acta 2007;376:213–219.

73. Laposata M. Fatty acid ethyl esters: nonoxidative metabolites of ethanol. Addict Biol 1988;3:5–14.

74. Hamamoto T, Yamada S, Hirayama C. Nonoxidative metabolism of ethanol in the pancreas: implication in alcoholic pancreatic damage. Biochem Pharmacol 1990;39:241–245.

75. Laposata EA, Lange LG. Presence of nonoxidative ethanol metabolism in human organs commonly damages by ethanol abuse. Science 1986;231:497–499.

76. Aleryani S, Kabakibi A, Cluette-Brown J, et al. Fatty acid ethyl ester synthase, an enzyme for nonoxidative ethanol metabolism, is present in serum after liver and pancreatic injury. Clin Chem 1996;42:24–27.

77. Lange LG, Bergmann SR, Sobel BE. Identification of fatty acid ethyl esters as a product of rabbit myocardial ethanol metabolism. J Biol Chem 1981;256:12968–12973.

78. Laposata M, Kabakibi A, Walden MP, et al. Differences in the fatty acid composition of fatty acid ethyl esters in organs and their secretions. Alcohol Clin Exp Res 2000; 24:1488–1491.

79. Laposata M, Hasaba A, Best CA, et al. Fatty acid ethyl esters: recent observations. Prostaglandins Leukot Essent Fatty Acids 2002;67:193–196.

80. Salem RO, Refaai MA, Cluette-Brown JE, et al. Fatty acid ethyl esters in liver and adipose tissues as postmortem markers for ethanol intake. Clin Chem 2001;47:722–725.

81. Refaai MA, Nguyen PN, Steffensen TS, et al. Liver and adipose tissue fatty acid ethyl esters obtained at autopsy are postmortem markers for premortem ethanol intake. Clin Chem 2002;48:77–83.

82. De Giovanni N, Donadio G, Chiarotti M. The reliability of fatty acid ethyl esters (FAEE) as biological markers for the diagnosis of alcohol abuse. J Anal Toxicol 2007; 31:93–97.

83. Auwärter V, Sporkert F, Hartwig S, et al. Fatty acid ethyl esters in hair as markers of alcohol consumption. Segmental hair analysis of alcoholics, social drinkers, and teetotallers. Clin Chem 2001;47:2114–2123.

84. Hartwig S, Auwärter V, Pragst F. Fatty Acid ethyl esters in scalp, pubic, axillary, beard and body hair as markers for alcohol misuse. Alcohol Alcohol 2003;38:163–167.

85. Auwärter V, Kiessling B, Pragst F. Squalene in hair – a natural reference substance for the improved interpretation of fatty acid ethyl ester concentrations with respect to alcohol misuse. Forensic Sci Int 2004;145:149–159.

86. Hartwig S, Auwärter V, Pragst F. Effect of hair care and hair cosmetics on the concentrations of fatty acid ethyl esters in hair as markers of chronically elevated alcohol consumption. Forensic Sci Int 2003;131:90–97.

87. De Giovanni N, Donadio G, Chiarotti M. Ethanol contamination leads to fatty acid ethyl esters in hair samples. J Anal Toxicol 2008;32:156–159.

88. Klein J, Karaskov T, Korent G. Fatty acid ethyl esters: a novel biologic marker for heavy in utero ethanol exposure: a case report. Ther Drug Monit 1999;21:644–646.

89. Caprara DL, Klein J, Koren G. Baseline measures of fatty acid ethyl esters in hair of neonates born to abstaining or mild social drinking mothers. Ther Drug Monit 2005; 27:811–815.

90. Wurst FM, Kempter C, Seidl S, et al. Ethyl glucuronide – a marker of alcohol consumption and a relapse marker with clinical and forensic implications. Alcohol Alcohol 1999;34:71–77.

91. Alt A, Janda I, Seidl S, et al. Determination of ethyl glucuronide in hair samples. Alcohol Alcohol 2000;35:313–314.

92. Wurst FM, Metzger J, WHO/ISBRA study on state and trait markers of alcohol use and dependence investigators. The ethanol conjugate ethyl glucuronide is a useful marker of recent alcohol consumption. Alcohol Clin Exp Res 2002;26:1114–1119.

93. Kamil IA. A new aspect of ethanol metabolism: isolation of ethyl-glucuronide. Biochem 1952;51:32–33.

94. Jaakonmaki PI, Know KL, Horning EC, et al. The characterization by gas-liquid chromatography of ethyl B-d-glucuronic acid as a metabolite of ethanol in rat and man. Eur J Pharmacol 1967;1:63–70.

95. Schmitt G, Aderjan R, Keller T, et al. Ethyl glucuronide, an unusual ethanol metabolite in humans, synthesis, analytical data and determination in serum or urine. J Anal Toxicol 1995;19:91–94.

96. Dahl H, Stephanson N, Beck O, et al. Comparison of urinary excretion characteristics of ethanol and ethyl glucuronide. J Anal Toxicol 2002;26:201–204.

97. Wurst FM, Kempter C, Metzger J, et al. Ethyl glucuronide: a marker of recent alcohol consumption with clinical and forensic implications. Alcohol 2000;20:111–116.

98. Sachs H. Drogennachweis in Haaren, in Proceedings of the Symposium "Das haar als spur – spur der haare" Lubecca (1993) 119–133.

99. Aderjan RE. Ethyl glucuronide, a non-volatile ethanol metabolite in human hair. Proceedings of the 1994 Joint TIAFT/SOFT International Meeting in Tampa (Florida).

100. Jurado C, Soriano T, Giménez MP, et al. Diagnosis of chronic alcohol consumption. Hair analysis for ethyl-glucuronide. Forensic Sci Int 2004;145:161–166

101. Politi L, Morini L, Leone F, et al. Ethyl glucuronide in hair: Is it a reliable marker of chronic high levels of alcohol consumption?. Addiction 2006;101:1408–1412.

102. Morini L, Politi L, Groppi A, et al. Determination of ethyl glucuronide in hair samples by liquid chromatography/electrospray tandem mass spectrometry. J Mass Spectrom 2006;41:34–42.

103. Nicholas PC, Kim D, Crews FT, et al. Proton nuclear magnetic resonance spectroscopic determination of ethanol-induced formation of ethyl glucuronide in liver. Anal Biochem 2006;358:185–191.

104. Appenzeller BM, Schuman M, Yegles M, et al. Ethyl glucuronide concentration in hair is not influenced by pigmentation. Alcohol Alcohol 2007;42:326–327.

105. Kintz P, Villain M, Vallet E, et al. Ethyl glucuronide: Unusual distribution between head hair and pubic hair. Forensic Sci Int 2008;176:87–90.

106. Pragst F, Yegles M. Alcohol markers in hair. In: Kintz P. (ed) Analytical and Practical Aspects of Drug Testing in Hair. Taylor & Francis, Boca Raton, FL, 2007, pp. 287–323.

107. Bendroth P, Kronstrand R, Helander A, et al. Comparison of ethyl glucuronide in hair with phosphatidylethanol in whole blood as post-mortem markers of alcohol abuse. Forensic Sci Int 2008;176:76–81.

108. Politi L, Zucchella A, Morini L, et al. Markers of chronic alcohol use in hair: comparison of ethyl glucuronide and cocaethylene in cocaine users. Forensic Sci Int 2007;172: 23–27.

109. Borucki K, Schreiner R, Dierkes J, et al. Detection of recent ethanol intake with new markers: comparison of fatty acid ethyl esters in serum and of ethyl glucuronide and the ratio of 5-hydroxytryptophol to 5-hydroxyindole acetic acid in urine. Alcohol Clin Exp Res 2005;29:781–787.

110. Yegles M, Labarthe A, Auwärter V, et al. Comparison of ethyl glucuronide and fatty acid ethyl ester concentrations

in hair of alcoholics, social drinkers and teetotallers. Forensic Sci Int 2004;145:167–173.

111. Pragst F, Yegles M. Determination of fatty acid ethyl esters (FAEE) and ethyl glucuronide (EtG) in hair: a promising way for retrospective detection of alcohol abuse during pregnancy?. Ther Drug Monit 2008;30:255–263.

112. Politi L, Leone F, Morini L, et al. Bioanalytical procedures for determination of conjugates or fatty acid esters of ethanol as markers of ethanol consumption: a review. Anal Biochem 2007;368:1–16.

113. Skopp G, Schmitt G, Pötsch L, et al. Ethyl glucuronide in human hair. Alcohol Alcohol 2000;35:283–285.

Chapter 38
Retinoic Acid Signalling in Neuropsychiatric Disease: Possible Markers and Treatment Agents

Sarah J. Bailey and Peter J. McCaffery

Abstract Retinoic acid is the transcriptionally active product of vitamin A, potentially controlling expression of several hundred genes through activation of specific nuclear receptors. The retinoic acid receptors (RARs) are expressed through much of the adult brain; in contrast the ligand for these receptors is much more restricted in its extent of expression. Non-liganded RARs likely control function through transcriptional repression whereas ligand activated receptors probably regulate more restricted functions including neuroplasticity and neurogenesis. Deregulation of retinoic acid signalling through application of excess levels of this ligand have been associated with depression while genetic and anatomical evidence has linked retinoic acid with the neurodevelopmental hypothesis of schizophrenia and autism. The RARs frequently act in conjunction with the second class of retinoid receptors, the retinoid X receptors (RXRs). These RXRs heterodimerize with other classes of nuclear receptors, including the thyroid hormone receptor which, like those for retinoic acid, are associated with schizophrenia and depression. It is possible that drugs that act on RXRs may be particularly effective in psychiatric disease by acting simultaneously on several nuclear receptor signalling pathways.

Keywords Depression · schizophrenia · autism · RXR · nuclear receptor · retinoic acid · vitamin A · serotonin · hippocampus · neurogenesis · dopamine receptor · epigenetic

Abbreviations ALDH: Aldehyde dehydrogenase; BDNF: Brain derived neurotrophic factor; CNS: Central nervous system; COMT: Catechol-*O*-methyltransferase; CRBP: Cellular retinol binding protein; CRABP: Cellular retinoic acid binding protein; CSF: Cerebrospinal fluid; DARPP-32: Dopamine and cyclic AMP-regulated phospho-protein-32; DSM-IV: Diagnostic and statistical manual of mental disorders; GABA: Gamma-aminobutyric acid; 5HIAA: 5-Hydroxy indole acetic acid; HPA: Hypothalamic–pituitary–adrenal; 5-HT: 5-Hydroxy-tryptamine; 5HTT: 5-Hydroxytryptamine transporter; 5-HT1A: 5-Hydroxytryptamine receptor 1A; NGF: Nerve growth factor; Nurr1: Nuclear receptor related 1; PI3K–PKB: Phosphatidylinositide 3-kinase-protein kinase B; RA: All-*trans* retinoic acid; RALDH: Retinaldehyde dehydrogenase; RAR: All-*trans* retinoic acid receptor; RARE: Retinoic acid response element; RBP4: Retinol binding protein; RXR: Retinoid X receptor; SNP: Single-nucleotide polymorphism; SSRI: Selective serotonin reuptake inhibitors; TPH2: Tryptophan hydroxylase 2; TTR: Transthyretin

Introduction

Overlapping Pathways in Psychiatric Disorders

Psychiatric diseases such as depression and schizophrenia have a moderate to strong genetic component but are very poorly understood mechanistically because, in each case, multiple genes are involved.[1] Various combinations of mutations of these genes will add to

S. J. Bailey
Department of Pharmacy & Pharmacology, University of Bath, Claverton Down, UK

P. J. McCaffery
University of Aberdeen, Institute of Medical Sciences, Foresterhill, Aberdeen, UK

the risk of the disease, although it is not clear what numbers of genes are involved, how heterogeneous they are between family groups who suffer these disorders, or whether mutations in particular disease associated genes are frequent or rare.[2,3] Although each disease can be defined in terms of symptoms listed in the Diagnostic and Statistical Manual of Mental Disorders (DSM-IV), there is a certain degree of overlap in symptoms; for instance depression can be one of the symptoms of schizophrenia. It perhaps is therefore not surprising that there is cross-over in the genes associated with psychiatric and other diseases of the mind. For instance, DISC1 is associated with schizophrenia, schizoaffective disorder, bipolar disorder, major depression and autism.[4] Similarly neuregulin, which guides neuronal development including synapse formation,[5] and catechol-*O*-methyltransferase (COMT), that degrades catecholamines,[6] are also linked with several disorders.

The overlap in pathways involved in different disorders also includes neurotransmitters. Dopamine and other monoamines are likely to be integral to schizophrenia and depression, as well as autism. The dopamine hypothesis of schizophrenia proposes hyperactive dopaminergic signalling in the disorder and is based on the findings that antipsychotic drugs, including chlorpromazine, block dopamine signalling whereas drugs such as amphetamines, that raise dopamine levels by blocking dopamine reuptake, aggravate schizophrenia. In contrast, a lack of dopaminergic reward pathway signalling is thought to contribute to the anhedonic symptoms of depression.[7] Indeed, many of the drugs used to treat depression raise monoamine levels either by blocking their reuptake or their degradation. Specific examples include the selective serotonin reuptake inhibitor fluoxetine and drugs that block the reuptake of either noradrenaline (reboxetine) or noradrenaline and dopamine (bupropion). Alternatively, the monoamine oxidase inhibitors, such as phenelzine and isocarboxazid, exert their anti-depressant effects by preventing the breakdown of monoamines.[8] The relationship of monoamines to autism is likely to be even more complex.[9]

Clearly, schizophrenia and depression are not entirely independent of each other.[10,11] A recent study on a very large number of individuals indicates that the group of disorders sharing genes responsible for the condition extends from the psychiatric diseases schizophrenia and depression to the childhood disorder autism.[12] One subset of genes in these polygenic disorders is overlapping, while a second set of genes results in the unique features of each disorder. The topic of this chapter, the retinoic acid (RA) signalling pathway, falls in the category of pathways shared between these disorders of the mind.

Retinoic Acid

RA is the active product of vitamin A and is the mediator of most of the functions of this vitamin. Its action to regulate cellular gene transcription is illustrated in Fig. 38.1. RA's precursor, retinol, is transported in the circulation bound to retinol binding protein (RBP4) and is transported into the cell facilitated by the RBP receptor Stra6.[13] Retinol is bound by cellular retinol binding protein (CRBP) and oxidized first to retinaldehyde and then to RA, which is bound to cellular retinoic acid binding protein (CRABP). RA can be released from the cell to activate nearby cells and can also enter the cell's own nucleus, binding to specific RA receptors of which there are three classes, RAR α, β and γ. The receptor docks to its corresponding RA response element (RARE) twinned as a dimer, most usually with the RXR nuclear receptor which is also divided into RXR α, β, and γ classes. The liganded RAR-RXR heterodimer activates gene transcription, potentially regulating a large pool of genes, numbering many hundred,[14] controlling both the developing and mature brain.[15,16] Receptors to which ligand have not bound can have the opposite action of repressing transcription (Fig. 38.1). Thus, RA is a locally synthesized transcriptional regulator present only in specific regions of the brain expressing the appropriate synthetic enzymes. These areas include the basal ganglia, olfactory bulbs, hippocampus and certain cortical regions and hindbrain nuclei.[17]

Retinoic Acid Receptor Signalling and Psychiatric Disorders

An understanding of the aetiology of schizophrenia, depression and autism based on the molecular mechanism of the drugs used to treat or exacerbate such disorders has been applied to chemical pathways in

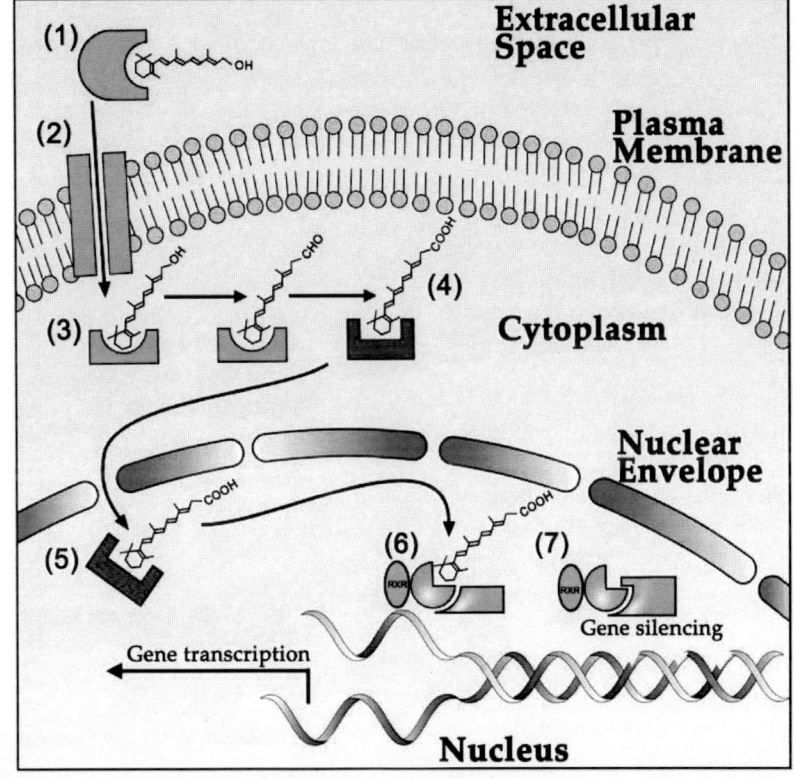

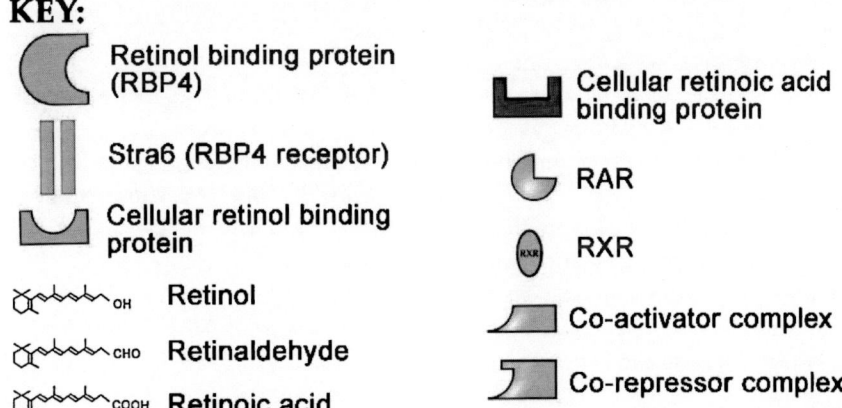

Fig. 38.1 The cellular RA signalling pathway. (1) The RA precursor, retinol, is carried in the plasma by RBP4. (2) This lipophilic molecule moves across the membrane in some cell types with the assistance of the membrane RBP4 receptor Stra6. (3) Retinol is then transferred to cellular retinol binding protein (CRBP). (4) Retinol undergoes a two-step oxidation, first catalyzed by a retinol dehydrogenase to generate retinaldehyde and then a retinaldehyde dehydrogenase to synthesize RA which is bound to cellular RA binding protein (CRABP). Alternatively RA can be generated in another nearby cell and diffuse directly into the target cell. (5) RA is transported into the nucleus probably by CRABP. (6) RA is released and binds to RAR and a complex of transcriptional cofactors, including co-activators, that allow gene transcription to take place. (7) When RAR is unbound to RA it can bind to a set of proteins including co-repressors that inhibit gene transcription

addition to that of the monoamine neurotransmitters. For instance, the finding that glucocorticoids can exacerbate depression provided evidence for the role of corticosteroid receptors, the hypothalamic-pitu-

itary-adrenal (HPA) axis and regulators of hippocampal neurogenesis in depression.[18,19] As described below, significant evidence has accumulated for a role for RA in schizophrenia, depression and autism.

A more complete understanding of the RA regulated pathways involved will potentially divulge elements of disease aetiology common between these disorders and identify potential biomarkers that are common and perhaps components that also distinguish these diseases.

Schizophrenia and Retinoic Acid Signalling

Schizophrenia is a devastating disease that shatters the sufferer's perception of reality and detaches them from society and normal social interaction. Schizophrenia has a worldwide lifetime prevalence rate of 0.4%[20] but despite this frequency the aetiology of schizophrenia is poorly defined. Unfortunately schizophrenia is a "complex" disorder with the involvement of a large number of genes combined with an environmental input. Some idea of the relative input of each can be estimated from twin studies and Sullivan et al.'s [21] meta-analysis of multiple studies found that heritability made a high (81%) contribution to schizophrenia but that environmental factors were also significant. The RA signalling pathway provides an interface for factors of both genetic (the receptors and genes modulating signalling) as well as environmental (dietary) sources. In addition, as described below, RA not only regulates the function of the adult brain but also guides its embryonic and postnatal development.[15] Thus genes that lie within the RA signalling pathway, if disrupted, could potentially influence the development of the brain. The neurodevelopmental hypothesis of schizophrenia proposes that the disease originates during development of the CNS, although the overt signs of schizophrenia are not evident until early adulthood, and this concept has gained widespread support.[22,23] Of environmental factors that act on the developing embryo and elevate the chance of schizophrenia later in life, nutrition is a prime candidate. Prenatal exposure to nutritional deficiency during famine results in an approximate doubling in the later risk of schizophrenia.[24,25]

Goodman[26,27] and LaMantia[28,29] originally proposed the concept of an association between RA and schizophrenia. It was propounded that "the minor physical anomalies,"[28] evident in schizophrenia, such as malformations in the face, limbs as well as heart, may betray the hallmark of an effect of RA on development, given that all these tissues are key targets of RA in the developing embryo. In the mouse RA is synthesized in limbic regions of the cortex associated with schizophrenia.[30] Endogenous RA regulates gene expression in GABAergic interneurons, a cell type repeatedly suggested to be altered in schizophrenia[31] in the cortex, olfactory bulb and amygdala.[29] The habenula is a further schizophrenia associated region in which RA has been shown to regulate gene expression.[29] Although there are a number of congenital malformations resulting from abnormal RA signalling that parallel defects in schizophrenia, such as enlarged ventricles, such changes result from mutation in very many other signalling systems and these changes are highly non-specific. Some of the best evidence of an association between RA and schizophrenia comes from potential changes in RA metabolism (synthesis and catabolism) in schizophrenia.

Alterations in RA Synthesis and Catabolism in Schizophrenia and DiGeorge Syndrome

Disruption of RA Synthesis in Schizophrenia

The last step of RA synthesis, the oxidation of retinaldehyde to RA, is performed by a retinaldehyde dehydrogenase which determines in which cells RA is synthesized[15,32] (Fig. 38.1). There are three such enzymes: ALDH1A1, ALDH1A2, and ALDH1A3 (also known as RALDH 1, 2 and 3). Microarray studies have identified ALDH1A1 (RALDH1 or AHD2) to be significantly decreased in the schizophrenic brain.[33,34] This enzyme is expressed in a subset of the midbrain dopaminergic neurons[35] and has been shown to decrease in expression in dopaminergic neurons of the ventral tegmental area, but not the substantia nigra in schizophrenia.[36] These dopaminergic neurons also express the nuclear receptor Nurr1, a receptor potentially associated with schizophrenia,[37,38] possibly suggesting a particular function of these RA synthesizing neurons to protect against schizophrenia. ALDH1A1 is expressed in these midbrain dopaminergic neurons both in the developing and mature brain, although it is unknown at which stage these neurons are lost. Interestingly inhibition of aldehyde dehydrogenases (ALDHs) by disulfiram in the adult, a chemical which potently inhibits ALDH1A1,[39,40] has frequently been reported to result in psychosis.[41–47] Disulfiram is used as an anti-alcoholic drug and in the early days of its use, when high doses were employed, psychosis resulted in between 2% up to 20% of cases.[48] It may be noted though that ALDH1A1 is not specific for retinaldehyde as a substrate; it may also act in the pathway to metabolize dopamine, although it is likely that the mitochondrial aldehyde dehydrogenases are more important for dopamine metabolism.[49] ALDH1A1 may also be involved in the oxidation of 3-deoxyglucosone, necessary to remove the product of a protein repair pathway.[50] Nevertheless ALDH1A1 is clearly important in the oxidation of retinaldehyde to RA[51,52] and, at least in regard to its role in the developing dopaminergic neurons, its function to synthesize RA may be critical.[53]

Retinoic Acid, 22q11 Deletion Syndrome and Schizophrenia

The 22q11 deletion syndrome (which includes the DiGeorge, velo-cardio-facial and catch-22 syndromes) is a haplodeficiency of a region of approximately 30 genes in which approximately 10% of patients show psychiatric disorders.[54,55] It is one of the highest genetic risk factors for the development of schizophrenia and the deletion has highlighted catechol-*O*-methyltransferase as a risk factor for schizophrenia.[56] Exposure of the developing brain to either excess RA[57] or decreased RA synthesis[58] results in many of the features of DiGeorge syndrome. Mouse mutant studies have shown that knockout of two critical genes in the 22q11 region, *Tbx1* and *Crkl*, severely disrupts RA homeostasis in the embryo with loss of both RA synthetic and catabolic enzymes leading to an overall increase in RA signalling.[59] It was proposed that RA may act as a modifier of the phenotype associated with 22q11 deletions. Given that Tbx1 is likely to be a strong contributor to the behavioural defects evident in 22q11.2 deletion syndrome[60] it would seem plausible that some of the downstream effects of haplodeficiency of *Tbx1* are mediated through developmental changes in RA metabolism.

Evidence for Changes in RA Receptors and Transthyretin in Schizophrenia

The suggested association between RA and schizophrenia has led several RA signalling genes to be examined for mutation in this disorder. Initial studies of the variation in the transthyretin (TTR) gene, which produces a protein product that carries both retinol bound to RBP4 as well as thyroid hormone in the circulation, did not show linkage with schizophrenia and the same study found no relationship of RPB4 or TTR plasma levels with the disorder as measured immunologically using radial-immunodiffusion.[61] This may be a problem though using antibody based detection because recent proteomic disease marker screening techniques, employing mass spectrometry, have found a significant change in TTR, decreasing in CSF and prefrontal cortex[62] and a decline in the CSF was also detected in the initial, prodromal state of psychosis.[63] A further study found decreasing TTR in the CSF of schizophrenics but rising 2 months after treatment with antipsychotics when measured in plasma.[64] Polymorphism of the RA receptors RAR and RXR α, β and γ have been examined, including exons and their splice junctions in 100 patients with schizophrenia; two uncommon missense mutations were found but there was no statistically significant evidence that structural variants played a major role in this disorder.[65] Similarly a search for mutations in RXRβ did not identify an association with schizophrenia[66] It is of note though in the Feng et al. study[65] that a RAR β structural variant was found in 3 of 100 schizophrenics and was absent from 92 patients without schizophrenia. The search for changes will continue with ever more sophisticated bioinformatics tools to examine, for instance, changes in gene enhancers. A suggestion that changes in receptor protein expression may occur in local brain regions comes from an analysis of RARα protein in the hippocampus, finding greater number of granule cells expressing this receptor in the schizophrenic versus control brain.[67] In contrast no change was evident in RARγ, RXRβ or RXRγ receptors. Although this was a small study with only 10 schizophrenic brains it is improbable that the effects were due to the influence of antipsychotic drugs as they have only a very marginal influence on RAR expression.[68]

Candidate Schizophrenia Genes Downstream of RA

Within the nuclear receptor superfamily of transcriptional regulators the RARs have probably the largest number of target genes, with over 500 genes regulated directly or indirectly.[14] With input into many signalling networks, in particular those controlling development of the CNS,[15] there are a number of potential pathways proposed as part of schizophrenia aetiology that RA may regulate. Several of these are listed below and some of the best candidates are described in Fig. 38.2a.

Reelin and Hippocampal Neurogenesis

Reelin is a secreted protein that regulates development of the brain, controlling neuronal migration and, in the adult, synaptic plasticity.[69] A series of reports have indicated that certain alleles of reelin and/or a decline in expression of this gene, contribute to schizophrenia (e.g. refs.[70–72]) as well as bipolar disorder and depression.[73] The epigenetic regulation of the reelin gene promoter has suggested this gene as part of an epigenetic input into schizophrenia[74] and the promoter has been reported to be hypermethylated in schizophrenia.[75] Reelin is regulated by RA which results in hypomethylation of the promoter, increasing reelin expression,[76] the opposite direction to that reported in schizophrenia. Thus any involvement of RA in epigenetic control of reelin in schizophrenia would imply a decrease in RA signalling.

One mechanism by which a decline in reelin may contribute to schizophrenia is via its influence on hippocampal neurogenesis.[77] Neurogenesis provides new granule neurons for the dentate gyrus and has been suggested to be contributory, although probably not a primary cause, of schizophrenia and depression, perhaps adding to the hippocampal component of the diseases.[78] The recent finding that DISC1, a gene with one of the tightest links to schizophrenia, as well as depression, regulates neurogenesis[79] suggests that this may also be one route that DISC1 acts to promote schizophrenia. RA has also been extensively described to depress hippocampal neurogenesis; either a deficiency[80] or excess[81] decreases the birth of hippocampal granule neurons, providing a route by which RA may have an impact on these psychiatric disorders (see also section on Depression and Retinoic Acid Signalling).

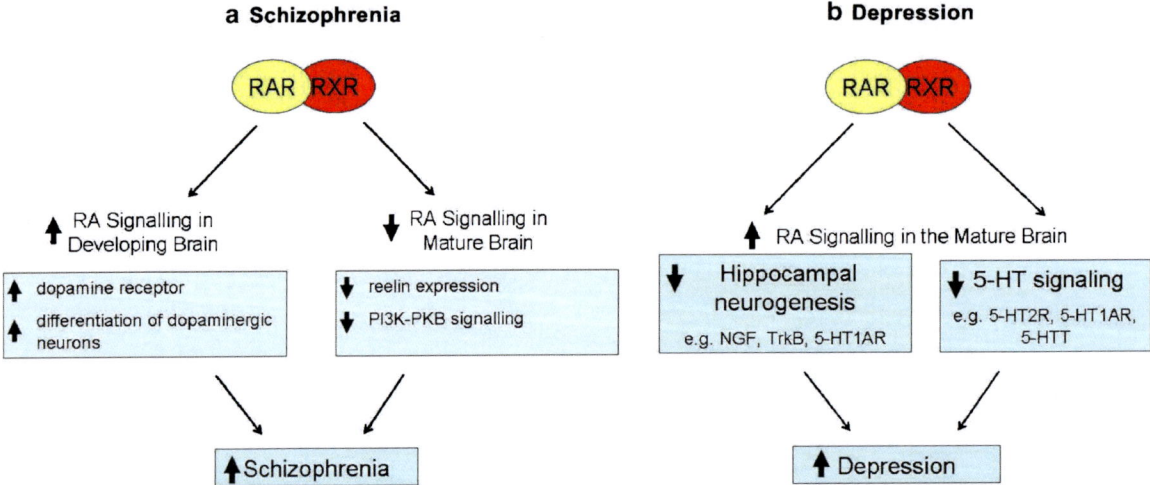

Fig. 38.2 Speculative RA signalling mechanisms promoting schizophrenia or depression. (**a**) In schizophrenia it is proposed that RA signalling in the developing embryo is abnormally high resulting in increased differentiation of dopaminergic neurons or increased expression of dopamine receptors. Possibly in response to high RA signalling in the embryo this pathway is attenuated postnatally leading to insufficient RA signalling in the adult resulting in changes in several downstream factors including reelin expression and PI3K-PKB signalling. (**b**) In depression it is proposed that a contributing factor is abnormally high levels of RA in the adult which either decrease birth (neurogenesis) of new granule cells in the hippocampus or disrupts regulation of the monoaminergic neurotransmitters with the specific involvement of the 5-hydroxytryptamine (5-HT) system

Dopamine Receptors

As already discussed (section on Overlapping Pathways in Psychiatric Disorders), the dopamine hypothesis of schizophrenia proposes that excess dopamine signalling is part of the aetiology of the disorder. Dopamine neurotransmission, in particular through the D2 dopamine receptor, is under regulation by RA during development.[82–85] Aberrant RA regulation during development may therefore result in abnormal dopamine receptor signalling. In the developing brain the ganglionic eminence is the germinal tissue in the basal forebrain that gives rise to the dopaminergic neurons of the striatum,[86] a key structure in the dopaminergic hypothesis for schizophrenia.[87] Indeed, a number of other cell types likely involved in the pathology of schizophrenia are born here, including the GABAergic interneurons of the cerebral cortex and hippocampus.[88,89] RA is crucial to the development of the ganglionic eminence and a RA synthesizing enzyme, ALDH1A3 (RALDH3), is strongly expressed in this region.[30,90] Among several functions, RA is suggested to be required for the expression of the proteins necessary to mediate dopamine transmission, including the D1 dopamine receptor, DARPP-32, the G_{olf} olfactory G protein, and adenylyl cyclase type V.[91] RA is probably a factor instructing striatal cells to become responsive to dopamine and, for instance, excess amounts of RA during development may result in a greater number of dopamine responsive neurons.

Retinoic Acid Inducible-1 Gene (RAI1)

RAI1 was initially identified as a RA inducible gene[92] and polymorphisms in a CAG-repeat region of this gene are associated with the severity of schizophrenia as well as the response of the patient to medication.[93,94] Mutations in the gene are associated with Smith-Magenis syndrome[95,96] a complex disorder with neurobehavioral disturbances. Gene dosage is critical and duplication of the chromosomal region containing this gene results in mental retardation and increased gene dosage has detrimental behavioural effects.[97] A second RA inducible gene, retinoic acid receptor responder 3, a tumour suppressor gene, was identified in a screen of blood leukocytes from schizophrenic and bipolar patients, although the link was only found with bipolar disorder.[98]

Signalling through Calcium Regulation and Other Pathways

A reduction in calcium binding proteins, including calbindin, parvalbumin and calretinin, has been correlated with schizophrenia[31] and calbindin is regulated by RA.[99] However, the decline in these binding proteins likely reflects changes in the number of GABAergic inhibitory interneurons that express these proteins. Neurogranin, in contrast, is a RA regulated gene[100,101] that may directly be linked with schizophrenia because single nucleotide polymorphisms in the gene are associated with this disorder.[102] Neurogranin is a calmodulin binding protein and controls the availability of the calcium sensor protein, calmodulin, by binding to calmodulin in the absence of calcium.

A decline in the phosphatidylinositide 3-kinase-protein kinase B (PI3K–PKB) signalling pathway has been proposed to be associated with schizophrenia[103,104] and, given that this pathway is integral to cell division, differentiation and a variety of cancers, a decrease in this pathway has been suggested as a possible explanation for the lower risk of certain types of cancers in schizophrenia.[104] PI3K–PKB is strongly promoted by RA and transduces the action of RA to induce differentiation in many cells types,[105–107] implying that a deficiency would lead to a decline in the PI3K–PKB pathway as a potential mechanism contributory to schizophrenia. However, activation of PI3K–PKB by RA is a short-term phenomenon and RA can even inhibit PKB with more prolonged exposure[106] while the parent compound to RA, retinol, can also inhibit PI3K, directly interacting with the protein.[108] The feedback loops involved in the regulation of PKB by retinoids need to be investigated further for RA and retinol can regulate PI3K–PKB in a complex manner and provides an intriguing route whereby retinoids may impact on schizophrenia pathways.

Depression and Retinoic Acid Signalling

Major depressive disorder affects 1 in 20 people during their lifetime and is one of the most prevalent medical illnesses.[109] The World Health Organization estimates that by 2020 depression will become the second most important cause of disability worldwide (after ischaemic heart disease). Depression, like schizophrenia, is a heterogeneous disorder characterized by a range of symptoms including emotional, psychomotor and cognitive processes. Depression is an umbrella term used to encompass short-lived episodes of depression, dysthymia, major depressive disorder and bipolar disorder. The primary symptoms for characterizing a major depressive disorder are persistent depressed mood or diminished interest or pleasure in every day activities, alongside disturbances in sleep, body weight, energy levels, psychomotor activity and suicidal thoughts (DSM-IV).

There are multiple inter-related ideas about the causes of depression.[7,110,111] The longest-standing is the monoamine theory of depression. As described in the section on Overlapping Pathways in Psychiatric Disorders, this is based largely on the effectiveness of antidepressants to increase synaptic levels of 5-hydroxytryptamine (5-HT, serotonin) and noradrenaline. As a consequence, deficits of 5-HT, noradrenaline and dopamine have all been implicated in the pathophysiology of depression. Numerous deficits in these monoaminergic transmitter systems have now been identified (reviewed by Manji et al.[110]). For 5-HT, for example, this includes reduced CSF levels of the metabolite 5-hydroxyindole acetic acid (5HIAA), the depressogenic effects of depleting tryptophan (the precursor for 5-HT) and reduced 5-HT1A receptor binding in patients.[112] There is also widespread interest in dysfunction of the hypothalamic-pituitary-adrenal axis (HPA) as a causative mechanism in depression. The HPA regulates cortisol production that acts on glucocorticoid receptors in the brain (see section on Retinoic Acid Receptor Signalling and Psychiatric Disorders). Hypercortisolaemia, nonsuppressive dexamethoasone test and blunted adrenocorticotrophic hormone responses to challenge have long been recognized in depressed patients. Glucocorticoid antagonists and cortisol synthesis inhibitors are being explored as possible antidepressant agents.[113] More recently the neurogenic theory of depression has been considered as antidepressants are known to increase hippocampal adult neurogenesis whereas stress produces the opposite effects.[114] These effects are thought to be mediated, at least in part, through brain derived neurotrophic factor (BDNF).[114,115] Antidepressant treatments of all classes can increase BDNF expression whereas stress reduces BDNF expression. Direct hippocampal infusions of BDNF protein can produce antidepressant effects. Other neurogenic factors and other mechanisms are also thought to be involved.

For example, the 5-HT1A receptor has been reported to be essential for the neurogenic and antidepressant behavioural effects of selective serotonin reuptake inhibitors (SSRI).[116] However, one weakness in this theory is that in situations where neurogenesis is experimentally reduced there is not always a concomitant increase in depression-related behaviours.[117]

Genetic studies have shown that mood disorders are substantially influenced by genetic components that are highly complex, polygenic and epistatic.[2,118] Compared with the general population, first-degree relatives of depressed individuals have a nearly three-fold increase in their risk of developing a major depressive disorder. Relatives of patients with bipolar disorder also have an increased risk of unipolar depression and affective disorders tend to coexist with anxiety in many families. A polymorphism in the 5-HT transporter (5HTT), a protein central to synaptic 5-HT function, is associated with altered expression and function and linked to negative emotionality traits including anxiety, depression and aggressiveness.[118] A functional SNP in the transcriptional control region of the gene for the 5-HT1A receptor is associated with depression and suicidality. SNPs in tryptophan hydroxylase 2 (TPH2), the enzyme involved in 5-HT synthesis, are also associated with major depressive disorder and suicide.[119] A SNP in BDNF (val/met) is also associated with bipolar disorder.[120] These genes are all potential biomarkers for depression.[121] The association of multiple genes with depression has lead to the idea that vulnerability to depression depends on the interaction of multiple genes with modest effect with each other and the environment.

Here we consider the evidence that the RA signalling pathway may contribute to susceptibility to depression. The RA receptors are widely distributed in the CNS and play a role in the developing, as well as the adult, brain[16,32] and the effects of RA signalling with regard to depression in the adult, and the potential mechanisms involved (Fig. 38.2b) are discussed.

An Association Between Retinoids and Depression

Hypervitaminosis A, either from the ingestion of supplements or liver with high vitamin A content, has long been associated with numerous side effects including psychiatric symptoms (reviewed by O'Donnell[122]). For example, depression and anxiety symptoms have been reported in patients with chronic vitamin A intoxication resulting from prolonged use of vitamin A.[123] Oral retinoid therapies, such as Accutane (13-*cis*-retinoic acid, 13-*cis*-RA), used for dermatological conditions have also been linked controversially with an increased risk of depression and suicidal behaviours.[124,125] Case report studies indicate that in about 5% of patients taking Accutane for the treatment of acne, depression, suicidality, psychosis, violence and aggression may develop[124,125] but the issue of causality remains controversial.[126,127] Interestingly, when high doses of Accutane are used (3 mg/kg/day) 25% of patients developed symptoms of depression.[128] Functional brain imaging studies of patients treated with Accutane show that brain metabolism is decreased in the orbitofrontal cortex, compared with patients treated with antibiotics.[129] While several of these patients had subtle changes in irritability or mood they were not clinically depressed as assessed by the Hamilton depression scale.[129] Acitretin is a retinoid used as a psoriasis treatment that carries a warning on the product insert that depression may be a side effect. However, the evidence for this is sparse.[130]

Retinoids and Animal Models of Depression

A complicating factor when considering whether patients taking Accutane may be depressed is the psycho-social impact of having severe acne. This can be obviated by studying the effects of retinoids in animal models of depression-related behaviour. In adolescent mice treated with 13-*cis*-RA, there is an increase in depression-related behaviour.[131] However, in adult rats, chronic administration of 13-*cis*-RA or RA did not alter depression-like behaviour in the forced swim test or in sucrose anhedonia.[132,133] This may represent a difference between species since 13-*cis*-RA administration has also been shown to impair learning and memory in mice[81] but not in rats.[134] This may also reflect a difference in the age of animals tested with adolescent animals perhaps being more vulnerable to the effects of 13-*cis*-RA than adult animals. Alternatively it may reflect a difference in the sensitivity of the behavioural paradigm to detect treatment effects.

In the resident-intruder paradigm, 13-*cis*-RA administration in adult rats reveals changes in aggression behaviours that are consistent with increased depression.[135] Further support for a role for retinoids in depression-related behaviours is provided by studies of the TTR null mouse. In this model, retinol transport would be impaired and depression-related behaviour is decreased.[136] This effect is opposite to that seen with 13-*cis*-RA administration.[131]

Candidate Depression Associated Genes Downstream of RA

The RAR/RXR receptor heterodimeric complex regulates gene transcription (Fig. 38.1) so the likely mechanisms by which retinoids could mediate changes in depression-related behaviour are via modification of gene expression. A large number of neuronal genes have been identified as being regulated by retinoids[16] and several biological pathways linked to depression-related behaviours may be influenced by retinoids.[137]

Recently, the ability of 13-*cis*-RA to upregulate 5-HT1A receptors and 5HTT has been demonstrated in cultured cells.[138] If such changes occur *in vivo*, then this could potentially lead to reduced neurotransmitter release (because of increased autoreceptor function) and reduced availability of 5-HT in the synapse (because of increased reuptake) and consequently impaired 5-HT neurotransmission. Although the effect of retinoid treatment on 5-HT release has not been directly studied *in vivo*, Ferguson et al.[139] have shown that the tissue content of 5-HT and 5-HIAA was not altered by 13-*cis*-RA administration, except in the striatum where it was elevated.

The monoamine hypothesis of depression suggests that impairments in noradrenaline or dopamine neurotransmission may also be important in the pathophysiology of depression. Regulation of noradrenaline levels could result from induction of expression of the noradrenaline reuptake transporter.[140] The TTR null mouse, that has reduced transport of retinol, has increased noradrenaline levels in limbic forebrain without a change in dopamine or 5-HT levels.[136] The expression of a range of dopamine signal transduction molecules is also regulated by retinoids including dopamine D1 and D5 receptors and the dopamine D2 receptor is one of the few neuronal genes with a verified

RARE.[82,84,91,141] This suggests that dopaminergic neurotransmission is amenable to modulation by retinoids (also see section on Candidate Schizophrenia Genes Downstream of RA). Ferguson et al.[139] have shown in rats that the tissue content of dopamine and its metabolites was not altered by 13-*cis*-RA treatment, except in the striatum, although dopamine release was not directly studied.

An alternative hypothesis is that retinoids may induce depression by impairing adult neurogenesis.[137] In mice, 3 weeks treatment with either RA or 13-*cis*-RA reduced cell proliferation and the number of new neurons surviving in the subgranular zone of the hippocampus.[81] Interestingly, in retinoid deficient mice, the number of new neurones surviving in the hippocampus is also reduced, although cell proliferation is not affected.[80] How can too much or too little retinoid cause the same effect? Establishing optimal concentrations of retinoids for biological function may be extremely important in determining the effects on neurogenesis. A concentration-dependent effect of retinoids on neuronal dendritic morphology has been demonstrated.[142,143] 13-*cis*-RA, acting via RXR and RAR, at low concentrations (10^{-8}M) promotes an increase in the number of hippocampal neurones in culture whereas high concentrations (10^{-4}M) reduces cell number.[143] Similarly, low concentrations of 13-*cis*-RA stimulate dendritic branching and outgrowth but this is reduced at higher concentrations.[143] Similar inhibitory effects of high concentrations of 13-*cis*-RA on dendritic morphology have been shown in 5-HT raphe neurons in slice culture.[142] So at high concentrations, retinoids may impair dendritic outgrowth, dendritic branching and reduce the survival of new neurons in the hippocampus leading to impaired hippocampal function. However, it should be noted that these effects are only detected at non-physiological concentrations of RA[142,143] and may reflect neuronal cell death.

The mechanisms regulating adult neurogenesis are complex with growth factors such as nerve growth factor (NGF) and BDNF, corticosteroids, and 5-HT all playing important roles.[114,115] Jacobs et al.[80] identified a number of retinoid-responsive genes in adult neural stem cells that suggest that the Wnt signalling pathway and lipid mobilization are important in the early stages of neuronal differentiation and that this is how retinoids are able to regulate the survival of new neurons. Interestingly, in cultured cells at least, NGF and TrkA/Trk B (NGF/BDNF receptors) have been shown to be induced by retinoids.[144–146] 5-HT and particularly the 5-HT1A receptor have been demonstrated to be required for the antidepressant-induced stimulatory effects on adult neurogenesis.[116] Activation of either NGF or 5-HT by retinoids might be anticipated to have a stimulatory effect on adult neurogenesis (i.e. an antidepressant response) but such information does highlight possible roles for retinoid-mediated regulation of adult neurogenesis.

Retinoids may also induce depression by an interaction with homocysteine levels and one-carbon metabolism. Increased homocysteine levels are associated with several psychiatric and neurodegenerative disorders.[147,148] High homocysteine interferes with one-carbon metabolism such that there is decreased DNA methylation capacity that may impair synthesis of neurotransmitters or membrane phospholipids. Homocysteine metabolism is intricately linked with vitamin B12 and folate (B9) levels and low serum folate is linked with high plasma homocysteine. Both low folate and low B12 levels are common in depression with as many as 56% of patients with affective disorders being folate deficient.[148] Both folate deficiency[149] and elevated homocysteine levels[150] have been reported in patients taking Accutane (13-*cis*-RA). Chanson et al.[151] have shown a subtle decrease in plasma folate levels, without a change in homocysteine levels, in 13-*cis*-RA treated healthy subjects. In rats, retinoids can perturb methyl group metabolism by increasing the abundance and activity of glycine-N-methyltransferase in a dose-dependent manner and this treatment results in a reduction of DNA methylation.[152,153] However, there is no direct evidence to date that a retinoid-induced change in folate or homocysteine levels is associated with an increase in depression.

Autism: A Third Disorder of the Mind Influenced by Nuclear Receptors

Autism is an early childhood disorder, usually occurring in the first 3 years of life, leading to deficits in communication, social awareness and interactions and symbolic or imaginative play and can include an increase in repetitive behaviour. However Delong has argued that there is some overlap in the clinical features of autism and major mood disorders and there is a high incidence of these disorders in family members of autistic patients.[154] Recent studies have indicated high levels of psychiatric illness in autism.[155] Certainly, a recent analysis of the genetic overlap between phenotypes studying a very large number of patients (1.5 million) identified significant genetic overlap between autism, bipolar disorder and schizophrenia.[12] As such, it would be predicted that if altered RA signalling contributes to schizophrenia and depression that it may also have some role to play in autism. Although there is a very strong genetic component to autism,[156] there is a clear contribution from environmental factors[157] and, as already discussed, RA lies at the intersection of genetic and dietary signal control.

London and Etzel[158] first proposed RA as an environmental modulator of autism[158] and as a factor that, in excess, produces abnormalities similar to those reported in the autistic hindbrain, including hypoplasia of the cerebellar vermis[159,160] and reduction in size of the hindbrain.[161–164] It has also been proposed that nuclear receptor ligands which regulate neurogenesis, such as RA, may contribute to macrocephaly evident in the early postnatal developmental stages of autism.[165]

The strongest evidence for an association between RA signalling and the developmental triggering of autism comes from the action of a drug that potently promotes the transcriptional activating capacity of nuclear receptors. Valproate is an anti-epileptic drug that is teratogenic with similarities to autism in its pathology.[166–170] Children prenatally exposed to valproate can show typical signs of autism such as impaired development in social interaction and communication and significantly restricted range of activities and interests. Several studies in animal models have further suggested that embryonic exposure to valproate can mimic the pathology of autism in the form of brain malformations,[171,172] monoamine concentration in blood and brain,[173] as well as behaviour.[174] Valproate's teratogenic effects are due to its ability to inhibit histone deacetylase.[175] Nuclear receptors, such as the RA receptors, activate transcription via the recruitment of histone acetylases into the transcriptional complex, opening chromatin into a conformation that allows transcription. Drugs such as valproate inhibit the histone deacetylase that would otherwise reverse this process thus promoting the action of nuclear receptors, with the RA signalling pathway as a main rate-limiting target of the histone deacetylase inhibitors.[176] These findings imply that over-stimulation of nuclear receptor signalling, in particular the RA signalling system, may lead to

abnormalities in the developing brain that can trigger autism in early postnatal life.

Conclusions and Future Directions

Treatments Based on Retinoic Acid

As described in the Introduction, RA signalling has been associated not with one but several psychiatric diseases and disorders of the mind. This might initially lower the interest in the involvement of RA in psychiatric disease or autism because abnormalities in this signal pathway may seem to have only a general effect on behaviour, simply mimicking these disorders but having no relationship to their actual aetiology. However it is evident that several factors, such as DISC1, are likely part of the aetiology of multiple syndromes, including schizophrenia, depression and autism[4] and such factors are still potential targets for the treatment of these disorders. Similarly, components of the RA signalling pathway are possible treatment targets. A summary of the predominant potential mechanisms by which this signalling pathway may impact schizophrenia and depression is shown in Fig. 38.2.

There are two developmental stages at which RA signalling may contribute to schizophrenia, depression or autism.

(i) A defect in signalling resulting in abnormal embryonic or postnatal development as part of the neurodevelopmental model of psychiatric disease
(ii) Abnormal signalling leading to interference with normal adult neural function

Needless to say, these are not mutually exclusive and defective RA signalling may lead to abnormal development and also later interfere with mature brain function. Only the latter is open to drug treatment for psychiatric illness but both are open to nutritional manipulation, i.e. during development or in the adult. The essential question is: in which way is RA signalling altered to contribute to schizophrenia, depression and autism? From what has been discussed in the chapter, there is insufficient evidence to make any strong conclusions on mechanisms involved. Some ideas though can be put forward. In the case of schizophrenia, the strongest associative evidence comes from the developmental changes in RA metabolism in the schizophrenia associated disorder 22q11 deletion syndrome, which leads to an overall increase in RA signalling.[59] Further evidence comes from the synthesis of RA in midbrain dopaminergic neurons[35] and the potent action of RA to promote expression of the dopamine receptors and dopaminergic signalling machinery.[82–85,91] It may be proposed then that excessive RA synthesis or signalling in the developing brain may contribute to schizophrenia by increasing the number of midbrain dopaminergic neurons, or neurons expressing dopamine receptors, as part of the neurodevelopmental and dopamine hypotheses of schizophrenia (Fig. 38.2a). This is impractical though to treat because an attempt to reduce embryonic levels of RA could have disastrous effects on development which is dependant in many ways on RA signalling, both in and outside the CNS.[177] It should be noted though that several features of the *adult* schizophrenic brain show markers of *reduced* RA signalling (Fig. 38.2a), including decreased reelin expression and activity of the PI3K–PKB signalling pathway while treatment with the anti-alcoholic drug disulfiram, that lowers RA levels by inhibiting its synthesis, induces psychosis.[41–48] It is possible that reduced RA signalling in the adult schizophrenic brain is in response to high RA signalling during development, leading to an attenuation of the RA signalling pathway in the adult brain. These changes will be regionalized to the limited regions of the brain in which RA is synthesized (see section on Retinoic Acid).

In the case of depression, the best associative evidence, although controversial, comes from the effects of RA to promote depression in a number of patients.[125] These effects are evident in about 5% of patients suggesting that there may be a vulnerable population of patients taking Accutane, perhaps because of altered endogenous retinoid-signalling or RA homeostasis. Studies using mouse (although not rat[133]) models have shown an increase in depression-related behaviour with RA treatment.[131] Further, the mouse knockout of the TTR gene, which results in a decreased delivery of retinol to cells, leads to a decline in depression-related behaviour.[136] The mechanism by which excess RA may have a depressogenic effect may be through RA upregulation of 5-HT1A receptors and 5HTT[137] or suppression of hippocampal neurogenesis.[81] Thus depression, like schizophrenia, may also result from excess RA signalling, but in the adult rather than developing brain. The third disorder discussed in this chapter, autism, may also

result from high RA signalling, but, as proposed for schizophrenia, is a developmental phenomenon. Such over-stimulation though may involve multiple nuclear receptor signalling pathways given that the best associative evidence comes from the embryonic effect of the histone deacetylase inhibitor valproate to promote autism and such inhibition of histone deacetylase activity can promote the activity of many nuclear receptor transcriptional activators, as well as RA.

It may then be suggested that over-stimulation of the RA signalling pathway can contribute to schizophrenia, depression and autism and that this is one of the factors common between the disorders, rather than distinguishing each. Even as a proposal though, this concept is highly simplistic and the effects of RA will depend on the interactions with the many other genes and environmental inputs into these complex disorders. There may even be cases of RA deficiency contributing to these diseases because, at least during development, some of the features of RA teratogenicity parallel those of RA deficiency.[15] Excess RA will trigger compensatory mechanisms to downregulate RA signalling and a prolonged decrease in RA signalling may contribute to these disorders, as we have suggested for schizophrenia. In this case, drugs that promote RA signalling may be therapeutic and two examples of this are discussed below.

Treatment of Schizophrenia with RXR Ligands

The RXR partners of the RARs were originally described as RA receptors themselves given that they bind with high affinity to the 9-*cis* isomer of RA.[178] However, it seems less likely that this is their endogenous ligand given that 9-*cis* RA is seldom found in tissues.[179] Artificial ligands though for the RXR can potentially stimulate the RA signalling pathway. Their action will be complex because they will influence all the nuclear receptor signalling pathways that use RXR as their heterodimeric receptor partner including the receptors for thyroid hormone, vitamin D and the peroxisome proliferation-activated receptor.[180] Their action is further complicated because, in some circumstances, ligand binding of both partners, e.g. 9-*cis*-RA for RXR and RA for RAR can promote activation of transcription but it is frequently not necessary and RXR often functions as an unliganded partner for the nuclear receptor to which it heterodimerizes.[180,181] For instance, 9-*cis*-RA should be more potent than all-trans RA given that it binds to both RAR and RXR, but this is not the case in a number of circumstances.[182] To even further complicate the picture, different synthetic ligands can have quite differing effects on the heterodimeric complex[181] so it is difficult to predict what effect each RXR ligand will have on RA or other signalling pathways, and each would need to be examined empirically for its action. However, an unequivocal function for RXR ligands with regard to RXR interaction with a heterodimeric partner is its induction of release of RXR from a tetrameric complex. In the absence of ligand, RXR exists as a complex of four RXRs and will tend not to interact with other nuclear receptors. The presence of RXR ligand will release RXR from this complex and allow it to interact with its other partners.[183,184] At least with reference to this action, RXR ligands will promote the action of heterodimeric receptors, i.e. RXR-RAR, RXR-thyroid hormone receptor, RXR-vitamin D receptor or RXR-Nurr1. How this may impact psychiatric disease is uncertain but there is evidence that RXR ligands have a positive effect on schizophrenia and depression.

Bexarotene (LGD1069, targretin) is a RXR agonist.[185] Lerner et al. (2008)[186] investigated the safety and efficacy of add-on oral bexarotene to ongoing antipsychotic treatment in chronic schizophrenia patients. A 6 week open label trial was conducted on 25 patients with chronic schizophrenia who received a low dose of Bexarotene (75 mg/day) augmentation. Significant improvement from baseline to endpoint was observed on total Positive and Negative Symptom Scale score, general psychopathology, and the dysphoric mood factor scores.[186] Although only a small study, these observations support a role for retinoids as treatment agents in schizophrenia.

Treatment of Schizophrenia with Epigenetic Modifiers

Epigenetic regulation of gene expression is an area of increasing interest in schizophrenia research. Such regulation takes place by covalent but reversible modification of either DNA itself or the packing proteins around it, modifying gene expression by altering the ability of proteins necessary for transcription to gain access to the requisite promoters or enhancers. The genes controlling epigenetic regulation are, of course, inherited, but

environmental influences can also modify the extent of epigenetic modulation of gene expression. The action of nuclear receptors as transcription factors is particularly prone to epigenetic modification. One such mechanism involves the modification of the histones by addition or subtraction of acetyl groups. DNA is compacted by its winding around the histones; addition of acetyl groups opens up this compaction (euchromatin) allowing transcription factors access to promoters whereas removal of these acetyl groups allows the DNA to collapse (heterochromatin) preventing transcription factors from reaching their DNA binding site. Drugs that inhibit the removal of acetyl groups (histone deacetylase inhibitors), such as trichostatin and valproate, promote transcription, and, in regards to their effects on cell proliferation, RA signalling is a key and rate limiting component of this type of control.[176] The action of the histone deacetylase inhibitor valproate on the developing embryo to promote autism may be through its enhancement of RA signalling leading to a teratogenic action on the developing brain (as described in the section on Autism: A Third Disorder of the Mind Influenced by Nuclear Receptors). In the case of syndromes resulting from deficient nuclear receptor signalling such histone deacetylase inhibitors may be beneficial and Shama[187] recently proposed these as psychotropic drugs.

Change in RA Signalling as a Putative Treatment for Depression

We propose that excessive RA signalling in the adult may contribute to depression and thus therapy would consist of treatment that reduces the strength of this signalling pathway. The difficulty presented is that balance of the RA signal needs to be maintained correctly and levels that are too low can be just as harmful as too high. For instance, as already described, disulfiram, which inhibits RA synthesis, can induce psychosis while a decline in hippocampal neurogenesis, proposed as one mechanism by which excess RA may result in depression[81] can also be induced by a deficiency of RA.[80] One simple route to attenuate RA signalling may be dietary reduction of vitamin A which, if plasma levels are monitored carefully, may produce a moderate decline in RA signalling, protected from large changes by homeostatic regulation of the system maintaining minimum circulatory levels under all but the most severe vitamin A depletion.

RA Signalling Components as Biomarkers

There is a need for biomarkers in psychiatric disorders that could be used clinically as an adjunct to diagnosis but also to help understand the molecular mechanisms underlying the particular disorder.[2,118,121,188,189] Biomarkers with overlap in multiple disorders have been identified including DISC1, BDNF and COMT.[188] Can components of the RA signalling pathway be added to this list? Analysis of association with schizophrenia of RA receptors or RBP4 is sparse or conflicting.[65,66] Similarly, with regard to depression or autism, no mechanistic association has yet been described with these signalling components. It is of note, too, that null mutation of the genes in mice for either the RA receptors or RBP4 have a relatively mild phenotype,[190,191] implying that each of these proteins is functionally backed up. In contrast knockout of the RA metabolic enzymes, either the synthetic enzymes ALDH1A2 or ALDH1A3, or the enzymes that degrade RA, CYP26A1 or CYP26B1, are lethal[192–196] and it is possible that a mutation in these genes that only reduces or increases activity of the enzymes may have a significant phenotype. Alternatively, given the importance of localized RA synthesis in discrete brain regions, changes in the distribution of these metabolic enzymes could have profound consequences for CNS function.

Thus, the enzymes in the metabolic pathways controlling RA synthesis and metabolism are good candidates as biomarkers, in particular ALDH1A1 in the midbrain dopaminergic neurons.[33,34,36] A focused effort is necessary to map the mutations in these enzymes to determine which modify their catalytic activity and which potentially map to disease. Further promising biomarkers are those that act on multiple nuclear receptor signalling pathways; TTR, which carries both thyroid hormone and retinol bound to RBP4 in the plasma is of particular note for reduction in expression as a marker for schizophrenia.[62–64] Similarly, low TTR has also been suggested as a marker of depression.[197] This is initially surprising given the link, at least in the mouse TTR null mutation model, of a decrease in depression related behaviour[136] but the decline in TTR in humans may be particularly associated with a subtype of depression; the strongest association being with depression with suicidal ideation.[197] It is of note though that individual biomarkers may not distinguish these disorders because of the putative similarities in RA signalling in schizophrenia (increase in the embryo with possible decrease in the adult), depression (increase in the adult) and autism (increase in the embryo).

In addition to the problem that these biomarkers may not distinguish disorders, changes in the steps that determine RA signalling are not restricted to one component but changes in many steps of the pathway (Fig. 38.1) could have the same result of raising or lowering the strength of the RA signal. The use of fast throughput techniques with bioinformatic analysis can define all elements of a pathway and provide a definition of the "normal versus abnormal" biological state. In addition to screening for gene mutations in the metabolic enzymes, changes in blood and CNS concentrations of RA, retinol and TTR could be quantified. The collective change of these values away from normal would become the biomarker, potentially even revealing subsets of schizophrenia, depression or autism, depending on which sets of these individual markers change and in which direction.

Acknowledgements We would like to thank Dr. Michelle Lane and Ann Goodman for their reading of the text and excellent suggestions. Thanks as well to Jemma Ransom for her reading of the section on schizophrenia.

References

1. Cannon TD, Keller MC. Endophenotypes in the genetic analyses of mental disorders. Annu Rev Clin Psychol 2006; 2: 267–90.
2. Levinson DF. The genetics of depression: a review. Biol Psychiatry 2006; 60: 84–92.
3. McClellan JM, Susser E, King MC. Schizophrenia: a common disease caused by multiple rare alleles. Br J Psychiatry 2007; 190: 194–9.
4. Hennah W, Thomson P, McQuillin A, et al. DISC1 association, heterogeneity and interplay in schizophrenia and bipolar disorder. Mol Psychiatry 2008. [epub ahead of print].
5. Blackwood DH, Pickard BJ, Thomson PA, et al. Are some genetic risk factors common to schizophrenia, bipolar disorder and depression? Evidence from DISC1, GRIK4 and NRG1. Neurotox Res 2007; 11: 73–83.
6. McClay JL, Fanous A, van den Oord EJ, et al. Catechol-O-methyltransferase and the clinical features of psychosis. Am J Med Genet B Neuropsychiatr Genet 2006; 141: 935–8.
7. Berton O, Nestler EJ. New approaches to antidepressant drug discovery: beyond monoamines. Nat Rev Neurosci 2006; 7: 137–51.
8. Berk M, Dodd S, Kauer-Sant'anna M, et al. Dopamine dysregulation syndrome: implications for a dopamine hypothesis of bipolar disorder. Acta Psychiatr Scand Suppl 2007: 41–9.

9. Lam KS, Aman MG, Arnold LE. Neurochemical correlates of autistic disorder: a review of the literature. Res Dev Disabil 2006; 27: 254–89.

10. Craddock N, Forty L. Genetics of affective (mood) disorders. Eur J Hum Genet 2006; 14: 660–8.

11. Hennah W, Thomson P, Peltonen L, et al. Genes and schizophrenia: beyond schizophrenia: the role of DISC1 in major mental illness. Schizophr Bull 2006; 32: 409–16.

12. Rzhetsky A, Wajngurt D, Park N, et al. Probing genetic overlap among complex human phenotypes. Proc Natl Acad Sci USA 2007; 104: 11694–9.

13. Kawaguchi R, Yu J, Honda J, et al. A membrane receptor for retinol binding protein mediates cellular uptake of vitamin A. Science 2007; 315: 820–5.

14. Balmer JE, Blomhoff R. Gene expression regulation by retinoic acid. J Lipid Res 2002; 43: 1773–808.

15. McCaffery PJ, Adams J, Maden M, et al. Too much of a good thing: retinoic acid as an endogenous regulator of neural differentiation and exogenous teratogen. Eur J Neurosci 2003; 18: 457–72.

16. Lane MA, Bailey SJ. Role of retinoid signalling in the adult brain. Prog Neurobiol 2005; 75: 275–93.

17. Wagner E, Luo T, Drager UC. Retinoic acid synthesis in the postnatal mouse brain marks distinct developmental stages and functional systems. Cereb Cortex 2002; 12: 1244–53.

18. Sapolsky RM. Glucocorticoids and hippocampal atrophy in neuropsychiatric disorders. Arch Gen Psychiatry 2000; 57: 925–35.

19. Holsboer F. The corticosteroid receptor hypothesis of depression. Neuropsychopharmacology 2000; 23: 477–501.

20. Bhugra D. The global prevalence of schizophrenia. PLoS Med 2005; 2: e151; quiz e75.

21. Sullivan PF, Kendler KS, Neale MC. Schizophrenia as a complex trait: evidence from a meta-analysis of twin studies. Arch Gen Psychiatry 2003; 60: 1187–92.

22. Jarskog LF, Miyamoto S, Lieberman JA. Schizophrenia: new pathological insights and therapies. Annu Rev Med 2007; 58: 49–61.

23. Lang UE, Puls I, Muller DJ, et al. Molecular mechanisms of schizophrenia. Cell Physiol Biochem 2007; 20: 687–702.

24. Susser ES, Lin SP. Schizophrenia after prenatal exposure to the Dutch Hunger Winter of 1944–1945. Arch Gen Psychiatry 1992; 49: 983–8.

25. St Clair D, Xu M, Wang P, et al. Rates of adult schizophrenia following prenatal exposure to the Chinese famine of 1959–1961. JAMA 2005; 294: 557–62.

26. Goodman AB. Congenital anomalies in relatives of schizophrenic probands may indicate a retinoid pathology. Schizophr Res 1996; 19: 163–70.

27. Goodman AB. Three independent lines of evidence suggest retinoids as causal to schizophrenia. Proc Natl Acad Sci USA 1998; 95: 7240–4.

28. LaMantia AS. Forebrain induction, retinoic acid, and vulnerability to schizophrenia: insights from molecular and genetic analysis in developing mice. Biol Psychiatry 1999; 46: 19–30.

29. Maynard TM, Sikich L, Lieberman JA, et al. Neural development, cell-cell signaling, and the "two-hit" hypothesis of schizophrenia. Schizophr Bull 2001; 27: 457–76.

30. Smith D, Wagner E, Koul O, et al. Retinoic acid synthesis for the developing telencephalon. Cereb Cortex 2001; 11: 894–905.

31. Reynolds GP, Abdul-Monim Z, Neill JC, et al. Calcium binding protein markers of GABA deficits in schizophrenia–postmortem studies and animal models. Neurotox Res 2004; 6: 57–61.

32. Mey J, McCaffery P. Retinoic acid signaling in the nervous system of adult vertebrates. Neuroscientist 2004; 10: 409–21.

33. Prabakaran S, Swatton JE, Ryan MM, et al. Mitochondrial dysfunction in schizophrenia: evidence for compromised brain metabolism and oxidative stress. Mol Psychiatry 2004; 9: 684–97, 43.

34. Goodman AB. Microarray results suggest altered transport and lowered synthesis of retinoic acid in schizophrenia. Mol Psychiatry 2005; 10: 620–1.

35. McCaffery P, Drager UC. High levels of a retinoic acid-generating dehydrogenase in the meso-telencephalic dopamine system. Proc Natl Acad Sci USA 1994; 91: 7772–6.

36. Galter D, Buervenich S, Carmine A, et al. ALDH1 mRNA: presence in human dopamine neurons and decreases in substantia nigra in Parkinson's disease and in the ventral tegmental area in schizophrenia. Neurobiol Dis 2003; 14: 637–47.

37. Buervenich S, Carmine A, Arvidsson M, et al. NURR1 mutations in cases of schizophrenia and manic-depressive disorder. Am J Med Genet 2000; 96: 808–13.

38. Chen YH, Tsai MT, Shaw CK, et al. Mutation analysis of the human NR4A2 gene, an essential gene for midbrain dopaminergic neurogenesis, in schizophrenic patients. Am J Med Genet 2001; 105: 753–7.

39. McCaffery P, Lee M-O, Wagner MA, et al. Asymmetrical retinoic acid synthesis in the dorso-ventral axis of the retina. Development 1992; 115: 371–82.

40. McCaffery P, Dräger UC. High levels of a retinoic-acid generating dehydrogenase in the meso-telencephalic dopamine system. Proc Natl Acad Sci USA 1994; 91: 7772–6.

41. Martensen-Larsen O. Psychotic phenomena provoked by tetraethylthiuram disulfide. Q J Stud Alcohol 1951; 12: 206–16.

42. Fiske D. "Psychotic reaction" to tetraethylthiuram disulfide (antabuse) therapy; report of a case. J Am Med Assoc 1952; 150: 1110–1.

43. Liddon SC, Satran R. Disulfiram (Antabuse) psychosis. Am J Psychiatry 1967; 123: 1284–9.

44. Scher JM. Psychotic reaction to disulfiram. JAMA 1967; 201: 1051.

45. Stuller S, Bell K, Read S, et al. Antabuse psychosis. Psychiatr J Univ Ott 1983; 8: 179–80.

46. Murthy KK. Psychosis during disulfiram therapy for alcoholism. J Indian Med Assoc 1997; 95: 80–1.

47. Ceylan ME, Turkcan A, Mutlu E, et al. Manic episode with psychotic symptoms associated with high dose of disulfiram: a case report. J Clin Psychopharmacol 2007; 27: 224–5.

48. Quail M, Karelse RH. Disulfiram psychosis. A case report. S Afr Med J 1980; 57: 551–2.

49. Tank AW, Deitrich RA, Weiner H. Effects of induction of rat liver cytosolic aldehyde dehydrogenase on the oxidation of biogenic aldehydes. Biochem Pharmacol 1986; 35: 4563–9.

50. Collard F, Vertommen D, Fortpied J, et al. Identification of 3-deoxyglucosone dehydrogenase as aldehyde dehydrogenase 1A1 (retinaldehyde dehydrogenase 1). Biochimie 2007; 89: 369–73.

51. McCaffery P, Wagner E, O'Neil J, et al. Dorsal and ventral retinoic territories defined by retinoic acid synthesis, breakdown and nuclear receptor expression. Mech Dev 1999; 85: 203–14.

52. Molotkov A, Duester G. Genetic evidence that retinaldehyde dehydrogenase Raldh1 (Aldh1a1) functions downstream of alcohol dehydrogenase Adh1 in metabolism of retinol to retinoic acid. J Biol Chem 2003; 278: 36085–90.

53. Jacobs FM, Smits SM, Noorlander CW, et al. Retinoic acid counteracts developmental defects in the substantia nigra caused by Pitx3 deficiency. Development 2007; 134: 2673–84.

54. Shprintzen RJ, Goldberg R, Golding-Kushner KJ, et al. Late-onset psychosis in the velo-cardio-facial syndrome. Am J Med Genet 1992; 42: 141–2.

55. Chow EW, Bassett AS, Weksberg R. Velo-cardio-facial syndrome and psychotic disorders: implications for psychiatric genetics. Am J Med Genet 1994; 54: 107–12.

56. Gothelf D, Eliez S, Thompson T, et al. COMT genotype predicts longitudinal cognitive decline and psychosis in 22q11.2 deletion syndrome. Nat Neurosci 2005; 8: 1500–2.

57. Coberly S, Lammer E, Alashari M. Retinoic acid embryopathy: case report and review of literature. Pediatr Pathol Lab Med 1996; 16: 823–36.

58. Vermot J, Niederreither K, Garnier JM, et al. Decreased embryonic retinoic acid synthesis results in a DiGeorge syndrome phenotype in newborn mice. Proc Natl Acad Sci USA 2003; 100: 1763–8.

59 Guris DL, Duester G, Papaioannou VE, et al. Dose-dependent interaction of Tbx1 and Crkl and locally aberrant RA signaling in a model of del22q11 syndrome. Dev Cell 2006; 10: 81–92.

60. Paylor R, Glaser B, Mupo A, et al. Tbx1 haploinsufficiency is linked to behavioral disorders in mice and humans: implications for 22q11 deletion syndrome. Proc Natl Acad Sci USA 2006; 103: 7729–34.

61. Ruano D, Macedo A, Soares MJ, et al. Transthyretin: no association between serum levels or gene variants and schizophrenia. J Psychiatr Res 2007; 41: 667–72.

62. Huang JT, Leweke FM, Oxley D, et al. Disease biomarkers in cerebrospinal fluid of patients with first-onset psychosis. PLoS Med 2006; 3: e428.

63. Huang JT, Leweke FM, Tsang TM, et al. CSF metabolic and proteomic profiles in patients prodromal for psychosis. PLoS ONE 2007; 2: e756.

64. Wan C, Yang Y, Li H, et al. Dysregulation of retinoid transporters expression in body fluids of schizophrenia patients. J Proteome Res 2006; 5: 3213–6.

65. Feng J, Chen J, Yan J, et al. Structural variants in the retinoid receptor genes in patients with schizophrenia and other psychiatric diseases. Am J Med Genet B Neuropsychiatr Genet 2005; 133: 50–3.

66. Ishiguro H, Okubo Y, Ohtsuki T, et al. Mutation analysis of the retinoid X receptor beta, nuclear-related receptor 1, and peroxisome proliferator-activated receptor alpha genes in schizophrenia and alcohol dependence: possible haplotype association of nuclear-related receptor 1 gene to alcohol dependence. Am J Med Genet 2002; 114: 15–23.

67. Rioux L, Arnold SE. The expression of retinoic acid receptor alpha is increased in the granule cells of the dentate gyrus in schizophrenia. Psychiatry Res 2005; 133: 13–21.

68. Langlois MC, Beaudry G, Zekki H, et al. Impact of antipsychotic drug administration on the expression of nuclear receptors in the neocortex and striatum of the rat brain. Neuroscience 2001; 106: 117–28.

69. Herz J, Chen Y. Reelin, lipoprotein receptors and synaptic plasticity. Nat Rev Neurosci 2006; 7: 850–9.

70. Impagnatiello F, Guidotti AR, Pesold C, et al. A decrease of reelin expression as a putative vulnerability factor in schizophrenia. Proc Natl Acad Sci USA 1998; 95: 15718–23.

71. Guidotti A, Auta J, Davis JM, et al. Decrease in reelin and glutamic acid decarboxylase67 (GAD67) expression in schizophrenia and bipolar disorder: a postmortem brain study. Arch Gen Psychiatry 2000; 57: 1061–9.

72. Wedenoja J, Loukola A, Tuulio-Henriksson A, et al. Replication of linkage on chromosome 7q22 and association of the regional Reelin gene with working memory in schizophrenia families. Mol Psychiatry 2007.

73. Fatemi SH, Earle JA, McMenomy T. Reduction in Reelin immunoreactivity in hippocampus of subjects with schizophrenia, bipolar disorder and major depression. Mol Psychiatry 2000; 5: 654–63, 571.

74. Tremolizzo L, Carboni G, Ruzicka WB, et al. An epigenetic mouse model for molecular and behavioral neuropathologies related to schizophrenia vulnerability. Proc Natl Acad Sci USA 2002; 99: 17095–100.

75. Grayson DR, Jia X, Chen Y, et al. Reelin promoter hypermethylation in schizophrenia. Proc Natl Acad Sci USA 2005; 102: 9341–6.

76. Chen Y, Sharma RP, Costa RH, et al. On the epigenetic regulation of the human reelin promoter. Nucleic Acids Res 2002; 30: 2930–9.

77. Reif A, Schmitt A, Fritzen S, et al. Neurogenesis and schizophrenia: dividing neurons in a divided mind? Eur Arch Psychiatry Clin Neurosci 2007; 257: 290–9.

78. Kempermann G, Krebs J, Fabel K. The contribution of failing adult hippocampal neurogenesis to psychiatric disorders. Curr Opin Psychiatry 2008; 21: 290–5.

79. Duan X, Chang JH, Ge S, et al. Disrupted-In-Schizophrenia 1 regulates integration of newly generated neurons in the adult brain. Cell 2007; 130: 1146–58.

80. Jacobs S, Lie DC, Decicco KL, et al. Retinoic acid is required early during adult neurogenesis in the dentate gyrus. Proc Natl Acad Sci USA 2006; 103: 3902–7.

81. Crandall J, Sakai Y, Zhang J, et al. 13-cis-retinoic acid suppresses hippocampal cell division and hippocampal-dependent learning in mice. Proc Natl Acad Sci USA 2004; 101: 5111–6.

82. Samad TA, Krezel W, Chambon P, et al. Regulation of dopaminergic pathways by retinoids: activation of the D2 receptor promoter by members of the retinoic acid receptor-retinoid X receptor family. Proc Natl Acad Sci USA 1997; 94: 14349–54.

83. Wolf G. Vitamin A functions in the regulation of the dopaminergic system in the brain and pituitary gland. Nutr Rev 1998; 56: 354–5.

84. Valdenaire O, Maus-Moatti M, Vincent JD, et al. Retinoic acid regulates the developmental expression of dopamine D2 receptor in rat striatal primary cultures. J Neurochem 1998; 71: 929–36.

85. Molotkova N, Molotkov A, Duester G. Role of retinoic acid during forebrain development begins late when Raldh3

generates retinoic acid in the ventral subventricular zone. Dev Biol 2007; 303: 601–10.

86. Deacon TW, Pakzaban P, Isacson O. The lateral ganglionic eminence is the origin of cells committed to striatal phenotypes: neural transplantation and developmental evidence. Brain Res 1994; 668: 211–9.

87. Meisenzahl EM, Schmitt GJ, Scheuerecker J, et al. The role of dopamine for the pathophysiology of schizophrenia. Int Rev Psychiatry 2007; 19: 337–45.

88. Pleasure SJ, Anderson S, Hevner R, et al. Cell migration from the ganglionic eminences is required for the development of hippocampal GABAergic interneurons. Neuron 2000; 28: 727–40.

89. Wonders C, Anderson SA. Cortical interneurons and their origins. Neuroscientist 2005; 11: 199–205.

90. Waclaw RR, Wang B, Campbell K. The homeobox gene Gsh2 is required for retinoid production in the embryonic mouse telencephalon. Development 2004; 131: 4013–20.

91. Wang HF, Liu FC. Regulation of multiple dopamine signal transduction molecules by retinoids in the developing striatum. Neuroscience 2005; 134: 97–105.

92. Imai Y, Suzuki Y, Matsui T, et al. Cloning of a retinoic acid-induced gene, GT1, in the embryonal carcinoma cell line P19: neuron-specific expression in the mouse brain. Brain Res Mol Brain Res 1995; 31: 1–9.

93. Joober R, Benkelfat C, Toulouse A, et al. Analysis of 14 CAG repeat-containing genes in schizophrenia. Am J Med Genet 1999; 88: 694–9.

94. Toulouse A, Rochefort D, Roussel J, et al. Molecular cloning and characterization of human RAI1, a gene associated with schizophrenia. Genomics 2003; 82: 162–71.

95. Seranski P, Hoff C, Radelof U, et al. RAI1 is a novel polyglutamine encoding gene that is deleted in Smith-Magenis syndrome patients. Gene 2001; 270: 69–76.

96. Slager RE, Newton TL, Vlangos CN, et al. Mutations in RAI1 associated with Smith-Magenis syndrome. Nat Genet 2003; 33: 466–8.

97. Girirajan S, Patel N, Slager RE, et al. How much is too much? Phenotypic consequences of Rai1 overexpression in mice. Eur J Hum Genet 2008; 16:941–54.

98. Middleton FA, Pato CN, Gentile KL, et al. Gene expression analysis of peripheral blood leukocytes from discordant sib-pairs with schizophrenia and bipolar disorder reveals points of convergence between genetic and functional genomic approaches. Am J Med Genet B Neuropsychiatr Genet 2005; 136: 12–25.

99. Wang YZ, Christakos S. Retinoic acid regulates the expression of the calcium binding protein, calbindin-D28K. Mol Endocrinol 1995; 9: 1510–21.

100. Iniguez MA, Morte B, Rodriguez-Pena A, et al. Characterization of the promoter region and flanking sequences of the neuron-specific gene RC3 (neurogranin). Brain Res Mol Brain Res 1994; 27: 205–14.

101. Enderlin V, Vallortigara J, Alfos S, et al. Retinoic acid reverses the PTU related decrease in neurogranin level in mice brain. J Physiol Biochem 2004; 60: 191–8.

102. Ruano D, Aulchenko YS, Macedo A, et al. Association of the gene encoding neurogranin with schizophrenia in males. J Psychiatr Res 2008; 42: 125–33.

103. Emamian ES, Hall D, Birnbaum MJ, et al. Convergent evidence for impaired AKT1-GSK3beta signaling in schizophrenia. Nat Genet 2004; 36: 131–7.

104. Kalkman HO. The role of the phosphatidylinositide 3-kinase-protein kinase B pathway in schizophrenia. Pharmacol Ther 2006; 110: 117–34.

105. Lopez-Carballo G, Moreno L, Masia S, et al. Activation of the phosphatidylinositol 3-kinase/Akt signaling pathway by retinoic acid is required for neural differentiation of SH-SY5Y human neuroblastoma cells. J Biol Chem 2002; 277: 25297–304.

106. Bastien J, Plassat JL, Payrastre B, et al. The phosphoinositide 3-kinase/Akt pathway is essential for the retinoic acid-induced differentiation of F9 cells. Oncogene 2006; 25: 2040–7.

107. Matkovic K, Brugnoli F, Bertagnolo V, et al. The role of the nuclear Akt activation and Akt inhibitors in all-trans-retinoic acid-differentiated HL-60 cells. Leukemia 2006; 20: 941–51.

108. Park EY, Wilder ET, Chipuk JE, et al. Retinol decreases phosphatidylinositol 3-kinase activity in colon cancer cells. Mol Carcinog 2008; 47: 264–74.

109. Peveler R, Carson A, Rodin G. ABC of psychological medicine: depression in medical patients. BMJ 2002; 325: 149–52.

110. Manji HK, Drevets WC, Charney DS. The cellular neurobiology of depression. Nat Med 2001; 7: 541–7.

111. Wong ML, Licinio J. Research and treatment approaches to depression. Nat Rev Neurosci 2001; 2: 343–51.

112. Neumeister A, Young T, Stastny J. Implications of genetic research on the role of the serotonin in depression: emphasis on the serotonin type 1A receptor and the serotonin transporter. Psychopharmacology (Berl) 2004; 174: 512–24.

113. Nemeroff CB. New directions in the development of antidepressants: the interface of neurobiology and psychiatry. Hum Psychopharmacol Clin Exp 2002; 17: S13–S6.

114. Duman RS, Monteggia LM. A neurotrophic model for stress-related mood disorders. Biol Psychiatry 2006; 59: 1116–27.

115. Martinowich K, Manji H, Lu B. New insights into BDNF function in depression and anxiety. Nat Neurosci 2007; 10: 1089–93.

116. Santarelli L, Saxe M, Gross C, et al. Requirement of hippocampal neurogenesis for the behavioral effects of antidepressants. Science 2003; 301: 805–9.

117. Vollmayr B, Mahlstedt M, Henn F. Neurogenesis and depression: what animal models tell us about the link. Eur Arch Psychiatry Clin Neurosci 2007; 257: 300–3.

118. Lesch KP. Gene-environment interaction and the genetics of depression. J Psychiatry Neurosci 2004; 29: 174–84.

119. Zill P, Baghai TC, Zwanzger P, et al. SNP and haplotype analysis of a novel tryptophan hydroxylase isoform (TPH2) gene provide evidence for association with major depression. Mol Psychiatry 2004; 9: 1030–6.

120. Sklar P, Gabriel SB, McInnis MG, et al. Family-based association study of 76 candidate genes in bipolar disorder: BDNF is a potential risk locus. Brain-derived neutrophic factor. Mol Psychiatry 2002; 7: 579–93.

121. Mossner R, Mikova O, Koutsilieri E, et al. Consensus paper of the WFSBP Task Force on Biological Markers: Biological Markers in Depression. World J Biol Psychiatry 2007; 8: 141–74.

122. O'Donnell J. Polar hysteria: an expression of hypervitaminosis A. Am J Ther 2004; 11: 507–16.

123. Muenter MD, Perry HO, Ludwig J. Chronic vitamin A intoxication in adults: Hepatic, neurologic and dermatologic complications. Am J Med 1971; 50: 129–36.

124. Hull PR, D'Arcy C. Acne, depression, and suicide. Dermatol Clin 2005; 23: 665–74.

125. Bremner JD, McCaffery P. The neurobiology of retinoic acid in affective disorders. Prog Neuropsychopharmacol Biol Psychiatry 2008; 32: 315–31.

126. Jacobs DG, Deutsch NL, Brewer M. Suicide, depression, and isotretinoin: is there a causal link? J Am Acad Dermatol 2001; 45: S168–75.

127. Wysowski DK, Pitts M, Beitz J. Depression and suicide in patients treated with isotretinoin. N Engl J Med 2001; 344: 460.

128. Meyskens FLJ. Short clinical reports. J Am Acad Dermatol 1982; 6: 732–3.

129. Bremner JD, Fani N, Ashraf A, et al. Functional brain imaging alterations in acne patients treated with isotretinoin. Am J Psychiatry 2005; 162: 983–91.

130. Starling J, 3rd, Koo J. Evidence based or theoretical concern? Pseudotumor cerebri and depression as acitretin side effects. J Drugs Dermatol 2005; 4: 690–6.

131. O'Reilly KC, Shumake J, Gonzalez-Lima F, et al. Chronic administration of 13-cis-retinoic acid increases depression-related behavior in mice. Neuropsychopharmacology 2006; 31: 1919–27.

132. Ferguson SA, Cisneros FJ, Gough B, et al. Chronic oral treatment with 13-cis-retinoic acid (isotretinoin) or all-trans-retinoic acid does not alter depression-like behaviors in rats. Toxicol Sci 2005; 87: 451–9.

133. Ferguson SA, Cisneros FJ, Hanig JP, et al. Oral treatment with ACCUTANE does not increase measures of anhedonia or depression in rats. Neurotoxicol Teratol 2007; 29: 642–51.

134. Ferguson SA, Berry KJ. Oral Accutane (13-cis-retinoic acid) has no effects on spatial learning and memory in male and female Sprague-Dawley rats. Neurotoxicol Teratol 2007; 29: 219–27.

135. Trent S, Mitchell PJ, Bailey SJ. Chronic treatment of rats with 13-cis retinoic acid changes aggressive behaviours in the resident-intruder paradigm From the Brighton Winter 2007 Meeting: Proceedings of the British Pharmacological Society at http://wwwpa2onlineorg/abstracts/Vol5Issue 2abst137Ppdf 2007

136. Sousa JC, Grandela C, Fernandez-Ruiz J, et al. Transthyretin is involved in depression-like behaviour and exploratory activity. J Neurochem 2004; 88: 1052–8.

137. O'Reilly K, Bailey SJ, Lane MA. Retinoid-mediated regulation of mood: possible cellular mechanisms. Exp Biol Med 2008; 233: 251–8.

138. O'Reilly KC, Trent S, Bailey SJ, et al. 13-Cis-retinoic acid alters intracellular serotonin, increases 5-HT1A receptor, and serotonin reuptake transporter levels in vitro. Exp Biol Med 2007; 232: 1195–203.

139. Ferguson SA, Cisneros FJ, Gough BJ, et al. Four weeks of oral isotretinoin treatment causes few signs of general toxicity in male and female Sprague-Dawley rats. Food Chem Toxicol 2005; 43: 1289–96.

140. Matsuoka I, Kumagai M, Kurihara K. Differential and coordinated regulation of expression of norepinephrine transporter in catecholaminergic cells in culture. Brain Res 1997; 776: 181–8.

141. Sodja C, Fang H, Dasgupta T, et al. Identification of functional dopamine receptors in human teratocarcinoma NT2 cells. Brain Res Mol Brain Res 2002; 99: 83–91.

142. Ishikawa J, Sutoh C, Ishikawa A, et al. 13-cis-retinoic acid alters the cellular morphology of slice-cultured serotonergic neurons in the rat. Eur J Neurosci 2008; 27: 2363–72.

143. Liu Y, Kagechika H, Ishikawa J, et al. Effects of retinoic acids on the dendritic morphology of cultured hippocampal neurons. J Neurochem 2008; 106:591–602.

144. Lucarelli E, Kaplan DR, Thiele CJ. Selective regulation of TrkA and TrkB receptors by retinoic acid and interferon-gamma in human neuroblastoma cell lines. J Biol Chem 1995; 270: 24725–31.

145. Salvatore AM, Cozzolino M, Gargano N, et al. Neuronal differentiation of P19 embryonal cells exhibits cell-specific regulation of neurotrophin receptors. Neuroreport 1995; 6: 873–7.

146. Scheibe RJ, Wagner JA. Retinoic acid regulates both expression of the nerve growth factor receptor and sensitivity to nerve growth factor. J Biol Chem 1992; 267: 17611–6.

147. Bottiglieri T. Homocysteine and folate metabolism in depression. Prog Neuropsychopharmacol Biol Psychiatry 2005; 29: 1103–12.

148. Coppen A, Bolander-Gouaille C. Treatment of depression: time to consider folic acid and vitamin B12. J Psychopharmacol 2005; 19: 59–65.

149. Jasim ZF, McKenna KE. Vitamin B12 and folate deficiency anaemia associated with isotretinoin treatment for acne. Clin Exp Dermatol 2006; 31: 599.

150. Schulpis KH, Karikas GA, Georgala S, et al. Elevated plasma homocysteine levels in patients on isotretinoin therapy for cystic acne. Int J Dermatol 2001; 40: 33–6.

151. Chanson A, Cardinault N, Rock E, et al. Decreased plasma folate concentration in young and elderly healthy subjects after a short-term supplementation with isotretinoin. J Eur Acad Dermatol Venereol 2008; 22: 94–100.

152. Ozias MK, Schalinske KL. All-trans-retinoic acid rapidly induces glycine N-methyltransferase in a dose-dependent manner and reduces circulating methionine and homocysteine levels in rats. J Nutr 2003; 133: 4090–4.

153. Rowling MJ, McMullen MH, Schalinske KL. Vitamin A and its derivatives induce hepatic glycine N-methyltransferase and hypomethylation of DNA in rats. J Nutr 2002; 132: 365–9.

154. DeLong R. Autism and familial major mood disorder: are they related? J Neuropsychiatry Clin Neurosci 2004; 16: 199–213.

155. Leyfer OT, Folstein SE, Bacalman S, et al. Comorbid psychiatric disorders in children with autism: interview development and rates of disorders. J Autism Dev Disord 2006; 36: 849–61.

156. Spence SJ. The genetics of autism. Semin Pediatr Neurol 2004; 11: 196–204.

157. Pardo CA, Eberhart CG. The neurobiology of autism. Brain Pathol 2007; 17: 434–47.

158. London E, Etzel RA. The environment as an etiologic factor in autism: a new direction for research. Environ Health Perspect 2000; 108: 401–4.

159. Hashimoto T, Tayama M, Miyazaki M, et al. Brainstem and cerebellar vermis involvement in autistic children. J Child Neurol 1993; 8: 149–53.

160. Makori N, Peterson PE, Hendrickx AG. 13-cis-retinoic acid causes patterning defects in the early embryonic rostral hindbrain and abnormal development of the cerebellum in the macaque. Teratology 2001; 63: 65–76.

161. Hashimoto T, Tayama M, Murakawa K, et al. Development of the brainstem and cerebellum in autistic patients. J Autism Dev Disord 1995; 25: 1–18.

162. Rodier PM, Ingram JL, Tisdale B, et al. Linking etiologies in humans and animal models: studies of autism. Reprod Toxicol 1997; 11: 417–22.

163. Holson RR, Cogan JE, Adams J. Gestational retinoic acid exposure in the rat: effects of sex, strain and exposure period. Neurotoxicol Teratol 2001; 23: 147–56.

164. Yamamoto M, Fujinuma M, Hirano S, et al. Retinoic acid influences the development of the inferior olivary nucleus in the rodent. Dev Biol 2005; 280: 421–33.

165. McCaffery P, Deutsch CK. Macrocephaly and the control of brain growth in autistic disorders. Prog Neurobiol 2005; 77: 38–56.

166. Christianson AL, Chesler N, Kromberg JG. Fetal valproate syndrome: clinical and neuro-developmental features in two sibling pairs. Dev Med Child Neurol 1994; 36: 361–9.

167. Williams PG, Hersh JH. A male with fetal valproate syndrome and autism. Dev Med Child Neurol 1997; 39: 632–4.

168. Moore SJ, Turnpenny P, Quinn A, et al. A clinical study of 57 children with fetal anticonvulsant syndromes. J Med Genet 2000; 37: 489–97.

169. Williams G, King J, Cunningham M, et al. Fetal valproate syndrome and autism: additional evidence of an association. Dev Med Child Neurol 2001; 43: 202–6.

170. Rasalam AD, Hailey H, Williams JH, et al. Characteristics of fetal anticonvulsant syndrome associated autistic disorder. Dev Med Child Neurol 2005; 47: 551–5.

171. Rodier PM, Ingram JL, Tisdale B, et al. Embryological origin for autism: developmental anomalies of the cranial nerve motor nuclei. J Comp Neurol 1996; 370: 247–61.

172. Ingram JL, Peckham SM, Tisdale B, et al. Prenatal exposure of rats to valproic acid reproduces the cerebellar anomalies associated with autism. Neurotoxicol Teratol 2000; 22: 319–24.

173. Narita N, Kato M, Tazoe M, et al. Increased monoamine concentration in the brain and blood of fetal thalidomide- and valproic acid-exposed rat: putative animal models for autism. Pediatr Res 2002; 52: 576–9.

174. Schneider T, Przewlocki R. Behavioral alterations in rats prenatally exposed to valproic acid: animal model of autism. Neuropsychopharmacology 2005; 30: 80–9.

175. Kultima K, Nystrom AM, Scholz B, et al. Valproic acid teratogenicity: a toxicogenomics approach. Environ Health Perspect 2004; 112: 1225–35.

176. Epping MT, Wang L, Plumb JA, et al. A functional genetic screen identifies retinoic acid signaling as a target of histone deacetylase inhibitors. Proc Natl Acad Sci USA 2007; 104: 17777–82.

177. Ross SA, McCaffery PJ, Drager UC, et al. Retinoids in embryonal development. Physiol Rev 2000; 80: 1021–54.

178. Heyman RA, Mangelsdorf DJ, Dyck JA, et al. 9-cis retinoic acid is a high affinity ligand for the retinoid X receptor. Cell 1992; 68: 397–406.

179. Wolf G. Is 9-cis-retinoic acid the endogenous ligand for the retinoic acid-X receptor? Nutr Rev 2006; 64: 532–8.

180. Mangelsdorf DJ, Evans RM. The RXR heterodimers and orphan receptors. Cell 1995; 83: 841–50.

181. Ahuja HS, Szanto A, Nagy L, et al. The retinoid X receptor and its ligands: versatile regulators of metabolic function, cell differentiation and cell death. J Biol Regul Homeost Agents 2003; 17: 29–45.

182. McCaffery P, Drager UC. Hot spots of retinoic acid synthesis in the developing spinal cord. Proc Natl Acad Sci USA 1994; 91: 7194–7.

183. Gampe RT, Jr., Montana VG, Lambert MH, et al. Structural basis for autorepression of retinoid X receptor by tetramer formation and the AF-2 helix. Genes Dev 2000; 14: 2229–41.

184. Egea PF, Rochel N, Birck C, et al. Effects of ligand binding on the association properties and conformation in solution of retinoic acid receptors RXR and RAR. J Mol Biol 2001; 307: 557–76.

185. oehm MF, Zhang L, Zhi L, et al. Design and synthesis of potent retinoid X receptor selective ligands that induce apoptosis in leukemia cells. J Med Chem 1995; 38: 3146–55.

186. Lerner V, Miodownik C, Gibel A, et al. Bexarotene as add-on to antipsychotic treatment in schizophrenia patients: a pilot open-label trial. Clin Neuropharmacol 2008; 31: 25–33.

187. Sharma RP. Schizophrenia, epigenetics and ligand-activated nuclear receptors: a framework for chromatin therapeutics. Schizophr Res 2005; 72: 79–90.

188. Craddock N, O'Donovan MC, Owen MJ. Genes for schizophrenia and bipolar disorder? Implications for psychiatric nosology. Schizophr Bull 2006; 32: 9–16.

189. Javitt DC, Spencer KM, Thaker GK, et al. Neurophysiological biomarkers for drug development in schizophrenia. Nat Rev Drug Discov 2008; 7: 68–83.

190. Mark M, Ghyselinck NB, Wendling O, et al. A genetic dissection of the retinoid signalling pathway in the mouse. Proc Nutr Soc 1999; 58: 609–13.

191. Quadro L, Blaner WS, Salchow DJ, et al. Impaired retinal function and vitamin A availability in mice lacking retinol-binding protein. EMBO J 1999; 18: 4633–44.

192. Niederreither K, Subbarayan V, Dolle P, et al. Embryonic retinoic acid synthesis is essential for early mouse post-implantation development. Nat Genet 1999; 21: 444–8.

193. Dupe V, Matt N, Garnier JM, et al. A newborn lethal defect due to inactivation of retinaldehyde dehydrogenase type 3 is prevented by maternal retinoic acid treatment. Proc Natl Acad Sci USA 2003; 100: 14036–41.

194. Abu-Abed S, Dolle P, Metzger D, et al. The retinoic acid-metabolizing enzyme, CYP26A1, is essential for normal hindbrain patterning, vertebral identity, and development of posterior structures. Genes Dev 2001; 15: 226–40.

195. Sakai Y, Meno C, Fujii H, et al. The retinoic acid-inactivating enzyme CYP26 is essential for establishing an uneven distribution of retinoic acid along the anterio-posterior axis within the mouse embryo. Genes Dev 2001; 15: 213–25.

196. Yashiro K, Zhao X, Uehara M, et al. Regulation of retinoic acid distribution is required for proximodistal patterning and outgrowth of the developing mouse limb. Dev Cell 2004; 6: 411–22.

197. Sullivan GM, Mann JJ, Oquendo MA, et al. Low cerebrospinal fluid transthyretin levels in depression: correlations with suicidal ideation and low serotonin function. Biol Psychiatry 2006; 60: 500–6.

Chapter 39
Abnormalities of Inositol Metabolism in Lymphocytes as Biomarkers for Bipolar Disorder

Galila Agam, Yuly Bersudsky, and Robert H. Belmaker

Abstract The phosphatidylinositol system is a key brain signal transduction system. Lithium has proven effects on this system particularly as an inhibitor of the enzyme inositol monophosphatase which may thereby reduce intracellular inositol and the responsiveness of this system. Therefore, we hypothesized that abnormalities in this system may represent genetic biomarkers for bipolar disorder. We studied inositol monophosphatase in lymphocyte derived cell lines of bipolar patients and found a marked reduction in activity especially in lithium responders. We then studied mRNA levels for IMPase in fresh lymphocytes of bipolar patients and found that these mRNA levels are significantly lower in bipolar patients. Inositol content in lymphocyte derived cell lines from bipolar patients was also significantly lower. These results suggest abnormalities of the system and different subgroups of bipolar patients may have different abnormalities. It is possible also that different combinations of abnormalities in these measures of the phosphatidylinositol system could define as biomarkers different subgroups of bipolar disorder patients.

Keywords Bipolar • inositol • inositol monophosphatase • inositol uptake

Abbreviations Li: Lithium; PI: Phosphoinositide; IMPase: Inositol monophosphatase; BPD: Bipolar disorder

G. Agam, Y. Bersudsky, and R. H. Belmaker
Faculty of Health Sciences, Ben Gurion University of the Negev and Beersheva Mental Health Center, Beersheva, Israel

Introduction

Intensive research has focused on the mechanism of Li's pharmacological actions and particularly on the phosphoinositide (PI) signal transduction cycle as the molecular substrate underlying this ion's clinical usefulness. Following the demonstration that therapeutic concentrations of Li uncompetitively inhibit inositol monophosphatase (IMPase) activity, it was proposed[1,2] that this ubiquitous enzyme which regenerates intracellular free inositol is a principal therapeutic target of Li.

IMPase in Bipolar Disorder Lymphocytes

The above considerations spurred us to examine the *in vitro* activity of IMPase in 106 transformed lymphoblastoid cell lines derived from control subjects and bipolar disorder (BPD) patients characterized by therapeutic response to Li.[3] Lymphocyte-derived cell lines were established and grown as previously described.[4] IMPase activity was assayed according to Agam and Livne.[5]

Cell lines from bipolar patients had significantly lower IMPase activity than cell lines from control subjects (p < 0.01, Fig. 39.1). When the bipolar patients were grouped according to clinical response to Li therapy the Li responders exhibited significantly lower IMPase activity compared to those patients with poor Li response. This difference in IMPase activity within the bipolar patient group suggests genetic heterogeneity in this complex disorder.

Considering the well-established *in vitro* inhibition of IMPase by therapeutic Li concentrations,[2] the results described[3] appear counterintuitive. If Li is an inhibitor

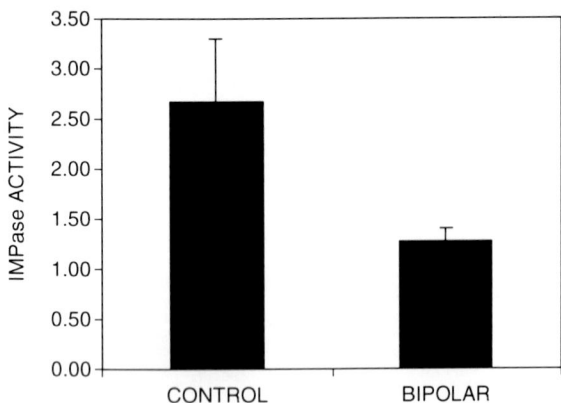

Fig. 39.1 Inositol monophosphatase activity in lymphoblastoid cell lines of BP patients (N = 77) and controls (N = 29)

of IMPase, and Li is a mood-stabilizing drug, one would expect elevated IMPase levels in drug-free BPD patients. The results suggest that the regulation of IMPase activity is more complicated and might be subject to transcriptional modulation. These considerations prompted us[6] to examine the effect of chronic Li treatment at therapeutic concentrations on IMPase mRNA levels for the predominant brain form of this enzyme.[7–9] Five days of 1 mM Li treatment representing a minimum of two generations of growth *in vitro* significantly raised by 40% IMPase mRNA levels.

IMPase mRNA Levels in Bipolar Disorder

We found[3] significantly lower IMPase activity in lymphoblastoid cell lines from BPD patients, but growth of these cell lines in culture requires extensive laboratory work and cannot be done as a real time rapid diagnostic test. We therefore planned to repeat our study in fresh lymphocytes. However, IMPase in fresh lymphocytes is inhibited *in vivo* by ongoing Li treatment and its pre-lithium activity cannot be evaluated. Therefore, we studied IMPase relative (to β-actin) mRNA levels of the gene (IMPA chr 8) coding for the predominant brain form of this enzyme.[8,9] The mRNA levels were measured in fresh lymphocytes of heterogeneous groups of bipolar patients and controls, including a limited number of drug-free bipolar patients.

Thirty-six non-hospitalized BPD patients were diagnosed according to DSM IV criteria, and recruited from the lithium clinic of the Beer Sheva Mental Health

Center. Eleven patients were treated with Li only whereas 20 were treated with Li and other drugs (such as carbamazepine, valproate, haloperidol, chlorpromazine, thioridazine, meclobemide, clonazepam and fluoxetine) or other drugs only. Five patients were drug-free. All patients were euthymic. Thirty-six control subjects (18 M, 18 F; average age± SD = 55.7 ± 15.7 years; range 19–86) were recruited from the Beer Sheva area and had no history of psychiatric illness. The control subjects included volunteers (n = 11) recruited from staff and students at the Soroka hospital and hospitalized non-psychiatric patients in the department of internal medicine (n = 25). Heparinized blood (10–20 ml) was withdrawn by venipuncture. Informed consent was obtained from all patients.

Isolation of lymphocytes, isolation of RNA from lymphocytes, and quantitative reverse transcriptase polymerase chain reaction (RT-PCR) have been previously described.[4]

IMPase relative mRNA levels were significantly lower in the drug-free (n = 5) BPD patients (p = 0.04) compared to the control subjects (n = 36, see Fig. 39.2). No significant effect of age or sex was observed on lymphocyte IMPase relative mRNA levels in either the patient or control groups. There was a nearly two-fold increase in IMPase relative mRNA levels in the drug-treated (Li only, Li and other drugs and other drugs only) patient cohort (n = 31) compared to the

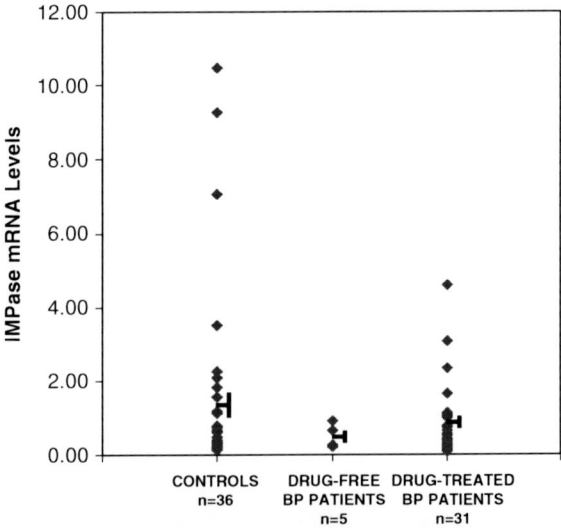

Fig. 39.2 Scatter plot of lymphocyte IMPase relative mRNA levels of control subjects (N = 36), drug-free (N = 5) and drug-treated BP patients (N = 31). Bars represent means ± SEM

drug-free patient group (t = 1.74, p = 0.049). For patients receiving solely Li there was a correlation between Li plasma concentrations and IMPase relative mRNA level (r = 0.54, p = 0.04, n = 11).

Inositol Levels in Lymphocytes of BPD

Inositol metabolism has been extensively studied in bipolar illness. Reduced post-mortem frontal cortex inositol levels were reported by Shimon et al.[10] and Davanzo et al.[11] although not replicated by Shapiro et al.[12] Soares et al.[13] reported reduced levels of phosphatidylinositol bisphosphate in platelet membrane from bipolar patients. Banks et al.[14] reported reduced inositol incorporation into membrane phosphoinositides of lymphoblastoid cell lines derived from bipolar patients. Shamir et al.[3] and Shaltiel et al.[15] reported decreased inositol monophosphatase activity in lymphocyte derived cell lines of bipolar patients. We then studied[16] the content of inositol in lymphocyte derived cell lines of bipolar patients and the capacity of the specific uptake site, the inositol transporter.

Thirty-seven non-hospitalized Type 1 bipolar patients (19 F, 18 M, age 50 ± 14 years, duration of illness 16 ± 10 years) from the Bipolar Clinic of the Beer-Sheva Mental Health Center were diagnosed according to DSM IV criteria. Forty control subjects with no history of psychiatric illness were recruited from hospital staff (21 F, 19 M, age 38 ± 8 years). Informed consent was obtained from all participating subjects. Twenty ml of venous blood was taken from each subject.

Lymphocyte-derived cell lines were established and grown as previously described.[3] Free inositol levels were analyzed as trimethylsilyl (TMS) derivatives by gas-liquid chromatography (GC), as previously described.[10, 12] Two cell lines (control and bipolar) were assayed on each experimental day. The examiner was blind to the identity of the cell lines. At the day of assay cells were suspended to 3.5–4.0 × 10^6 cells/ml in low inositol Basal Medium Eagle (BME, Gibco) supplemented with 10% fetal calf serum. Two ml cell suspension aliquots of each cell line were distributed and appropriate aliquots of stock inositol solution were added to reach concentrations in the range of 10–3,000 μM. Tubes were preincubated for 30 min at 37° in a humidified incubator (95% O_2 + 5% CO_2).

High affinity (Na$^+$ *myo*-inositol transporter, SMIT) and low affinity Km and Vmax of inositol uptake were

Table 12.1 Inositol content and uptake in lymphocyte-derived cell lines of bipolar patients

	Control	Bipolar	p
Inositol content	99 ± 41	82 ± 29	<0.05
N	40	36	
Inositol uptake			
SMIT Km	215 ± 128	182 ± 135	N.S.
SMIT Vmax	450 ± 212	487 ± 183	N.S.
Low-affinity Km	2,152 ± 1477	1,891 ± 1567	N.S.
Low-affinity Vmax	2,244 ± 1020	2,233 ± 724	N.S.
n	26	37	–

Results are means ± SD.

Inositol content in nmol/10^8 cells; Km in μM; Vmax in nmol/min × mg protein.

measured as described.[16] Samples in duplicates were assayed for total cell protein. The values of the kinetic parameters for each cell line were deduced by plotting the best-fit Eadie-Hofstee plot for each experiment.

Replicate reliability was studied by six repeated assays of the same cell line during several passages. Coefficients of variance of the four parameters were 18% for the high affinity Km and Vmax, 38% for the low affinity Km and 30% for the low affinity Vmax.

Table 39.1 presents inositol content and Km and Vmax of inositol uptake for cell lines derived from bipolar patients and control subjects. Inositol content in lymphocyte-derived cell lines from bipolar patients is 17% lower than in control subjects (p < 0.05). There was no significant difference between the two groups in the kinetic parameters of inositol uptake though there was a trend for lower Km values of both the high- and low-affinity processes.

Conclusions and Future Directions

A previous preliminary study in a small group of post-mortem BPD patient brains found no reduction in IMPase activity, but drug treatment and agonal state were not controlled.[10,17] However, in lymphoblastoid cell lines devoid of differing environmental and humoral variables, our study[3] showed significantly lower IMPase activity in BP patients, particularly in Li-responders. One could speculate that BPD Li responders start with genetically determined low IMPase activity and chronic inhibition of this enzyme by Li therapy leads to transcriptional up-regulation of mRNA levels which may lead to normalized enzyme levels. A possible molecular mechanism for this Li effect is based on the presence of two adjacent consensus sequences in the 3′ UTR of the human IMPase (IMPA) gene.[8] These two sequences are homologous to a nine base pair motif in the promoter region of some yeast genes that represents a binding site for a regulatory factor. The factor is a complex of two proteins that together control the transcription of several phospholipid biosynthetic enzymes.[18] This complex formation is favored in the absence of inositol. Li treatment, by inhibiting IMPase, results in low inositol levels, especially in Li-responding BP patients. The drop in inositol could, as in yeast, release the suppression of transcription and thus normalize enzyme translation and activity.

There was a corresponding reduction in lymphocyte IMPase relative mRNA levels in drug-free bipolar patients.[6] These results suggest that low IMPase transcription and translation are possibly trait markers of BPD.

Biochemical abnormalities in a complex disorder such as bipolar illness could occur at several steps in a metabolic system. Different patients could have different abnormalities and more than one abnormality may be required to cause the system as a whole to malfunction. The phosphatidylinositol signaling system, so important for communication among neurons in the brain, could malfunction if inositol uptake is insufficient, if inositol formation is inadequate, or if inositol conversion into second messengers is defective. It is possible that more than one deviation from normal levels is required to overcome redundant back-up mechanisms. Thus, we have undertaken a multi-dimensional survey of parameters of inositol metabolism in bipolar patients, using both post-mortem tissue and peripheral tissue from living patients.

The significant 17% reduction in inositol content in lymphocyte derived cell lines from bipolar patients is less than the 40% reduction in post-mortem cortical inositol previously reported by our group.[10] Li reduces cellular inositol, but these cells have been grown for at least five generations without lithium in vitro.

Our finding of reduced IMPase activity in lymphocyte derived cell lines from bipolar patients[3,15] could conceivably also be a cause of low inositol content, though blockade of IMPase by 50% might only be expected to affect the PI cycle under conditions of over stimulation.

Cellular endophenotypes are positioned much closer to genes and are considerably easier to map to specific chromosomal loci.[19] Our results showing that low lymphocyte IMPase mRNA levels and activity and low brain and lymphocyte inositol levels partially characterizes BPD patients suggest that this measure may be one such useful cellular endophenotype.

Further progress in understanding PI metabolism in bipolar disorder will depend on multiple measurements of this system in the same sample of patients, allowing for study of non-linear interactions between different components of the system and of heterogeneity among bipolar patients for possible abnormalities at different points in the same system.

In order for a diagnostic biomarker to be useful, certain criteria need to be met. These criteria include the following (see Chapter 1 by Ritsner and Gottesman, this volume):

1. The biomarker should reflect some basic pathophysiological process, and detect a fundamental feature of the disease with high sensitivity and specificity. With regard to inositol content, inositol monophosphatase activity and inositol monophosphatase mRNA in lymphocyte, these measures clearly reflect basic physiological processes important in brain but we are not yet completely certain that they are involved in bipolar disorder.

2. The biomarker should be specific for the disease compared with related disorders. Inositol monophosphatase was reported not to be reduced in schizophrenia,[15] but the specificity vis a vie other disorders has not been investigated.

3. The biomarker should not reflect symptomatology of disorder. The markers discussed in this chapter have not yet been investigated during the depressive vs. the manic phase of bipolar disorder and have only been studied in lymphocyte related cell lines in the euthymic state.

4. The biomarker can be measured repeatedly over time and should be reproducible. While some in vitro replicability studies have been done and the assay is replicable in vitro, true repeat measurements and replicability over the life course have not been done.

5. The biomarker should be measured in noninvasive and easy-to-perform tests that can be done at the bedside or in the outpatient setting. The measures discussed in this chapter involved blood testing and are therefore relatively non invasive and not harmful; however, they involved complicated laboratory assays that are not generally available in the average laboratory today.

They are expensive as well compared to other laboratory tests.

It seems unlikely given the small effect size and great overlap in all of the above measures that any one of them has potential to be developed as a single test for bipolar disorder. However, a multivariate approach to the phosphatidylinositol system where ratios of these three markers might be diagnostic could be a potentially exciting future area of research.

References

1. Berridge MJ, Downes CP, Hanley MR. Neural and developmental actions of lithium: a unifying hypothesis. Cell 1989;59(3):411–419.
2. Hallcher LM, Sherman WR. The effects of lithium ion and other agents on the activity of myo-inositol-1-phosphatase from bovine brain. J Biol Chem 1980;255(22):10896–901.
3. Shamir A, Ebstein RP, Nemanov L, Zohar A, Belmaker RH, Agam G. Inositol monophosphatase in immortalized lymphoblastoid cell lines indicates susceptibility to bipolar disorder and response to lithium therapy. Mol Psychiatry 1998; 3(6):481–2.
4. Ebstein RP, Bennett ER, Hadjez J, Silver H, Yedgar S, Lerer B. Cyclic AMP second messenger signal generation in EBV-transformed lymphoblastoid cells from schizophrenic patients. J Psychiatr Res 1990;24(2):121–127.
5. Agam G, Livne A. Inositol-1-phosphatase of human erythrocytes is inhibited by therapeutic lithium concentrations. Psychiatry Res 1989;27(2):217–224.
6. Nemanov L, Ebstein RP, Belmaker RH, Osher Y, Agam G. Effect of bipolar disorder on lymphocyte inositol monophosphatase mRNA levels. Int J Neuropsychopharmacol 1999;2(1):25–29.
7. McAllister G, Whiting P, Hammond EA, et al. cDNA cloning of human and rat brain myo-inositol monophosphatase Expression and characterization of the human recombinant enzyme. Biochem J 1992;284 (Pt 3):749–754.
8. Sjoholt G, Molven A, Lovlie R, Wilcox A, Sikela JM, Steen VM. Genomic structure and chromosomal localization of a human myo-inositol monophosphatase gene (IMPA). Genomics 1997;45(1):113–122.
9. Yoshikawa T, Turner G, Esterling LE, Sanders AR, Detera-Wadleigh SD. A novel human myo-inositol monophosphatase gene, IMP.18p, maps to a susceptibility region for bipolar disorder. Mol Psychiatry 1997;2(5):393–397.
10. Shimon H, Agam G, Belmaker RH, Hyde TM, Kleinman JE. Reduced frontal cortex inositol levels in postmortem brain of suicide victims and patients with bipolar disorder. Am J Psychiatry 1997;154(8):1148–1150.
11. Davanzo P, Thomas MA, Yue K, et al. Decreased anterior cingulate myo-inositol/creatine spectroscopy resonance with lithium treatment in children with bipolar disorder. Neuropsychopharmacology 2001;24(4):359–369.
12. Shapiro J, Belmaker RH, Biegon A, Seker A, Agam G. Scyllo-inositol in post-mortem brain of bipolar, unipolar and schizophrenic patients. J Neural Transm 2000;107(5):603–607.
13. Soares JC, Dippold CS, Wells KF, Frank E, Kupfer DJ, Mallinger AG. Increased platelet membrane phosphatidylinositol-4,5-bisphosphate in drug-free depressed bipolar patients. Neurosci Lett 2001;299(1–2):150–152.
14. Banks RE, Aiton JF, Cramb G, Naylor GJ. Incorporation of inositol into the phosphoinositides of lymphoblastoid cell lines established from bipolar manic-depressive patients. J Affect Disord 1990;19(1):1–8.
15. Shaltiel G, Shamir A, Nemanov L, et al. Inositol monophosphatase activity in brain and lymphocyte-derived cell lines of bipolar patients. World J Biol Psychiatry 2001;2(2):95–98.
16. Belmaker RH, Shapiro J, Vainer E, Nemanov L, Ebstein RP, Agam G. Reduced inositol content in lymphocyte-derived cell lines from bipolar patients. Bipolar Disord 2002;4(1):67–69.
17. Agam G, Shimon H. Evidence of the role of inositol in bipolar disorder and antibipolar treatment. In: Manji HK, Bowden CL, Belmaker RH (eds) Mechanisms of Action of Antibipolar Treatment. Washington DC: American Psychiatric Press; 2000:31–45.
18. Schuller HJ, Hahn A, Troster F, Schutz A, Schweizer E. Coordinate genetic control of yeast fatty acid synthase genes FAS1 and FAS2 by an upstream activation site common to genes involved in membrane lipid biosynthesis. Embo J 1992;11(1):107–114.
19. Brzustowicz LM, Gardner JP, Hopp L, et al. Linkage analysis using platelet-activating factor Ca2 + response in transformed lymphoblasts. Hypertension 1997;29(1 Pt 2):158–164.

Contents to Volumes 1, 2, and 4

Volume 2

Volume 4

Contributors to Volumes 1, 2, and 4

Volume 1

Trygve E. Bakken M.Sc., Scripps Genomic Medicine and Scripps Translational Science Institute; Medical Scientist Training Program and Graduate Program in Neurosciences, University of California, San Diego, CA, USA
E-mail: tbakken@ucsd.edu

Patricia E.G. Bestelmeyer, Ph.D., Post-Doc Centre for Cognitive Neuroimaging Department of Psychology, Glasgow, UK
E-mail: p.bestelmeyer@psy.gla.ac.uk

Cinnamon S. Bloss, Ph.D., Research Scientist, Scripps Genomic Medicine and Scripps Translational Science Institute, Scripps Health and The Scripps Research Institute, La Jolla, CA, USA
E-mail: cbloss@scripps.edu

David L. Braff, M.D., Professor, Department of Psychiatry, University of California, San Diego, CA, USA
E-mail: DBraff@ucsd.edu

Jessica A. Burket, B.S., Mental Health Service Line, Department of Veterans Affairs Medical Center, Washington, USA

Yue Chen, Ph.D., Director, Visual Psychophysiology Laboratory, McLean Hospital, Department of Psychiatry, Harvard Medical School, Belmont, MA, USA
E-mail: ychen@mclean.harvard.edu

Stephen I. Deutsch, M.D., Ph.D., Mental Health Service Line, Department of Veterans Affairs Medical Center, Washington, Department of Psychiatry, Georgetown University School of Medicine, Washington, USA
E-mail: Stephen.Deutsch@va.gov

Ayman H. Fanous, M.D., Mental Health Service Line, Department of Veterans Affairs Medical Center, Washington, Department of Psychiatry, Georgetown University School of Medicine, Washington, USA

Brooke L. Gaskins, B.A., Mental Health Service Line, Department of Veterans Affairs Medical Center, Washington, USA

Irving I. Gottesman Professor, Departments of Psychiatry and Psychology,
University of Minnesota, Minneapolis, MN, USA
E-mail: gotte003@umn.edu

Michael F. Green, Ph.D., Professor, Semel Institute, University of California,
Los Angeles, CA, USA
E-mail: mgreen@ucla.edu

Tiffany A. Greenwood, Ph.D., Assistant Adjunct Professor of Psychiatry,
Department of Psychiatry, University of California, San Diego, CA, USA
E-mail: tgreenwood@ucsd.edu

Raquel E. Gur, M.D., Ph.D., The Karl and Linda Rickels Professor and Vice Chair
for Research Development, Departments of Psychiatry, Neurology and Radiology,
Director, Neuropsychiatry Section, University of Pennsylvania Medical Center,
Philadelphia, PA, USA
E-mail: raquel@upenn.edu

William P. Horan, Ph.D., VA Greater Los Angeles Healthcare system &
University of California, Los Angeles, CA, USA
E-mail: horan@ucla.edu

Assen Jablensky, M.D., D. Med.Sci., Professor of Psychiatry,
School of Psychiatry and Clinical Neurosciences, The University of Western
Australia, Director, Centre for Clinical Research in Neuropsychiatry, Australia
E-mail: assen@cyllene.uwa.edu.au

Holger Jahn University of Hamburg, Department of Psychiatry,
Hamburg, Germany

Alexander H. Joyner M.Eng., Scripps Genomic Medicine and Scripps
Translational Science Institute; Graduate Program in Biomedical Sciences,
University of California, San Diego, CA, USA
E-mail: ajoyner@ucsd.edu

Fergus Kane, Ph.D. student at the section of General Psychiatry,
Department of Psychiatry, Institute of Psychiatry, London, UK

Liezl Koen, M.B. Ch.B., M.Med. (Psych), Department of Psychiatry,
University of Stellenbosch, South Africa
E-mail: liezlk@sun.ac.za

Eugenia Kravariti, M.A., M.Sc., Ph.D., Lecturer, NIHR Biomedical Research
Centre for Mental Health, South London and Maudsley NHS Foundation Trust and
Institute of Psychiatry, King's College London, UK
E-mail: e.kravariti@iop.kcl.ac.uk

Jukka M. Leppänen, Ph.D., Assistant Professor, Department of Psychology,
University of Tempere, Tempere, Finland
E-mail: jukka.leppanen@ufa.fi

Ryan J. Van Lieshout The Offord Centre for Child Studies, McMaster Children's
Hospital and Department of Psychiatry and Behavioural Neurosciences, McMaster
University, Hamilton, Ontario, Canada
E-mail: rjv@prontomail.com

Carolina Lopez Eating Disorders Research Unit, Department of Academic Psychiatry, King's College London, UK
E-mail: c.lopez@iop.kcl.ac.uk

Angus W. MacDonald, III, Ph.D., Associate Professor, Departments of Psychology and Psychiatry, University of Minnesota, Minneapolis, Minnesota, USA
E-mail: angus@umn.edu

John Mastropaolo, Ph.D. Mental Health Service Line, Department of Veterans Affairs Medical Center, Washington, USA

Ryan McBain McLean Hospital, Belmont, MA, USA

Harald Mischak Mosaiques Diagnostics and Therapeutics AG, Hannover, Germany

Vijay A. Mittal, Ph.D., Postdoctoral Scholar, Department of Psychology, University of California Los Angeles, USA
E-mail: vmittal@mednet.ucla.edu

Robin M. Murray, Professor of Psychiatry, Institute of Psychiatry, King's College London, UK
E-mail: robin.murray@iop.kcl.ac.uk

Dana J.H. Niehaus, M.B. Ch.B., M.Med. (Psych.), D.Med. (Psych.), FC Psych., Department of Psychiatry, University of Stellenbosch, South Africa
E-mail: djhn@sun.ac.za

Daniel Norton McLean Hospital, Belmont, MA, USA

Jonna Perälä, M.D., Researcher, National Public Health Institute, Department of Mental Health and Alcohol Research, Helsinki, Finland
E-mail: jonna.perala@ktl.fi

Thomas J. Raedler, M.D., Associate Professor, Department of Psychiatry, Faculty of Medicine, University of Calgary, Calgary, Alberta, Canada
E-mail: Thomas.Raedler@albertahealthservices.ca;
Thomas.Raedler@Calgaryhealthregion.ca

Michael S. Ritsner, M.D., Ph.D., Associate Professor of Psychiatry and Head of Cognitive and Psychobiology Research Laboratory, The Rappaport Faculty of Medicine, Technion – Israel Institute of Technology, Haifa and Chair, Acute Department, Sha'ar Menashe Mental Health Center, Hadera, Israel
E-mail: ritsner@sm.health.gov.il

Marion Roberts, Eating Disorders Research Unit, Department of Academic Psychiatry, Institute of Psychiatry, King's College London, 5th Floor Bermondsey Wing, Guy's Hospital, London, SE1 9RT ddi. 0207 188 0181
E-mail: marion.roberts@iop.kcl.ac.uk; www.eatingresearch.com

Richard B. Rosse, M.D., Mental Health Service Line, Department of Veterans Affairs Medical Center, Washington, Department of Psychiatry, Georgetown University School of Medicine, Washington, USA

Renata Schoeman, M.B. Ch.B., M.Soc. Sc., M.Med. (Psych.), FC Psych., Department of Psychiatry, University of Stellenbosch, South Africa
E-mail: bibitica@mweb.co.za

Nicholas J. Schork, Ph.D., Director of Research, Scripps Genomic Medicine; Director of Biostatistics and Bioinformatics, The Scripps Translational Science Institute; Professor, Molecular and Experimental Medicine, Scripps Health and The Scripps Research Institute, CA, USA
E-mail: nschork@scripps.edu

Barbara L. Schwartz, Ph.D., Mental Health Service Line, Department of Veterans Affairs Medical Center, Washington, Department of Psychiatry, Georgetown University School of Medicine, Washington, USA

Jaana Suvisaari, M.D., Ph.D., Academy research fellow, National Public Health Institute Department of Mental Health and Alcohol Research, Helsinki, Finland
E-mail: jaana.suvisaari@ktl.fi

Peter Szatmari Department of Psychiatry and Behavioural Neurosciences, McMaster University, Hamilton, Ontario, Canada
E-mail: szatmar@mcmaster.ca

Janet Treasure Psychological Medicine Department, King's College London, Institute of Psychiatry, London, UK
E-mail: Janet.Treasure@iop.kcl.ac.uk

Annamari Tuulio-Henriksson, Ph.D., Senior Researcher, National Public Health Institute Department of Mental Health and Alcohol Research, Helsinki, Finland
E-mail: annamari.tuulio-henriksson@ktl.fi

Elaine F. Walker, Ph.D., Samuel Candler Dobbs Professor of Psychology and Neuroscience, Department of Psychology, Emory University, USA
E-mail: psyefw@emory.edu

Abraham Weizman, M.D., Professor of Psychiatry, Research Unit, Geha Mental Health Center and the Laboratory of Biological Psychiatry at Felsenstein Medical Research Center, Sackler Faculty of Medicine, Tel-Aviv University, Ramat-Aviv, Tel-Aviv
E-mail: Israel. aweizman@clalit.org.il

Klaus Wiedemann University of Hamburg, Department of Psychiatry, Hamburg, Germany

Odette de Wilde, Ph.D., Academic Medical Center, University of Amsterdam, Department of Psychiatry, The Netherlands
E-mail: o.dewilde@amc.uva.nl

Georg Winterer, M.D., Ph.D., Associate Professor, Department of Psychiatry, Heinrich-Heine University, Duesseldorf, and Institute of Neurosciences and Biophysics, Juelich Research Centre, Juelich, Germany
E-mail: georg.winterer@uni-duesseldorf.de

Volume 2

Caleb M. Adler, M.D., Associate Professor of Psychiatry, Co-Director, Division of Bipolar Disorders Research, Department of Psychiatry, University of Cincinnati College of Medicine, Cincinnati, OH, USA
E-mail: adlercb@uc.edu

Deanna M. Barch, Ph.D., Professor, Departments of Psychology, Psychiatry and Radiology, Washington University, St. Louis, MO, USA
E-mail: dbarch@wustl.edu

Stephan Bender Senior scientist and commissionary Head of the joint Neurophysiological Laboratory of the Psychiatric, Psychosomatic and Child and Adolescent Psychiatric Hospital of the University of Heidelberg, Germany
E-mail: Stephan.Bender@med.uni-heidelberg.de

Mona K. Beyer, M.D., Ph.D., Department of Radiology, Stavanger University Hospital, Stavanger, Norway; The Norwegian Centre for Movement Disorders, Stavanger University Hospital, Stavanger, Norway

Michael A. Cerullo, M.D., Assistant Professor of Psychiatry,Division of Bipolar Disorders Research, Department of Psychiatry,University of Cincinnati College of Medicine, Cincinnati, OH, USA
E-mail: michael.cerullo@uc.edu

John G. Csernansky, M.D., Lizzie Gilman Professor and Chairman, Department of Psychiatry and Behavioral Sciences, Northwestern University Feinberg School of Medicine, Chicago, IL, USA
E-mail: igc@nonthwestern.edu

Turi O. Dalaker, M.D., Buffalo Neuroimaging Analysis Center, Department of Neurology, State University of New York at Buffalo, Buffalo, NY, USA; Department of Radiology, Stavanger University Hospital, Stavanger, Norway; The Norwegian Centre for Movement Disorders, Stavanger University Hospital, Stavanger, Norway
E-mail: datu@sus.no

Melissa P. DelBello, M.D., M.S., Vice-Chair for Clinical Research, Department of Psychiatry; Associate Professor of Psychiatry and Pediatrics, Division of Bipolar Disorders Research, University of Cincinnati College of Medicine, Cincinnati, OH, USA
E-mail: delbelmp@email.uc.edu

David E. Fleck, Ph.D., Assistant Professor of Psychiatry, Division of Bipolar Disorders Research, Department of Psychiatry,University of Cincinnati College of Medicine, Cincinnati, OH, USA
E-mail: david.fleck@uc.edu

Shabnam Hakimi, B.A., Clinical Brain Disorders Branch, Genes, Cognition, and Psychosis Program, National Institute of Mental Health, National Institutes of Health, Bethesda, MD, USA

David Linden Professor of Biological Psychiatry, Wales Institute of Cognitive Neuroscience and North Wales Clinical School, School of Psychology, University of Wales Bangor, Bangor, UK
E-mail: d.linden@bangor.ac.uk

Valentina Lorenzetti Ph.D. candidate, Melbourne Neuropsychiatry Centre, Department of Psychiatry, The University of Melbourne and Melbourne Health, Australia
E-mail: vlor@unimelb.edu.au

Dan I. Lubman, Ph.D., FRANZCP, FAChAM; Associate Professor, ORYGEN
Research Centre, Department of Psychiatry, University of Melbourne, Victoria,
Australia
E-mail: dan.lubman@mh.org.au

Frank P. MacMaster, Ph.D., Psychiatry and Behavioral Neurosciences,
Wayne State University School of Medicine Detroit, MI, USA
E-mail: fmacmast@med.wayne.edu

Daniel Mamah, M.D., M.P.E., Instructor, Department of psychiatry, Washington
University School of Medicine St. Louis; President, Eastern Missouri Psychiatric
Society, USA
E-mail: mamahd@psychiatry.wustl.edu

Chong Mei Sian Consultant, Department of Geriatric Medicine,
Tan Tock Seng Hospital, 11 Jalan Tan Tock Seng, Singapore
E-mail: Mei_Sian_Chong@ttsh.com.sg

Andreas Meyer-Lindenberg, M.D., Ph.D., Director of the Central Institute of
Mental Health, Professor of Psychiatry and Psychotherapy, Faculty of Clinical
Medicine Mannheim, University of Heidelberg, Germany
E-mail: a.meyer-lindenberg@zi-mannheim.de

Jayasree J. Nandagopal, MD, Assistant Professor of Psychiatry, Division of
Bipolar Disorders Research, Department of Psychiatry, University of Cincinnati
College of Medicine, Cincinnati, OH, USA
E-mail: jayasree.nandagopal@uc.edu

Nick C. Patel, Pharm.D., Ph.D., Clinical Pharmacist, Lifesynch; and Clinical
Assistant Professor & Health Behavior, Medical College of Georgia; USA
E-mail: npatel5@lifesynch.com

Armin Raznahan, MBBS, MRCPCH, MRCPsych., Medical Research Council
Clinical Research Training Fellow, Institute of Psychiatry,
King's College London, UK
E-mail: Armin.Raznahan@iop.kcl.ac.uk

Franz Resch Professor, Director of the Child and Adolescent Psychiatric
Hospital of the University of Heidelberg, Germany
E-mail: Franz.Resch@med.uni-heidelberg.de

David R. Rosenberg, M.D., Psychiatry and Behavioral Neurosciences,
Wayne State University School of Medicine, Children's Hospital of Michigan,
Detroit, MI, USA
E-mail: drosen@med.wayne.edu

Lim Wee Shiong Consultant, Department of Geriatric Medicine, Tan Tock Seng
Hospital, 11 Jalan Tan Tock Seng, Singapore

Nadia Solowij, Ph.D., Senior Lecturer, School of Psychology and Illawarra
Institute for Mental Health, University of Wollongong, Australia, Affiliated
Scientist, Schizophrenia Research Institute, Sydney, Australia
E-mail: nadia@uow.edu.au

Milena Stosic, M.D., Buffalo Neuroimaging Analysis Center, Department of Neurology, State University of New York at Buffalo, Buffalo, NY, USA

Stephen M. Strakowski, MD, The Stanley and Mickey Kaplan Professor and Chair of Psychiatry Professor of Psychology and Biomedical Engineering Director, Center for Imaging Research University of Cincinnati College of Medicine, Cincinnati, OH, USA
E-mail: stephen.strakowski@uc.edu

Heike Tost, M.D., Ph.D., Post-Doctoral Research Fellow, Clinical Brain Disorders Branch, Genes, Cognition, and Psychosis Program, National Institute of Mental Health, National Institutes of Health, Bethesda, MD, USA
E-mail: tosth@mail.nih.gov

Matthias Weisbrod, Professor, Director of the SRH Psychiatric Hospital Karlsbad-Langensteinbach; Head of the Section for Experimental Psychopathology of the University of Heidelberg, Germany
E-mail: Matthias.Weisbrod@kkl.srh.de

Murat Yücel, Ph.D., MAPS; Senior Lecturer and Clinical Neuropsychologist, Melbourne Neuropsychiatry Centre and ORYGEN Research Centre, Department of Psychiatry, University of Melbourne and Melbourne Health, National Neuroscience Facility, Melbourne, Australia
E-mail: murat@unimelb.edu.au

Robert Zivadinov, M.D., Ph.D., Buffalo Neuroimaging Analysis Center, Department of Neurology, State University of New York at Buffalo, Buffalo, NY, USA
E-mail: rzivadinov@bnac.net

Volume 4

Danielle M. Andrade, M.D., M.Sc., Assistant Professor, Department of Medicine, University of Toronto, Division of Neurology – Epilepsy Program UHN – Toronto, Western Hospital, Toronto, Canada
E-mail: Danielle.Andrade@uhn.on.ca

Ramón Cacabelos Professor and Chairman EuroEspes Biomedical Research Center, Institute for CNS Disorders and Genomic Medicin, EuroEspes Chair of Biotechnology and Genomics, Camilo José Cela University, Coruña, Spain
E-mail: rcacabelos@euroespes.com

Gursharan Chana, Ph.D., Department of Psychiatry, University of California, San Diego, La Jolla, CA, 92093-0603, USA

Rebecca Dang Johns Hopkins University, Baltimore, MD, USA

Juergen Deckert, M.D., Full Professor and Chairman, Department of Psychiatry, University of Wuerzburg, Germany
E-mail: Deckert_J@klinik.uni-wuerzburg.de

Chantal Depondt, M.D., Ph.D., Department of Neurology, Hôpital Erasme, Université Libre de Bruxelles, Belgium
E-mail: cdepondt@ulb.ac.be

Katharina Domschke, M.D., M.A., Head of Group "Genetics of Affective Disorders", Department of Psychiatry, University of Muenster, Germany
E-mail: katharina.domschke@ukmuenster.de

Ian P. Everall, M.D., Ph.D, FRCPsych., FRCPath., Department Of Psychiatry, University of California, San Diego, CA, USA

Stephen J. Glatt, Ph.D., Department of Psychiatry and Behavioral Sciences, State University of New York, Syracuse, NY 13210, USA

Marco A. Grados, M.D., M.P.H., Assistant Professor, Johns Hopkins University School of Medicine, Baltimore, MD, USA
E-mail: mjgrados@jhmi.edu

Susumu Higuchi, M.D., Ph.D., National Hospital Organization, Kurihama Alcoholism Center, Kanagawa, Japan
E-mail: h-susumu@db3.so-net.ne.jp

Yasue Horiuchi, Ph.D., Department of Medical Genetics, Graduate School of Comprehensive Human Sciences, University of Tsukuba, Ibaraki, Japan
E-mail: yasu-horiuchi06@ob.md.tsukuba.ac.jp

Hiroki Ishiguro, M.D., Ph.D., Assistant Professor of Department of Medical Genetics, Graduate School of Comprehensive Human Sciences, University of Tsukuba, Ibaraki, Japan
E-mail: hishigur@md.tsukuba.ac.jp

Dorothée Kasteleijn-Nolst Trenité Professor Medical Genetics, University of Utrecht; Department of Medical Genetics, University Medical Centre, Utrecht, the Netherlands; Professor Neuroscience, University of Rome "Sapienza", Department of Neuroscience, Rome, Italy
E-mail: Dorothee.Kasteleijn@uniroma1.it; D.kasteleijn@umcutrecht.nl

Minori Koga, Ph.D., Department of Medical Genetics, Graduate School of Comprehensive Human Sciences, University of Tsukuba, Ibaraki, Japan
E-mail: taq-pol@tmail.plala.or.jp

Janet Kwok, B.Sc., Department Of Psychiatry, University of California, San Diego, La Jolla, CA, 92093-0603, USA

Daniel Lévesque Associate Professor, Senior Scientist, Faculty of Pharmacy University of Montreal, Canada
E-mail: daniel.levesque.1@umontreal.ca

Dick Lindhout Department of Medical Genetics, University Medical Centre Utrecht, Utrecht, The Netherlands

Laura Mandelli, Psy.D., Assistant Professor of Psychiatry, Institute of Psychiatry, University of Bologna, Italy
E-mail: laura.mandelli@unibo.it

Alessandra Nivoli, M.D., Associate Professor of Psychiatry, Institute of Psychiatry, University of Sassari, Via Luna e Sole 55, 07100, Sassari, Italy
E-mail: alessandro.serretti@unibo.it

Emmanuel S. Onaivi, Ph.D., Associate Professor of Department of Biology, William Paterson University, Wayne, NJ, USA
E-mail: OnaiviE@wpunj.edu

Dalila Pinto, Ph.D., Research Fellow. Department of Medical Genetics. University Medical Center Utrecht, Utrecht, the Netherlands; and Genetics and Genome Biology, The Center for Applied Genetics, The Hospital of Sick Children, Toronto, Canada.
E-mail: d.c.s.pinto@gmail.com

Michael S. Ritsner, M.D., Ph.D., Associate Professor of Psychiatry and Head of Cognitive and Psychobiology Research Laboratory, The Rappaport Faculty of Medicine, Technion – Israel Institute of Technology, Haifa and Chair, Acute Department, Sha'ar Menashe Mental Health Center, Hadera, Israel
E-mail: ritsner@sm.health.gov.il

Claude Rouillard, Ph.D., Professor, Department of Medicine, Faculty of Medicine, Laval University and Neuroscience Research Centre, Laval University Hospital Research Centre, Québec City, Québec, Canada
E-mail: claude.rouillard@neurosciences.ulaval.ca

Alessandro Serretti, M.D., Associate Professor of Psychiatry, Institute of Psychiatry, University of Bologna, Italy
E-mail: alessandro.serretti@unibo.it

Ehud Susser, M.D., Senior Psychiatrist, Sha'ar Menashe Mental Health Center, Hadera, and the Rappaport Faculty of Medicine, Technion – Israel Institute of Technology, Haifa, Israel
E-mail: udiwudi@yahoo.com

Ming T. Tsuang, M.D., Ph.D., D.Sc., Behavioral Genomics Endowed Chair and University Professor, University of California; Distinguished Professor of Psychiatry and Director, Center for Behavioral Genomics, Department of Psychiatry, University of California, San Diego, CA; Director, Harvard Institute of Psychiatric Epidemiology and Genetics, Harvard Medical School and Harvard School of Public Health, USA
E-mail: mtsuang@ucsd.edu

Index